T0281539

PROTON-ANTIPROTON
COLLIDER PHYSICS

ADVANCED SERIES ON DIRECTIONS IN HIGH ENERGY PHYSICS

Cover Artwork by courtesy of Los Alamos National Laboratory.
"This work was performed by the University of California, Los Alamos National Laboratory, under the auspices of the United States Department of Energy."

Advanced Series on
Directions in High Energy Physics-Vol. 4

PROTON-ANTIPROTON COLLIDER PHYSICS

Editors

G. Altarelli
L. Di Lella

World Scientific
Singapore • New Jersey • London • Hong Kong

Published by

World Scientific Publishing Co. Pte. Ltd.
P O Box 128, Farrer Road, Singapore 9128

USA office: World Scientific Publishing Co., Inc.
687 Hartwell Street, Teaneck, NJ 07666, USA

UK office: World Scientific Publishing Co. Pte. Ltd.
73 Lynton Mead, Totteridge, London N20 8DH, England

Library of Congress Cataloging-in-Publication Data

Proton-antiproton collider physics.
    (Advanced series on directions in high energy physics; vol. 4)
1. Proton-antiproton colliders-Switerland-Geneva.   2. Proton-antiproton
interactions.   3. European Organization for Nuclear Research.
I. Altarelli, G.   II. DiLella, L.   III. Series.
QC787.P73P76   1989   539.7'3   88-33939
ISBN 9971-50-562-2
ISBN 9971-50-563-0 (pbk.)

PROTON-ANTIPROTON COLLIDER PHYSICS

Copyright © 1989 by World Scientific Publishing Co. Pte. Ltd.

All rights reserved. This book, or parts thereof, may not be reproduced in any form or by any means, electronic or mechanical, including photocopying, recording or any information storage and retrieval system now known or to be invented, without written permission from the Publisher.

ISBN 9971-50-562-2
      9971-50-563-0 (pbk)
ISSN 0218-0324

Printed in Singapore by JBW Printers & Binders Pte. Ltd.

PREFACE

Carlo Rubbia
CERN, Geneva, Switzerland

The Proton–Antiproton Collider: A Facility or an Experiment?

One of the most unique features of the CERN Proton–Antiproton Collider is that it was conceived as an *experiment to observe the charged and neutral Intermediate Vector Bosons,* rather than as a general-purpose accelerator facility. As amply illustrated in this book, these goals have been significantly extended, and the findings of the several teams that have worked enthusiastically on the project — often with remarkable skill and ingenuity — have provided a rather exhaustive picture of the basic phenomenology of hadron collisions in the energy domain of $\sqrt{s} \approx 0.6$ TeV.

The luminosity of the CERN Collider, now limited by the number of available antiprotons, is being increased about tenfold with a new, more sophisticated and powerful accumulation scheme, the Antiproton Collector (ACOL). It is still possible to further improve the luminosity by adding a superconducting low-beta section. I believe that the ultimate luminosity of the Collider will then be in the vicinity of 10^{31} cm^{-2} s^{-1}, where other limitations associated with the detector technology, the triggering, and the multiple interactions start to become dominant. It is expected that in the near future the TeV I complex now in operation in Batavia (USA), and largely inspired by our project, would extend the data on this phenomenology by about a factor of 3 in $\sqrt{s}$. Therefore, the results described in this book will be further perfected, and a phase of 'systematic' measurements will follow the 'exploratory' phase.

The utilization of proton–antiproton collisions, rather than the more straightforward proton–proton collisions pioneered at CERN with the Intersecting Storage Rings (ISR), was, for us, dictated by the reason of necessity, namely that we had available only one magnetic structure capable of storing a hadron beam at sufficiently high energy and therefore the two counter-rotating colliding beams had to share the same vacuum chamber. After having briefly considered H$^-$ ions — much too fragile to withstand acceleration and storage — the only choice was to make use of antiprotons, following an old suggestion by Budker and in analogy with e$^+$e$^-$ colliders.

Although proton–antiproton colliders bear some resemblance to the e$^+$e$^-$ colliders that have dominated, uncontested, the field of leptonic colliders — it is likely that future, dedicated, higher-energy colliders will once again rely on protons-against-protons, and that the CERN p$\bar{\text{p}}$ Collider, together with its bigger 'brother' the Fermilab TeV I, will, in our field, remain unique examples of their kind. This is because each mass scale of high-energy phenomena is associated with its own magnitude of the cross-sections and drops as M^{-2}. Therefore, at higher energies the luminosity becomes a more essential parameter. Two separate rings, although more costly, provide us with a higher stored current and hence higher luminosities. This would still be true if one were to find ways of constructing much more prolific sources of antiprotons, and it has to do with the general dynamics of two beams sharing the same vacuum chamber.

It should be added, however, that the use of particle–antiparticle collisions in the case of the production and decay of the W particles permits—at least at energies relatively close to the production threshold—the direction of the incoming fermion and antifermion to be identified. Because of helicity arguments, the distribution of the decay electrons exhibits one of the most salient features of weak interactions, namely a remarkable charge asymmetry along the direction of the incoming beams. If, instead, proton–proton collisions were used, evidence of such an asymmetry would have been either absent or much more difficult to find. The demonstration of the existence of such an asymmetry has played a fundamental role in associating the phenomenology of W particles observed at the CERN Collider with the expectations of the weak interaction, and it has, for instance, made it possible to demonstrate that the W has spin 1, and that both the production and the decay have maximal V–A mixing.

It has often been argued that since both the quark and the antiquark that are needed in the fusion process to produce a W or a Z are coming from the 'valence', for equal luminosities and for energies near threshold, the proton–antiproton reaction is more prolific of Intermediate Vector Bosons. However, this point should not be over-emphasized, since proton–proton reactions—even if less efficient—can naturally lead to much larger luminosities because of the much greater availability of proton beams, and therefore the initial advantage in the cross-sections can easily be off-set.

There are, however, a number of very important experiments of a specialized nature, and at low energies, in which the use of intense and highly concentrated antiprotons is of fundamental importance. Therefore the technology of intense antiproton sources is an extremely promising field. This was indeed already promptly realized during the early days of the CERN Collider project, and has led to the exciting physics that the low-energy antiproton accumulator LEAR is now producing.

Another unique feature of the Collider was the necessity to operate the Super Proton Synchrotron (SPS) complex alternately as an accelerator and as a storage ring. Ours were experiments that needed to alternate and to compete for time with the other, fixed-target programmes at the main CERN accelerator. However, if the SPS settings could easily be changed from accelerator to storage ring, for the heavy and complex detectors it has been quite another matter, since they had to be staged periodically in and out of the main accelerator tunnel.

Finding the space for the detectors has not been an easy task. The SPS has six long straight sections, of which, however, four were already committed (one for injection, one for the two extraction lines, one for RF acceleration, and one for the beam dump). For some time we were afraid that only one interaction point, namely LSS5, and therefore only one general-purpose experiment, namely UA1, could be suitable since LSS5 was at a relatively shallow depth. The other point, LSS4, was more than sixty metres underground. Fortunately, this problem was eventually overcome by the ingenuity of the experimenters who conceived UA2—a compact but effective non-magnetic device—and of the civil engineers who were interested in experimenting with the idea of novel, deep-underground solutions, with implications for the design of LEP.

In essence, experiments at the CERN pp̄ Collider have been successful because of a relatively narrow window, bracketed on one side by the mass scale and hence the cross-section of the electroweak phenomena, and on the other side by an achievable luminosity in the range of 10^{29} to 10^{30} cm^{-2} s^{-1}, sufficient to produce a modest but convincing rate of W and Z events. This scenario is unlikely to repeat itself at much higher energies. Even if it has brought an immense amount of know-how to the physics of high-energy collisions and of the associated detector and accelerator technologies, the Collider programme will remain a separate and self-contained chapter amongst the key experiments in the field of elementary particle physics.

The Birth of a Project

At the time when the CERN p$\bar{\text{p}}$ Collider project was taking shape, a number of extremely wild extrapolations had to be made, with full awareness of the associated risks involved. The most obvious was the number of antiprotons needed and the rate at which they had to be accumulated. But, as we shall see, there were other, perhaps more sticky, issues.

Phase-space collapse for the antiprotons was clearly a formidable task, with a required compression factor of order 10^8-10^9 in the six-dimensional phase space. We had at that time *a priori* two possible alternative methods: one based on electron cooling, which had just been demonstrated in Novosibirsk, and the other based on stochastic cooling, the validity of which had been shown at the ISR. We hesitated for quite some time over the final choice. I believe, in retrospect, that either scheme may have eventually worked. Also, hybrid schemes in which an initial stochastic cooling was to be followed by electron cooling were widely considered. In the end, the choice was made rationally — in favour of the extensive use of several stochastic devices acting in the various components of the phase space — on the basis of experience accumulated with the Initial Cooling Experiment (ICE), which had been constructed using the old g − 2 muon ring and brought into operation in less than one year! Stochastic cooling was clearly in the lead because of its flexibility, elegance, and conceptual simplicity. Electron cooling could provide a sufficient cooling speed, but only at such low antiproton energies that the production yield was infinitesimal. Therefore in order to use it, one had to decelerate the antiprotons from about 3.0 GeV (the production maximum at the CERN PS) to about 100–200 MeV, cool them, and then bring them back to high energies. Also, the electron beam needed to cool the antiprotons involved a current of several tens of amperes at energies in excess of 100 kV in a single-pass arrangement, from a very low emittance cathode to a collector. The antiprotons had to be maintained in the 'electron cooling bath' for a considerable fraction ($\geqslant$ 10%) of the circumference of the antiproton storage ring. The handling of such a large amount of power in the beam is by no means trivial. For instance, electrons had to be slowed down by an efficient energy-recovery scheme before hitting the collector.

In retrospect, the choice of the all-stochastic-cooling scheme for the Antiproton Accumulator (AA) was the correct one, and the experience gained with ICE made us confident that it would indeed closely fulfil our expectations. Also, the AA was built remarkably quickly. Indeed, the full complex was made operational for protons in less than two years from the date of approval. A few months later it accepted antiprotons.

Meanwhile, other preoccupations were becoming more pressing. During a test run, the ISR had been operated at a limit in the beam-tune shift which, if equalled at the Collider, would have given us the required luminosity. However, whilst the ISR had a d.c. beam, what we needed in order to operate the Collider were few and very tight bunches. The CERN p$\bar{\text{p}}$ Collider was the first example of a storage ring in which bunched protons and antiprotons collided head-on. It was, of course, well known that e^+e^- colliders could work very well with bunched beams: however, they had had a strong, continuous phase-space damping due to synchrotron radiation, which now was completely absent. Furthermore, since antiprotons are scarce, the Collider had to be operated under conditions of relatively large beam–beam interactions, which was not the case for the continuous proton beams of the ISR. Looking back, I believe that probably one of the most remarkable features of the Collider was that it operated at such a large beam–beam tune shift, which in turn meant a high luminosity.

In order to explain more clearly what one had to expect, let us assume that the beam–beam force, impressed on the particle at each beam–beam crossing, can be approximated as a periodic succession of extremely non-linear potential kicks. They are, of course, expected to excite the continuum of resonances of the storage ring, which has, in principle, the density of rational numbers. Reduced to bare essentials, let us consider a weak antiproton beam colliding head-on with a strongly bunched proton beam. The increment — due to the angular kick $\Delta x'$ — of the action invariant

$$W = \gamma x^2 + 2\alpha xx' + \beta x'^2$$

of an antiproton is

$$\Delta W = \beta(\Delta x') + 2(\alpha x + \beta x')\Delta x' .$$

This can be expressed in terms of the 'tune shift' ΔQ as: $\Delta x' = 4\pi\Delta Qx/\beta$. If we now assume that the successive kicks are randomized, the second term of ΔW averages to zero, and we get

$$\Delta W/W = \frac{1}{2}(4\pi\Delta Q)^2 .$$

In order to achieve the design luminosity, we need $\Delta Q \approx 0.003$, leading to $(\Delta W/W) = 7.1 \times 10^{-4}$. This is a very large number indeed, giving an e-fold increase of W in only $1/(7.1 \times 10^{-4}) = 1.41 \times 10^3$ kicks!

Therefore the only reason why the antiproton motion may remain stable over approximately 10^{10} beam–beam traversals is because these strong kicks are not random but periodic, and the beam has a long 'memory' that allows them to be added *coherently* rather than at random. Then, off-resonance, the effects of these kicks will cancel on the average, giving an overall zero-amplitude growth.

These beam–beam effects were very difficult, if not impossible, to evaluate theoretically, since this *a priori* purely deterministic problem can exhibit stochastic behaviour and irreversible diffusion-like characteristics. Strongly non-linear lenses can simulate only poorly the effects of the beam–beam interactions, known to excite resonance up to $10-12^{th}$ order.

A possible experiment was, then, to lessen the damping effects of the synchrotron radiation by reducing the energy of an existing e^+e^- storage ring and extrapolating to infinite damping time — which is the case with the CERN p$\bar{\text{p}}$ Collider.

The measurement was performed at SPEAR (Stanford). Although computer simulations over an enormous number of turns showed no loss of emittance, the experimental result had aggravated, at least for some time, the general concern about the viability of the Collider scheme. Reducing the energy of SPEAR resulted in a smaller value of the maximum allowable tune shift, interpreted as being due to the reduced synchrotron radiation damping.

Equating the required beam lifetime for the p$\bar{\text{p}}$ Collider (where damping is absent) with the extrapolated damping time of an e^+e^- collider gives a maximum allowed tune shift $\Delta Q = 10^{-5}-10^{-6}$ — which is catastrophically low. This bleak prediction was not confirmed by experience at the Collider, where $\Delta Q = 0.003$ per crossing, and six crossings are routinely achieved with a beam luminosity lifetime approaching one day.

What, then, is the reason for such a striking contradiction between the experiments with protons and those with electrons? The difference is caused by the presence of synchrotron radiation in the latter. The emission of synchrotron photons is a major source of quick randomization between crossings, and leads to a rapid deterioration of the beam emittance. Fortunately, the same phenomenon also provides us with an effective damping mechanism. The CERN p$\bar{\text{p}}$ Collider does not contradict the SPEAR result, because at the Collider *both* the randomizing *and* the damping mechanisms are absent. This unusually favourable combination of effects has ensured that proton–antiproton colliders have become viable devices.

The three parts of the Collider project — namely the accumulator, the modification of the SPS and the PS, and the main experiments — were approved by the CERN Council in June 1978, for the total, anticipated sum of 137 million Swiss francs. The first collisions were observed three years later, in early July 1981. Even if the antiproton current was, at that time, infinitesimal, the few people who had stayed on that night could witness, in the first hours of the morning and after many hours of unsuccessful acceleration and storage attempts, the final proof of the stability of the scheme: an antiproton bunch safely surviving the multiple passages through an intense counter-rotating proton bunch.

The Ups and Downs of Hadron Colliders

In the late seventies, colliding-beam reactions were classified as 'clean' and 'dirty'. Proton–proton collisions and a fortiori also proton–antiproton collisions were 'dirty'. Electron–positrons, instead, were 'clean'. The adjectives were coined after the spectacular nature of the successes of the e^+e^- machines in the systematic studies of the J/ψ and the Υ, and the cleanliness of the initial quantum numbers 1^{--} of the photon propagator on the one hand, and, in spite of the much higher centre-of-mass energy, the apparent difficulty of the ISR to produce evidence of comparable discovery potentials, on the other hand. Indeed, both charm and beauty were within the kinematical reach of the ISR. For a luminosity of 5×10^{31} cm^{-2} s^{-1} — typical of the ISR and the charm-associated cross-section of 0.5 mb — there were as many as 25,000 charm pairs produced each second in each of the interaction points! In spite of this, only very few events were observed, very painfully and very late. Because of a number of experimental problems, the discovery of both the J/ψ and of the Υ was missed. They were found, instead, in proton–proton collisions with fixed-target experiments and with much lower centre-of-mass energies, the first at the AGS (Brookhaven) and the second at Fermilab (Batavia).

The naïve explanation for these problems was ordinarily blamed on the 'complexity' of the hadron collisions. Dick Feynman used to say: 'What will one ever learn colliding Swiss(!) watches against Swiss watches?'

The success of the collider experiments is due, in a major way, to the fact that the detectors were at last well matched to the task of analysing the full complexity of hadronic events. This new detection technique represents enormous progress from the technology used in the early ISR experiments, and no doubt it is a result of the experience gained there.

The initial ISR philosophy was the one of dividing up the solid angle into a large number of small slices, each assigned to a separate experimental team, much like secondary beams from an accelerator used to be assigned to separate experiments. The Split Field Magnet (SFM), a 4π-detector facility, was chosen under the assumption that little or no physics could occur around 90°, and therefore large angles have since been effectively cut out by a highly non-uniform and weak analysing field. The trigger problem was in general neglected, at least in the design of the early detectors.

The UA1/2 detectors were therefore the first of a new kind. Had they operated at the ISR from the start, fundamental discoveries such as the J/ψ and Υ would not have been missed. The new key elements were: a truly 4π detection geometry (for instance, 97% of the mass of UA1 is calorimetrically active); the extensive use of calorimetry for both hadrons and electromagnetic showers; and (at least for UA1) a huge magnetic field volume (100 m^3) equipped with a bubble-chamber-like redundant image detector (50–100 points per track).

Both detectors were, however, designed as *experiments* having in mind, almost exclusively, the goal of discovering the W and the Z. One must recognize that the risk of failure was considerable. The leptonic signature for the Z was obvious, but the rate of events was marginal. For the W, the rate was an order of magnitude higher, but the signature was more elusive because of the single charged lepton and the neutrino. However, for the first time global calorimetry has made it possible to measure the *missing energy,* namely to identify at least the transverse components of the neutrino emission in the W decay from the overall momentum imbalance. To this effect, the detector had to be fully sensitive and *hermetic* in order that no appreciable energy release would escape detection.

One of the most gratifying surprises of the CERN p$\bar{p}$ Collider has been the observation of jets through the so-called 'Lego' plot, where they appear as clusters called 'Manhattan towers'. Again, calorimetry was essential in order to perform quantitative physics. Collider experiments have transformed 'jettology' into an almost exact science, bypassing fragmentation and looking directly at the elementary hadronic constituents (quarks and gluons) through the energy flow recorded over the solid angle. Jet-dominated events — in spite of the tremendous number of particles produced — most often exhibit a remarkable simplicity, and just like in the case of e^+e^- collisions they can be directly related to the underlying Feynman diagrams. These detectors will be further improved in order to become even more fragmentation-independent when the energy response to electromagnetic (π^0) and hadronic ($\pi^{\pm}$) components of the cascade will be matched, using for instance the energy gain from fission to correct for losses in the hadronic cascade. The improved UA1 calorimeter has in this way achieved a $\pi^0/\pi^{\pm}$ ratio of 1 within the per cent level over the full energy range. Another important feature would be 'particle identification', namely the ability to distinguish a quark jet from a gluon jet. Even if statistically the two phenomena are different, with a gluon-initiated jet being 'softer' than the one from a quark, no detector has so far been able to provide such a refined separation on individual jets.

There is no doubt that these very important advances in detector instrumentation have made it possible to produce, with hadronic collisions, high-quality physics results that are in a class with those traditionally obtained with e^+e^-. The new generation of colliders — the Large Hadron Collider (LHC) at CERN and the Superconducting Super Collider (SSC) in the US — will greatly profit from the detector developments for the proton–antiproton colliders, in the same way that we have built up, out of the ISR, the basic experience that was necessary to produce the results described in this book.

CONTENTS

THE CERN p̄p COLLIDER

Lyndon Evans, Eifionydd Jones and Heribert Koziol

CERN
European Organization for Nuclear Research
CH-1211 Geneva 23
Switzerland

CONTENTS

1. INTRODUCTION

Antiprotons, produced at an accelerator by bombardment of a target with a high energy proton beam, were first observed at the Bevatron at Berkeley, California, USA, in the autumn of 1955. Seven years later at CERN, proposals were discussed for colliding protons and antiprotons in the ISR (Intersecting Storage Rings) then being designed, but luminosities were discouragingly low. The densities of the antiproton beams were far too poor and a further four years had to elapse before the ideas of Budker[1] and O'Neill[2] on electron cooling emerged in 1966. In that same year, Rubbia first proposed[2] the use of the projected 300 GeV machine (to be built years later as the SPS, the CERN Super Proton Synchrotron), as a proton-antiproton collider but using the scheme of electron cooling for the concentration of the antiprotons. For many years, the idea of electron cooling remained in its infancy; the technology was complicated and little understood outside of Budker's home institute in Novosibirsk. A proposal to investigate these techniques by converting CESAR, the small CERN electron storage ring (which was an electron model for the proposed proton-proton collider, ISR) attracted little interest and was quickly abandoned in favour of more pressing projects.

However, increasing the density of particle beams had long been a major preoccupation of the community of accelerator builders and the next step came in 1968 when Simon van der Meer thought up a method based on the fact that all beams are particulate and that therefore, on a microscopic level, the density within a given sample volume will be subject to statistical fluctuations. Experimental confirmation of both techniques came in 1974, electron cooling at Novosibirsk[3] and stochastic cooling at the CERN ISR[4] in Geneva. Rubbia took up again the idea of converting the SPS into a proton-antiproton (p̄p) collider. The CERN Direction, persuaded by Rubbia's enthusiasm and forcefulness, finally accepted the idea and the Initial Cooling Experiment (ICE) was set up to verify in greater detail both cooling methods. Stochastic cooling of both the transverse and longitudinal beam emittances was so successful in ICE that beam lifetimes, at 2 GeV/c, were extended from a few hours without cooling to several days.

The feasibility of effective antiproton accumulation thus demonstrated, the p̄p project was quickly launched.

An interesting aside is that at that time, 1978, notwithstanding theoretical expectation, the experimentally proven lower limit to the lifetime of the antiproton was only about 120 µs. An experiment in ICE pushed this limit to 80 h at 2.1 GeV/c (32 h at rest) and some people heaved a secret sigh of relief because the design of the Antiproton Accumulator (AA) was already underway. Construction of the AA started in 1979 and the first proton-antiproton collisions at a centre-of-mass energy of 540 GeV were observed in the SPS on the 10th July 1981. The first "W data" were taken in 1982 and the discovery of both the W and Z was proclaimed in 1983.

2.   *ANTIPROTON PRODUCTION AND ACCUMULATION*

A brief outline of the whole scheme is illustrated in Fig. 1.

3

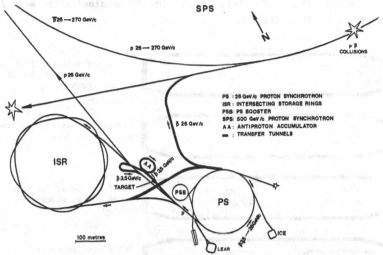

Fig. 1 - Layout of accelerators and storage rings involved in the p̄
project. pp̄ collisions first occurred in the ISR; the Low
Energy Antiproton Ring, LEAR, was added in 1983.

The CERN Proton Synchrotron (PS) first accelerates protons to 26 GeV/c
and concentrates them into one quarter of its circumference to match the
burst length to the smaller circumference of the AA. The protons are then
ejected through a transfer line towards the AA and focused on the antiproton
production target. The emerging burst of antiprotons at 3.58 GeV/c, selected
via a spectrometer, is injected into the AA near the outer edge of its
vacuum chamber.

The antiprotons remain around the injection orbit for almost 2 s while
they are being subjected to fast stochastic cooling to reduce their momentum
spread. They are next trapped by a radiofrequency system and, after removal
of the shutters separating the injection regions of the kicker and cooling
devices from the accumulation region, they are moved inwards, into the
stacking region. The injection region has thus been cleared for the next
burst of antiprotons due to arrive 2.4 s (the PS cycle time) after the pre-
ceding one. This process is repeated during the whole accumulation period.
The antiprotons in the stack are subjected continuously to six further
stochastic cooling systems which slowly create a dense core of antiprotons
around the accumulation orbit near to the inner edge of the vacuum chamber.
The accumulation process takes about a day and a typical performance for the

4

AA would be 6×10^9 $\bar{p}$/h. The most intense stack, obtained after many days, contained 5.2×10^{11} $\bar{p}$ (Fig. 2).

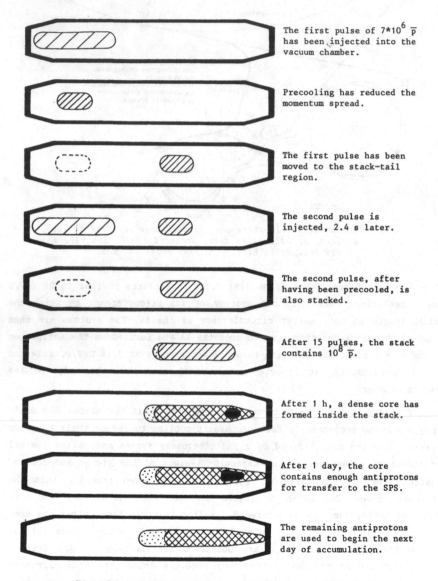

The first pulse of $7*10^6$ $\bar{p}$ has been injected into the vacuum chamber.

Precooling has reduced the momentum spread.

The first pulse has been moved to the stack-tail region.

The second pulse is injected, 2.4 s later.

The second pulse, after having been precooled, is also stacked.

After 15 pulses, the stack contains 10^8 $\bar{p}$.

After 1 h, a dense core has formed inside the stack.

After 1 day, the core contains enough antiprotons for transfer to the SPS.

The remaining antiprotons are used to begin the next day of accumulation.

Fig. 2 - *Schematic sequence illustrating cooling and accumulation of a stack in the AA.*

The dense core can now be tapped for bunches of antiprotons which, after suitable shaping with an rf system, are ejected from the AA and sent along a loop to the PS. There, they are accelerated to 26 GeV/c and, just before ejection, again shaped to make them acceptable to the injection conditions obtaining to the SPS. Three antiproton bunches are consecutively transferred to the SPS, every 2.4 s. Just before the antiproton transfer, the PS will have already accelerated and transferred three proton bunches circulating in the opposite direction to the antiprotons. These are sitting parked at 26 GeV/c, waiting in the SPS for the arrival of the three counter-rotating antiproton bunches. Upon the arrival of the latter, the accelerating procedure of the SPS is set into motion, both beams are brought to 315 GeV and then allowed to circulate for many hours. The phasing of the bunches is adjusted such that collisions between protons and antiprotons occur at those regions along the SPS circumference which are surrounded by the detectors, set up to examine the results of such collisions. With one "fill" of the collider, the experiments last typically for a day. During this time the AA continues to accumulate, topping up its stack with more antiprotons to be ready for the next day's transfer.

Typical luminosities achieved are a few 10^{29} cm^{-2}s^{-1} and the luminosity lifetime, at the beginning of a store, is about 18 h with around $1.5 \cdot 10^{11}$ p/bunch and $1.5 \cdot 10^{10}$ $\bar{p}$/bunch.

So much for an outline of the scheme. Below, the separate systems are discussed in more detail. However, if the reader has in mind the building of such a device he is advised to assimilate in depth the proceedings of the CERN Accelerator School (CAS) on Antiprotons for Colliding Beam Facilities held at CERN during October 1983 [5].

2.1 Production of Antiprotons[6,7]

A proton loses energy as it traverses and interacts with any material. When the proton energy is sufficiently high, the dominant energy-loss mechanism is particle production. The probability that a proton interacts in a given target material is defined in terms of a "total" cross section σ_{tot}. The definition is simply understood in terms of the following. Let N_{in} be the number of incident protons and n the number of target atoms seen by them per cm^2, where

$$n = N_A \varrho L / A \qquad (1)$$

N_A is Avogadro's number, i.e. atoms per mole, A the atomic weight, ϱ the density of the material in g cm^{-3} and L the length of the target in cm. Let N_{out} be the number of outgoing protons which have not interacted.

σ_{tot} is then defined by

$$N_{out} = N_{in} \exp(-n\sigma_{tot}) \tag{2}$$

or

$$\sigma_{tot} = (1/n) \ln(N_{in}/N_{out}) \tag{3}$$

The commonly used unit of total cross section is the barn where

$$1 \text{ b} = 10^{-24} \text{ cm}^2$$

or

$$1 \text{ nb} = 10^{-33} \text{ cm}^2$$

In all target materials there are essentially two types of interaction:

(i) Elastic scattering - no particle production involved
(ii) Inelastic scattering - involving particle production

Now consider only inelastic scattering and define an "absorption" cross section σ_{abs} for this type of interaction, as was done above for the total cross section:

$$\sigma_{abs} = (1/n) \ln(N_{in}/N'_{out}) \tag{4}$$

where now N'_{out} is the number of outgoing protons which have not interacted inelastically (not forgetting that N'_{out} includes the protons which have interacted elastically).

Targets are generally long and thin so that a useful concept is the absorption length λ_{abs} defined by

$$N'_{out} = N_{in} \exp(-L/\lambda_{abs}) \tag{5}$$

where L is the target length, hence

$$\lambda_{abs} = \frac{A}{N \varrho \sigma_{abs}} \tag{6}$$

The absorption cross section depends upon the target material and usually varies roughly as

$$\sigma_{abs} \propto A^{2/3} \qquad (7)$$

To design an antiproton accumulator we need to know the rate at which antiprotons can be produced. Two obvious variables need to be considered:

- the momentum of the produced antiproton = p(GeV/c)
- its production angle = θ (rad).

If the probability to produce a particle (say an antiproton) of momentum p at the angle θ into the solid angle dΩ and inside the momentum spread dp, is defined as

$$\frac{d^2\sigma(p,\theta)}{d\Omega dp} \quad cm^2 \cdot (GeV/c)^{-1} \cdot sr^{-1} , \qquad (8)$$

a related function may be defined, which is the "double differential" antiproton production "yield":

$$\frac{d^2N(p,\theta)}{d\Omega \, dp} = \frac{1}{\sigma_{abs}} \frac{d^2\sigma(p,\theta)}{d\Omega \, dp} \ (GeV/c)^{-1} \cdot sr^{-1} \qquad (9)$$

which is the number of particles produced per interacting proton, per unit solid angle and per unit momentum spread dp.

Such quantities have been measured experimentally and their values can be found in the literature. One of the most reliable sources summarising such measurements is believed to be a publication by Eichten et al.[8] wherein an invariant production density W(p,θ) is quoted, from which values of the double differential yield $d^2N/d\Omega dp$ may be estimated.

The yield is obviously a function of the momentum of the produced antiprotons and we should seek its maximum. This is observed to occur when an antiproton is produced at rest in the centre-of-mass system of the incoming proton and the stationary target nucleon with which it collides.

Since the protons, incident on the target, have a momentum of 26 GeV/c, the production yield has a maximum (albeit rather flat) around an antiproton momentum of 3.5 GeV/c; the AA was chosen to have a momentum of 3.58 GeV/c on its injection orbit for a number of reasons, including the above.

Figure 3 is a compilation[9] of the measured yields for antiprotons of 4 GeV/c per interacting proton of 24 GeV/c and for various target materials.

8

The yield peaks at forward production angles. It is estimated from Fig. 3 that the yield per interacting proton may be roughly written as

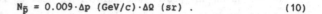

$$N_{\bar{p}} = 0.009 \cdot \Delta p \ (GeV/c) \cdot \Delta\Omega \ (sr) \ . \tag{10}$$

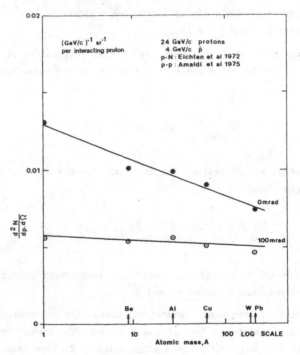

Fig. 3 - Measured antiproton production yield
for various target materials[9].

The AA ring was designed to accept a total momentum spread of 1.5%, so that $\Delta p = 0.015 \times 3.58$ GeV/c: if θ_{max} is the maximum angle (assumed small or near zero) that can be accepted by the beam transport system and matched into the acceptance of the AA ring, then $\Delta\Omega = \pi\theta_{max}^2$.

Some antiprotons will be reabsorbed by the target nuclei and many of them will be elastically (Coulomb) scattered by the electric charge of the nuclei. However, since most antiprotons will be found at the largest angles, it turns out that as long as the target is thin, this Coulomb scattering has a negligible effect on the yield. As for reabsorption, those antiprotons produced at smaller angles spend a longer time traversing the target than those at larger angles, so that the reduction in yield with production angle (see Fig. 3) is largely compensated for by less reabsorption. Hence we can write with good accuracy for a copper target 3 mm in diameter and 110 mm long (the AA target dimension):

$$N_{\bar{p}} = 0.009 \times 0.015 \times 3.58 \times \pi \times \theta^2_{max} \quad \text{per interacting proton.} \quad (11)$$

The number of protons which will interact with a target, N_p, is from definitions given above

$$N_p = N_{in}[1 - \exp(-L/\lambda_{abs})] \quad (12)$$

$\lambda_{abs} = 140$ mm for copper, so that in a target of 110 mm length only two thirds of the protons interact. Then the yield, Y, per incident proton, which is the more usual parameter quoted, is given as

$$Y = \frac{2}{3} N_{\bar{p}} \approx 10^{-3} \theta^2_{max} . \quad (13)$$

This implies that, since we measure Y to be around 6×10^{-7} in the AA ring, the maximum angles accepted are around 50 mrad.

Antiprotons emerging from the target at such angles are not easily focused by conventional iron-cored quadrupoles having reasonably dimensioned apertures. A more suitable lens would have a focal length of around 50 cm or less and preferably cylindrical symmetry. That chosen for the AA, for a number of reasons[7], was a current-sheet lens in the form of a magnetic horn, similar to the one described in 1962 by S. van der Meer[10]. It is highly non-linear, with the azimuthal magnetic field varying as the inverse of the radius. The exact profile of the inner conductor (Fig. 4) - the current sheet - is chosen by computer modelling such that it concentrates upon focusing the larger production angles (where most of the antiprotons are to be found) while maintaining a depth of focus equal to an appreciable fraction of the target length.

A lens having a short focal length will have a small depth of focus; hence targets should be made of materials which have a relatively short proton absorption length, λ_{abs}, in order to stay within the depth of focus of the device placed after it. Tungsten and copper were chosen for the AA. The

a)

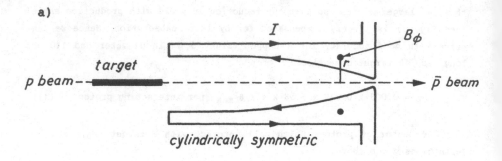

b)

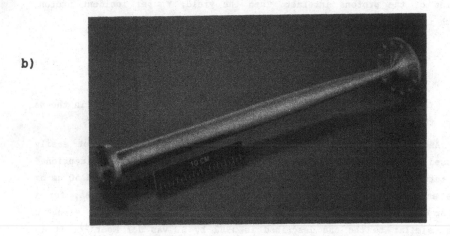

Fig. 4 - a) The flow of electrical current in a magnetic horn.
b) The inner conductor, made from aluminium alloy.

targets have to be cooled because an average of 1.1 kW is dissipated within
them by the 26 GeV/c beam of 10^{13} protons per 2.4 s (the proton burst is 0.4
µs long and the energy in the beam is around 40 kJ per pulse of which about
one third is dissipated in the target and the rest absorbed in a dump located straight ahead of the target). These numbers imply that the temperatures
in the tungsten target rise by 1350°C and in the copper target by 400°C at
each pulse. To avoid oxidation and disintegration the targets are enclosed
in sealed containers. Figure 5 shows a typical target assembly[11].

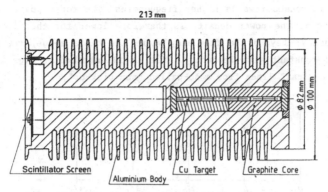

Fig. 5 - *A target assembly, as it was used for
the first several years of operation.*

2.2 Stochastic Cooling

There are many excellent texts on stochastic cooling theory[12, 13] and
experiment[13]. Here we will only give a brief introduction to the ideas relevant to the AA.

"Cooling" is the compaction of density in a particle beam. Indeed, in
the AA the density in 6-dimensional phase space (x, p_x ; y, p_y; z, p_z) is
increased by a factor 10^9. This is not done by compressing the phase space
volume, that is forbidden by Liouville's theorem, but by re-ordering the
particles by moving them into the empty volume that lies between them and
squeezing out this empty volume to the peripheries of the phase space.

A circulating beam of particles is "grainy", not only in phase space,
but also in real (x, y, z) space. This "graininess" shows up in the noise-
effect described by Schottky in the case of electrons travelling in a conductor or a thermionic vacuum valve. A mono-energetic beam of particles
circulating in a ring and passing through a perfect detector which gave out
a signal whose amplitude relates to the circulating charge, would yield a
spectrum of lines, when plotting amplitude versus frequency, at the revolution frequency f_0 and all harmonics, nf_0.

Natural beams are never monoenergetic; hence each spectral line becomes
a practically continuous family of lines, the envelope of which is referred
to as the "Schottky band", illustrated in Fig. 6.

The width of these bands is proportional to the harmonic number and therefore increases towards higher frequencies. The total power is the same for each band. The power density is therefore lower for the wider bands at high frequency; beyond the frequency where the bands merge, their combined density is constant with frequency. This is illustrated in Figs 6 and 7 for the "longitudinal" lines from a so-called "sum pick-up".

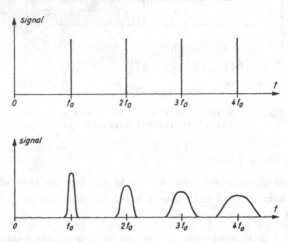

Fig. 6 - *The line spectrum from a hypothetical mono-energetic beam and, below, the Schottky bands (exaggerated in width) of a real beam with momentum spread.*

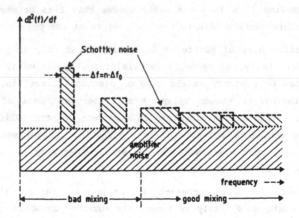

Fig. 7 - *Amplifier- and Schottky-noise, which determine the incoherent heating. The amplifier-noise has a continuous spectrum, the Schottky-noise has a band spectrum (the width of the bands is exaggerated)[14]*

If the detector is sensitive to the transverse position of the particles, a "difference pick-up", then the Schottky signals are modulated at the
frequencies of any transverse oscillation. The transverse motion (say in x)
of a single particle of revolution frequency f_0, around an ideal orbit at x
= 0, can be described by

$$x = a \cdot \cos[2\pi Q f_0 t + \varphi] \tag{14}$$

where Q is the number of so-called "betatron oscillations" per revolution
(the "tune"). Betatron oscillations are inherent in all beams in synchrotrons and have a spread in amplitude as well as a spread in frequency.
Transverse cooling aims at reducing just these amplitudes, while "longitudinal" (momentum) cooling aims at reducing the frequency (or momentum) spread.
The betatron oscillations create Schottky bands centered at the frequencies
$(m-Q)f_0$ and $(m+Q)f_0$, where m is called the "mode".

The Schottky signals are at the base of the techniques used in stochastic cooling. They are the natural voice of particle beams from which
knowledge concerning particle density both in longitudinal and transverse
phase space can be drawn. The "graininess" of the beam is revealed by these
signals; information about the individual particle's (or small samples of
particles) position in phase space can thus be treated and directed at
"correcting" the position so that it lines up with its fellows in an orderly
manner; it is "cooled".

A stochastic cooling system therefore consists of a detector (or a
pick-up) that acquires the Schottky signals from the particles, and a
corrector (or kicker) that pushes the very same sample of particles in the
appropriate direction.

Such a system is illustrated for the cooling of horizontal betatron
oscillations (transverse phase space) in Fig. 8. Betatron oscillations are
executed around some closed orbit. At each passage of a single particle (in
practice a sample of particles is treatable in this manner), the difference
(or transverse) pick-up provides a short signal that is proportional to the
distance of the particle from the centre of the pick-up. This is amplified
and applied to the kicker which will deflect the particle. The distance
between pick-up and kicker must equal an odd number of quarter betatron
wavelengths, in order that maximum positional error at the pick-up transforms to a maximum angular error at the kicker. If the gain of the system is
chosen correctly, any oscillation will then be cancelled. The signal must

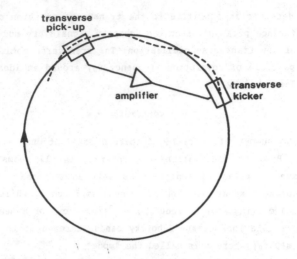

<u>Fig. 8</u> - Cooling of the horizontal betatron oscillation
of a single particle.

arrive at the kicker at the same time as the particle, hence the signal path
must cut off a bend in the particle's trajectory because cables and amplifiers have a delaying action on the pick-up signal. This correcting effect
is coherent in just such a manner as the workings of an injection kicker for
transferring beams into synchrotrons. However, the particle is not alone in
the beam and even with the fastest electronics, signals from neighbouring
particles will overlap and "noisily" perturb the coherent corrections. The
perturbing effect is a "noise" these neighbours create, producing kicks at
random phase with respect to the "corrected" particle. This leads to an
increase of the mean square of the amplitude in a stochastic, incoherent
manner - the beam is "heated".

Now, whereas the coherent, cooling, correction is dependent linearly on
the gain of the amplification system, the incoherent, heating, effect is
dependent upon the square of the gain. Hence, correct choice of gain will
always allow the cooling effect to predominate.

Each particle will cool itself with its own coherent signal. This requires that the phase of the correcting signal must be correct at the kicker
so that it exerts its influence always in the right direction. The other
particles produce an incoherent heating effect, at each Schottky line, proportional to the noise power density around that line[15]. Hence, only those

other particles with frequencies very near to those of the "corrected" particle will make any appreciable contribution to its heating. Of course, any
power from thermal noise (in amplifiers, etc.) must be added to the Schottky
band's power density.

It was explained above that the Schottky bands are well separated at
low frequencies, with high spectral power-density. This means that, for the
coherent effect to dominate over the incoherent, the gain has to be sufficiently low at the bottom of the frequency range and increasing towards the
top of the frequency range as the power density decreases. This effect is
also related to what is termed "mixing". The coherent displacement signal
from a particular sample of particles can be eliminated, ideally, by a
single perfectly phased well chosen coherent kick. However, next time round
through the pick-up, assuming all particles in the sample travel at the same
velocity, there is no longer any coherent signal left on the sample and yet
its transverse dimensions may still be large in spite of its being now perfectly "centered" in the pick-up. In order to reduce further the transverse
size of the beam, the sample must be "renewed", such that on average it has
again an off-centre charge distribution. These renewed samples can be created by mixing particles of the original sample (through different particle
velocities due to momentum spread) with others in the beam. If by the time
the original coherently corrected "sample" arrives again at the pick-up, it
contains a completely different set of particles with a correspondingly different offset, then the "mixing" is said to be perfect. This happens at the
higher frequencies where also the Schottky bands begin to overlap. Here the
gain of the system should be independent of frequency. In general, both
optimum gain and optimum cooling rate per Schottky line are inversely proportional to the density (dN/df) around that line, rather than to the total
number of particles N.

Longitudinal cooling, which reduces the momentum spread and hence the
revolution frequency spread in a beam, may be carried out in two ways. One
method, called "Palmer cooling"[16] uses a pick-up sensitive to horizontal
position placed at a point along the circumference of the machine where the
dispersion has been arranged to be large. The particle position hence
depends strongly upon momentum and the central position corresponds to the
momentum towards which all particles are driven by a longitudinal kicker
(either accelerating or decelerating particles). The kicker however needs to
be placed in a region of zero dispersion, for transverse heating to be

avoided. Otherwise, each sudden change in momentum would mean that a beta-tron oscillation is induced about the now-changed momentum orbit. Alternatively, the kicker may be split into two halves, placed in regions of identical dispersion and such that they are half a betatron oscillation apart: the transverse effects of the two kicker-halves will then cancel.

The second way to cool longitudinally, and used extensively in the AA, is the "Thorndahl" or "filter" method[17]. This uses a sum pick-up whose signal is subsequently filtered by means of shorted transmission lines.

Figure 9 illustrates that such filters cause a change in the sign of the phase at the middle of each Schottky band when the delay-length of the transmission line corresponds to exactly half the orbit length. The revolution frequencies for particles of different momenta being different, the particles are thus pushed from either side of the bands towards the centre. The momentum spread is therefore reduced.

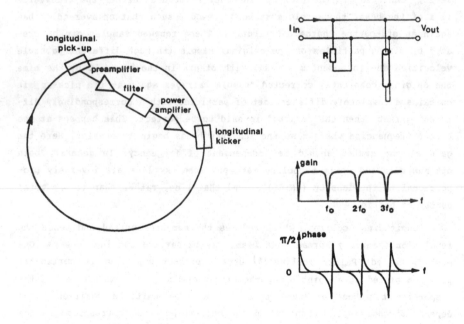

Fig. 9 - a) Cooling of the momentum spread using the 'Thorndahl'
(filter) method. b) Simple transmission line filter.
c) Amplitude and phase response vs. frequency.

The filter method is particularly advantageous for low beam intensity because the attenuation at the centre frequency can be attained after the preamplifier (the signal strength in the Palmer method is determined by the geometrical position of the beam in the pick-ups, hence before the preamplifier). The signal-to-noise ratio is therefore better and low beam intensities can be cooled faster. However, the Schottky bands must be well separated for practical filters to be usable; this means that they have to work with "bad mixing".

The design of pick-ups and kickers is a highly skilled activity calling upon the latest artistry of mechanical and microwave engineering. There are many structures[18] in use but the three dominant ones, all used in the AA, consist of quarter-wavelength coupling-loops or antennae, ferrite single-turn current-transformers for sum pick-ups and kickers and slot-coupled transmission lines. All three are illustrated in Fig. 10 and their practical uses are described in the next section.

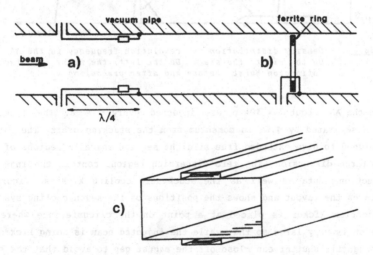

Fig. 10 - The three structures currently used for pick-ups and kickers. a) λ/4 directional coupler. b) Ferrite ring current transformer. c) Slotted transmission line.

3.   THE ANTIPROTON ACCUMULATOR (AA)[19]

In the following, and at the risk of somewhat annoying readers not intimately familiar with accelerator physics, we have to resort to an increased use of notions common in this field.

18

3.1 Accumulator Performance

Figure 11 illustrates how the particle density in the AA depends upon the revolution frequency (or momentum). The stack, on the right, consists of those particles already accumulated, and the low density injected beam is represented, before and after precooling, on the left.

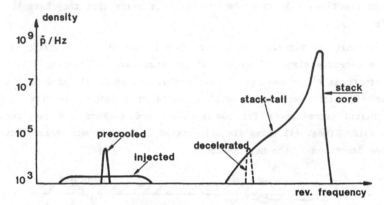

Fig. 11 - *Density distribution vs. revolution frequency in the AA. On the right, the stack. On the left, the newly injected antiproton burst, before and after precooling.*

In the AA, about 7×10^6 $\bar{p}$ are injected every 2.4 s. The injection orbit is separated by 4.6% in momentum from the stacking orbit. The lattice was designed to have extended free straight section space in regions of zero and of large dispersion. The zero dispersion regions contain the injection and ejection septa as well as the stochastic cooling kickers. Figure 12 illustrates the layout and shows the positions of the seven cooling systems. The injection kicker is placed at a point on the circumference where the dispersion is very large so that while the injected beam is being kicked, an electromagnetic shutter can close off the magnet gap to avoid that the stack also gets affected. The precooling systems, both pick-ups and kickers, are placed in similar dispersive regions and also have shutters.

Precooling, using the Thorndahl filter method, is the first to be applied. It cools the injected beam from its initial momentum spread of 1.5% down to 0.2% in 2 s. The precooling structures consist of many transformer-like ferrite rectangles (200 each for pick-ups and kickers) surrounding the injected beam[21]. One vertical ferrite side of each pick-up is formed by

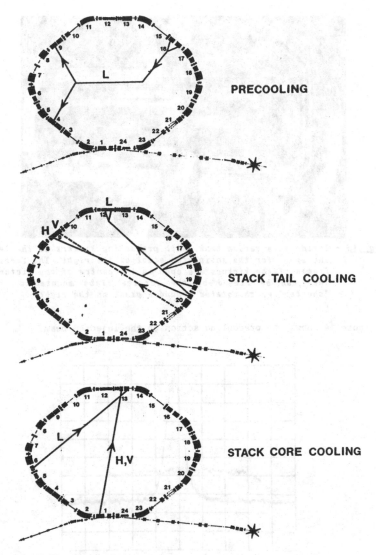

Fig. 12 - *The 7 cooling systems of the AA. For clarity, the 3 sets are shown separately, H: horizontal, V: vertical, L: longitudinal (momentum).*

a movable shutter which is opened when precooling is finished so that the beam can be moved into the tail of the stack. The precooling kickers are identical to the pick-ups except for the addition of water cooling (some kW of power in the band 150 to 500 MHz are used for the precooling process) and are shown in Fig. 13.

Fig. 13 - *Inside of a vacuum tank, with precooling kickers at the left and space for the antiproton stack at the right. The ferrite frames of the kickers are open in the centre of the picture. They can be closed by the ferrite slabs mounted on the "shutter" which rotates around a pivot on the right.*

Figure 14 shows the precooling action on the injected beam.

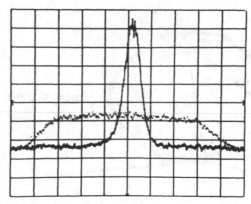

Fig. 14 - *Precooling in the AA of 6 × 10⁶ p̄ in 2 s. Longitudinal Schottky band at the 170th harmonic (314 MHz), before and after cooling.*

After 2 s the precooled beam is bunched by a radiofrequency system and decelerated towards the low-frequency tail of the stack (Fig. 11), where it is debunched and deposited. This process, including the opening and closing of the injection and the precooling shutters, takes a further 0.4 s. The injection orbit is now free to receive the next burst of antiprotons from the target. The particles deposited in the tail have to be cooled longitudinally well into the stack in the next 2.4 s to make room for the following set. This is done by means of the stack-tail stochastic cooling system using the 250 to 500 MHz band.

The particles have to be pushed up against the density gradient which is almost exponential and the gain should vary inversely with the density. The gain dependence on frequency is achieved by positioning coupling-loop pick-ups in regions of large dispersion. Two sets of pick-ups are used, each at different radial positions over the tail and each with its own preamplifier and gain adjustment. With this, fast cooling can be obtained at the lowest density of the tail where particles are deposited, while slower cooling occurs near the dense core because here particles remain for many hours. The kickers of the stack-tail system are placed in a zero-dispersion region to prevent them from exciting betatron oscillations. The core itself is also stochastically cooled both longitudinally and transversely by the Palmer method with another system using slotted-wave-guide pick-ups and kickers[23] working over a band reaching from 1 to 2 GHz at up to 20 W power. These are shown in the layout in Fig. 12. A general view of the AA is shown in Fig. 15 where some of the transmission lines transporting signals from pick-ups to kickers can be seen.

Accumulation follows the above procedure for typically a day, after which time some $2 \cdot 10^{11}$ $\bar{p}$ are found in the stack, cooled down to transverse emittances of about 2π mm.mrad from the initial 100π mm.mrad in both horizontal and vertical planes. The stack momentum spread is as narrow as 0.3%. This allows three bunches of antiprotons, in 2.4 s intervals, to be unstacked by the rf system, moved out to the injection orbit, and ejected with an ejection kicker (moved into position just before ejection is to occur) towards the PS machine. The latter accelerates each bunch from 3.58 GeV/c to 26 GeV/c and then ejects it to the SPS which is waiting, already filled with three proton bunches circulating in the opposite direction, for the antiproton bunches to arrive. Protons and antiprotons are then accelerated together to the storage energy of 315 GeV for each beam.

Fig. 15 – *View of the AA before it was buried under a concrete shield.*

3.2 Accumulator Limitations

The first and most fundamental limitation encountered in the accumulation of antiprotons is the stochastic cooling rate. For an ideal system
this can approximately be written in terms of the rate at which the root
mean square amplitude is reduced:

$$\frac{1}{\tau} = \frac{W}{2N} \quad (s^{-1}) \tag{15}$$

with W the bandwidth in Hz and N the number of particles.

The goal of the AA is to stack 10^{12} $\bar{p}$. A cooling band from 1 to 2 GHz
would thus give a cooling time of 30 min. In practice cooling times are of
the order of an hour even with the present stack intensities of a few 10^{11}
$\bar{p}$. However, cooling rate is but one of the limitations encountered in
practice.

An accumulator's performance may be characterised in terms of accumulation rate, stack core emittance and intensity-related instabilities[24].

3.2.1 <u>Accumulation rate limitations</u>. The average accumulation rate in the AA is about 6 × 10^9 $\bar{p}$/h. This is determined by the:

i)   intensity of the primary 26 GeV/c proton beam,

ii) configuration of target and magnetic horn,

iii) effective transverse acceptances of the accumulator ring,

iv) precooling efficiency,

v)   RF capture and deposit efficiency,

vi) stack-tail efficiency,

vii) stack-core loss-rate,

viii) unstacking and ejection efficiency,

ix) availability (or reliability).

The primary proton beam intensity depends upon the Booster and PS main ring performance. 10^{13} p per pulse every 2.4 s is easily achieved and a factor of 1.5 to 2 improvement is within reach. In most antiproton collider runs the standard target and horn configuration was used. This consisted of a 3 mm diameter copper target, 120 mm in length, encased in a graphite and aluminium alloy air cooled container. The magnetic horn is pulsed with 160 kA through its aluminium alloy inner sheath. Higher density targets and even pulsed current targets have been tested along with a collector lens consisting of a current-carrying lithium rod[25]. Improvements in antiproton yield were obtained well above the AA operational value of 6 × 10^{-7} $\bar{p}$/proton on target; more development work is however needed before the excellent reliability of the copper-target plus aluminium-horn configuration, with a working life in excess of one year, can be obtained.

The design acceptance of the AA is 100π mm.mrad in both horizontal and vertical planes. Much closed orbit corrections and shimming of quadrupoles resulted finally in values of A_H = 98π and A_V = 78π mm.mrad across the 6% momentum acceptance. However, there is still a depopulation of this phase space at the larger amplitudes caused by linear and nonlinear coupling. Partial cures are attained by choosing a tune for accumulation away from the diagonal on the Q_H, Q_V plot and adding a skew quadrupole and two sextupoles in the zero-dispersion regions of the ring. These latter additions improved accumulation by 10 to 20%.

Typical precooling efficiencies are around 75%, mainly limited by thermal noise from the ferrite material of the pick-ups.

The rf capture and deposit efficiency approaches 90% and corresponds closely to that expected from measurements of momentum spread after precooling.

The stack-tail system is probably the most difficult and consequently the least efficient, mainly because of the extreme demands on the gain profile. Although its efficiency can be as high as 90% at low core densities ($<5 \times 10^{10}$ $\bar{p}$) it drops to well below 50% with a stack of 5×10^{11} $\bar{p}$.

Another culprit for particle loss from the stack-tail is thought to be ions. Trapped in the stack-core, they excite, on high-order resonances, the larger amplitude particles before these have been sufficiently cooled transversely to the smaller amplitudes of the core.

The stack core loss rate, when the beam is left to coast with the cooling systems off, is normally dominated by single Coulomb and nuclear scattering on the residual gas nuclei. The stainless steel vacuum chambers of the accumulator are baked at up to 200°C resulting in a residual gas (90% H_2) pressure of 5×10^{-11} Torr. The lifetime is then about 3000 h. Even at the record stack intensity of 5.23×10^{11} $\bar{p}$ the loss rate due to residual gas scattering is only a few percent of the stacking rate. However, during the accumulation process the actual loss rate is substantially increased to an appreciable fraction of the stacking rate as the stack intensity approaches a few 10^{11} $\bar{p}$. There are many contributary effects, mostly having to do with the behaviour of the stack-core and its cooling systems - these are described below - but nevertheless, with core emittances below 4π mm.mrad, antiprotons can be unstacked and ejected with efficiencies as high as 95%.

The availability of the AA is high, as much as 95% of the scheduled time is achieved. If the accidental loss of stacks is accounted for as downtime by converting the number of antiprotons lost into stacking time, the availability is reduced to 85%. As a measure of this reliability it is worth mentioning that the longest continuously circulating antiproton stack was kept for 999 h, or 42 days, in autumn 1985.

3.2.2 <u>Limitations on the stack-core emittance</u>. The stack-core is designed to be cooled to less than 2π mm.mrad in both transverse planes, and to contain up to 6×10^{11} $\bar{p}$ inside a momentum spread of $\approx 0.3\%$.

Many factors are known to influence the core emittances:

i)   Common mode overlap.

ii)   Asymmetry in the stack-tail momentum kicker.

iii)   Residual dispersion at the location of the stack-tail kicker.

iv)   Noise in the stack-tail momentum cooling.

v)   Limitations in the core cooling system.

vi)   Coulomb scattering on the residual gas.

vii)   Intrabeam scattering.

viii) Ion-induced nonlinear resonances.

ix)   Ion-antiproton coherent instabilities.

x)   Charged, solid microparticles (dust).

xi)   rf overlap knock-out.

The first five effects are all present during stacking and are essentially due to hardware weaknesses which could all be cured to some extent by improved knowledge of the systems coupled with engineering ingenuity.

Coulomb scattering was discussed above and is not a problem with pressures below 5×10^{-11} Torr.

Intrabeam scattering is probably the most fundamental of all the limitations and results from Coulomb scattering between individual particles. This heats the beam transversely, drawing energy from the longitudinal momentum. It is a well known effect and good agreement has been obtained between theory and experiment[26].

Factors (viii) to (x) are peculiar to stacked antiproton beams.

Excessive transverse heating rates, well above what can be accounted for by intrabeam scattering, have been observed with high intensity antiproton stacks - this effect is attributed to heating by ion-driven nonlinear resonances. The ions are produced by interaction of the beam with the residual gas molecules and may be trapped in the negative potential of the beam if the ion-clearing system is inadequate. Transverse heating can be reduced or avoided by careful choice of tune and reduction of the chromaticity so as to avoid nonlinear resonance lines at least up to 10th order. There are signs that even resonances of up to 34th order can be driven by ions when stack intensities reach 4×10^{11} $\bar{p}$ [24].

The trapped ions can also produce coherent instabilities[27]. These were observed to limit the minimum to which the vertical emittance can be cooled for intensities above 1.5×10^{11} $\bar{p}$. Attempts to damp these instabilities with active feedback systems are partially successful but better clearing of ions out of the trapping potentials of the beam is best.

A most interesting fast emittance-growth effect, which is possibly unique to the AA, is that provoked by the charging-up of solid (dust) micro-particles[28]. The dust can be charged up by ionisation and thermionic emission caused by the beam, eventually producing very intense electric fields. The dust particles then can get trapped in the potential well of the beam and their electric fields scatter the antiprotons. The emittance can quickly double in times of the order of a few minutes. The dust particles can usually be eliminated by slowly cycling the clearing electrode fields so that the dust particles move out of their traps and are picked up on a clearing electrode - a form of vacuum cleaning unimagined by Mr. Hoover! A well chosen distribution of clearing electrodes to eliminate all possible traps is the best solution to this limitation.

Limitation (xi), "rf overlap knock-out", was first observed at the ISR[29]. It is caused in the process of unstacking by higher harmonics of the bunch spectrum overlapping the betatron side-band frequencies of the stack-core. It is cured by keeping the rf voltage of the unstacking bucket low, while its frequency is changing, so as to increase the bunch length and thereby reduce the high harmonic content of its spectrum.

3.2.3 <u>Intensity limitations due to instabilities</u>. Some of the limitations linked to beam intensity have already been touched upon above. Further limitations are:

(1) Longitudinal instabilities due to

    i) Stack-tail momentum cooling system interacting with the stack core.

    ii) Precooling system interacting with the stack core.

    iii) Radiofrequency cavity and other longitudinal impedances.

2) Transverse coherent instabilities due to

    i) Resistive wall effects.

    ii) Ion-antiproton effects.

3) Transverse incoherent instabilities due to

    i) Ion-induced nonlinear resonances.

    ii) Trapping of solid microparticles.

When the core is sufficiently intense, in spite of all attempts to reduce coupling to the sensitive pick-ups of the precooling and stack-tail systems, these latter systems heat and widen the core momentum distribution. The heating can be reduced by lowering the gain of the system but this affects the stacking rate and is undesirable. Care with shielding and ground connections of coaxial lines between the structures and the vacuum tanks and feedthroughs has reduced these effects.

The longitudinal coupling impedances of all structures mounted in the AA ring were kept to a minimum by means of suitable screens or damping resistors. The rf cavity was equipped with an active feedback system so that its impedance as seen by the beam was kept below 200 Ω. The overall longitudinal coupling impedance was measured to be 500 Ω, low enough to avoid longitudinal instabilities, even with the most intense stacks envisaged, 10^{12} $\bar{p}$. However, stacks have been lost due to this effect when the rf cavity feedback system was inadvertently switched off whilst the cavity was tuned to the revolution frequency.

The transverse resistive wall instabilities are kept at bay by means of a bidirectional transverse feedback damping system[30] and have never been a problem.

The other instabilities have already been discussed.

Most of the limitations to the performance of the AA are now well understood and even though they cannot all be eliminated, the accumulation rate and stack intensity have reached levels which allow colliding beams with respectable luminosities to be achieved in the SPS collider.

4. THE COLLIDER

The Super Proton Synchrotron (SPS) was commissioned in the summer of 1976. At the time of its conception, only the most visionary could imagine that one day it would serve as a colliding beam device. It was designed as a fixed-target machine, accelerating protons, provided at 10 GeV/c by its injector, the CERN Proton Synchrotron (PS), up to a maximum energy of 450 GeV. The particles are spilled out of the machine over a few seconds and transported into two experimental areas (Fig. 16). The acceleration cycle repeats every 14.4 s.

The SPS is a "separated-function" synchrotron, i.e. the bending of the particles in a circular trajectory is made by dipole magnets and the focusing is made independently with quadrupole magnets. It consists of 108 FODO periods each 64 m long (a FODO period consists of a horizontally focusing

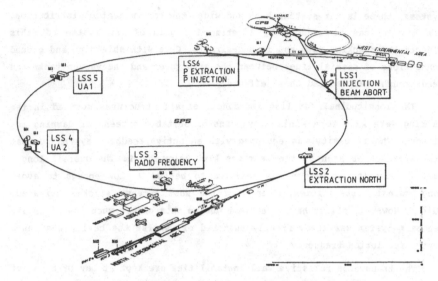

Fig. 16 - *CERN's accelerator complex, with the SPS
prominently in the foreground.*

quadrupole followed by 4 dipole bending magnets, a horizontally defocusing
quadrupole and another 4 dipole magnets), giving a total machine circumference of 6.9 km. In order to accommodate major systems (injection, extraction, acceleration and beam abort), 6 long straight sections (LSS) were
created by leaving out the bending magnets over two lattice periods at each
of six azimuthally symmetric locations around the ring. LSS1 was allocated
the injection system, extraction to the North Experimental Area is in LSS2,
acceleration in LSS3, beam abort in LSS4. LSS5 was reserved for "future
development", the only veiled indication in the design report[31] of a
possible future use as a collider. LSS6 was used for extraction to the West
experimental area.

4.1 Modification of the SPS for Colliding Beam Operation

The requirements of a storage ring, in which beams must circulate for
many hours, are much more demanding than those of a pulsed synchrotron. The
first obvious improvement required was to the vacuum system. The design
vacuum of 2×10^{-7} Torr, perfectly adequate for pulsed acceleration, needed
improving by almost three orders of magnitude for transverse emittance

blow-up of the beams over many hours of storage to be negligible compared with other effects.

It was decided that two major detectors would be built: UA1 (for Underground Area 1) located in LSS5 and UA2, for which space was made by moving the beam abort system from LSS4 into the injection straight section LSS1. At the location of these detectors the SPS tunnel had to be enlarged into huge experimental halls. In addition, the machine lattice was substantially modified in order to provide "low-beta insertions"[32] at the experimental collision points. In these insertions additional quadrupoles reduce the cross-sectional areas of the beams at the collision points by about a factor of 70 compared to their values in the regular lattice.

The design injection momentum for the SPS as an accelerator was 10 GeV/c, later increased to 14 GeV/c. Around 22 GeV/c, "transition" must be crossed. There, the revolution frequency of the particles is independent of their momentum, an intrinsically unstable situation. Transition crossing under normal conditions poses no great problems. However, the bunch intensity for the collider is at least a factor of 20 higher than for the "fixed-target" machine. Early experiments[33] with such dense single bunches revealed two classic instabilities in the region of transition, the "negative mass" instability in the longitudinal plane and the "head-tail" instability in the transverse plane. These instabilities conspired to limit the bunch intensity to about a factor of five less than required for the collider. The solution to this problem was to inject both protons and antiprotons into the SPS at 26 GeV/c, above the transition momentum. This led to the use of the PS as an intermediate accelerator between the accumulator and the SPS. For the protons this required an upgrading of the injection line TT10. For the antiprotons a completely new beam line (TT70) was built.

4.2 Operation of the Collider Complex

In order to achieve colliding beams in the SPS under the best possible conditions, all three machines in the injector chain must be carefully optimised. At the present time (1987) the SPS operates with three (soon to be increased to six) bunches of protons and antiprotons per beam. The intensity of each of the proton bunches being about a factor of 20 higher than normal for fixed-target physics, the production of such intense bunches ($1.5 \cdot 10^{10}$ p/bunch) in the PS and their subsequent capture and acceleration in the SPS is not trivial.

Normally, bunches on the 26 GeV/c ejection flat-top in the PS are about 16 ns long, impossible to capture in the 5 ns buckets provided by the SPS 200 MHz radiofrequency system. They are thus compressed by using a technique of bunch rotation[34] just before extraction from the PS. In this way, dense single bunches of up to 2×10^{11} p/bunch with a longitudinal emittance of 0.5 eV.s (a length of 4 ns and a momentum spread of $\pm 3 \times 10^{-3}$) are produced and transported to the SPS.

In order to economise on precious antiprotons, adjustment of the anti-proton injection line and fine tuning of the two machines is all done with protons. A single proton bunch circulating at 26 GeV/c in the SPS can be extracted using the antiproton injection equipment and transported back to the PS along the antiproton line TT70 (Fig. 1). In this way the whole system is adjusted so that the antiprotons will finally follow exactly the trajectory of the protons in the reverse direction. In a similar manner the transfer between AA and PS is adjusted by sending protons at 3.58 GeV/c from the PS backwards along the antiproton transfer channel to the AA where they are injected onto the antiproton extraction orbit.

Finally, when the whole system is ready, three proton bunches are injected into the SPS at 2.4 s intervals and equally spaced around its circumference. Three dense antiproton bunches are then extracted from the AA stack, also at 2.4 s intervals, using a precisely defined radiofrequency program to capture antiprotons within a longitudinal emittance of 0.5 eV.s, the same as that of the protons. They are then transported to the PS and accelerated to 26 GeV/c, circulating in the opposite sense to the proton direction. After the bunch compression manipulation, identical to that of the protons, the antiproton bunches are ejected from the PS and transported to the SPS where they are injected at precisely timed instances, so that collisions will occur at the centres of the detectors.

After this, the magnetic field rises and the two beams are accelerated to 315 GeV. During the first second of the high-energy flat-top the low-beta insertions squeeze the beam to its final small size. This involves simultaneously modifying the currents in 30 independently powered quadrupoles in such a way that the machine optics remains stable throughout the squeezing process. The SPS then passes into storage for 15 to 20 h of physics data-taking whilst the PS/AA complex resumes accumulation in preparation for the next SPS fill.

4.3 Collider Performance

The most frequently used figure of merit for a storage ring is the luminosity L, defined as the event rate per unit of cross section and most often quoted in units of $cm^{-2} \cdot s^{-1}$. The luminosity for head-on collisions between bunched beams is given by

$$L = \frac{2MN_p \ N_{\bar{p}} \ f \ \beta \ \gamma}{\pi [\beta_H^* \beta_V^* (E_{Hp} + E_{H\bar{p}})(E_{Vp} + E_{V\bar{p}})]^{1/2}} \tag{16}$$

where M is the number of bunches per beam, N_p and $N_{\bar{p}}$ the numbers of parti-cles per bunch, f the revolution frequency (43.4 kHz), $\beta_{H,V}^*$ the amplitude-or beta- function values at the collision point and $E_{H,V}$ the normalised emittances defined in terms of beam size as $E/\pi = 4\beta\gamma\sigma^2/\beta^*$ rad.m for a Gaussian bunch.

A better measure of the productivity of a collider is the integrated luminosity L_I, which determines the number of interesting events observable by a physics experiment in a given time interval. It depends both on the peak luminosity L_0 and on the luminosity lifetime T_L. For an exponential luminosity decay the integrated luminosity for a run of duration T is

$$L_I = L_0 T_L (1 - e^{-T/T_L}) . \tag{17}$$

For the SPS collider, the integrated luminosity is normally quoted in "inverse nanobarns" (1 nb^{-1} = 10^{33} cm^{-2}).

The first physics run took place at the end of 1981, when 0.2 nb^{-1} of integrated luminosity was obtained over the whole period of more than a month. In successive runs over the years 1982 to 1986 the machine perform-ance was improved steadily. Table 1 summarises the evolution of the main collider parameters between 1982 and 1985. The 1986 run was prematurely terminated because of a major fault with the UA1 detector.

Until 1983 the energy was limited to 273 GeV due to heating of the magnet coils. The addition of further water cooling allowed the energy to be pushed to 315 GeV in 1984.

The β^* values at the experimental insertions have been progressively squeezed from their design values of β_H^* = 2 m, β_V^* = 1 m to their present values of 1.0 m $\ast$ 0.5 m. The highest initial luminosity, at the beginning of

a store, achieved to date is 3.9×10^{29} cm-2·s-1, with routine values generally in excess of 2×10^{29} cm-2·s-1. The luminosity lifetime in normal operation is more than 24 h, averaged over a long store.

Table 1

SPS Collider operation, 1982-1985

Operational features	1982	1983	1984	1985
Beam Energy (GeV)	273	273	315	315
β_H^* (m)	1.5	1.3	1.0	1.0
β_V^* (m)	0.75	0.65	0.5	0.5
Integrated luminosity (nb-1)				
average per store	0.5	2.1	5.3	8.2
average per day	0.4	1.8	5.1	5.8
per year	28	153	395	655
Luminosity (10^{29} cm-2·s-1)				
peak	0.5	1.7	3.5	3.9
average per store	0.1	0.5	1.0	1.3
Hours scheduled	1750	2064	2136	2688
Hours realised	748	889	1065	1358
Number of stores	56	72	77	80
Average store duration (h)	13	12	15	17
% stores terminated by faults	41	40	32	18

Although the collider is limited to 630 GeV centre-of-mass energy in
its high-luminosity mode of operation, the machine has been operated for a
short period of time at the maximum achievable energy of 910 GeV c.m. in the
so-called "pulsed storage" mode[35]. Protons and antiprotons are injected
at 26 GeV/c as usual but instead of going into d.c. storage at 315 GeV beam
energy, the machine was cycled between 455 GeV and 100 GeV in order to limit
the mean power dissipation to an acceptable value.The low-beta insertions
could not be used due to limitations in quadrupole gradient, so the initial
luminosity was only about 3×10^{26} cm-2·s-1. In addition, the luminosity
lifetime was much shorter than in d.c. storage, of the order of a few hours,
due to unavoidable small tune variations during ramping and to radiofrequency noise. Nevertheless, in total about half a million events at 910 GeV
c.m. were recorded by the two experiments UA1 and UA5.

4.4 Performance Limitations

The luminosity is governed by the number of bunches per beam, the number of particles in each bunch and their emittance, and by the horizontal and vertical β^* values at the interaction points. Although the initial transverse emittances and antiproton bunch intensities are defined by the injector chain, the proton bunch intensity is limited in the SPS itself by a fast longitudinal instability, the "microwave instability"[36].

The luminosity lifetime is mainly limited by multiple Coulomb scattering between particles in the same bunch (intrabeam scattering) and by the effect of the global electromagnetic field of the particles in the opposing beam (beam-beam interaction). These phenomena are discussed briefly in the following chapters, mainly in terms of their physical manifestation. For a detailed description, the reader is referred to the literature[37,38].

4.4.1 Microwave instability.
Above a threshold around 9×10^{10} p/bunch, a fast longitudinal emittance blow-up occurs immediately after injection. Certain parts of the SPS vacuum chamber behave like cavities and are excited by the beam. The electromagnetic fields generated in this way act back on the beam, causing self-bunching at very high frequency and a blow-up of the momentum spread.

The effect can easily be observed by measuring the bunch length as a function of intensity, keeping all parameters of the injected beam constant and varying the intensity by vertically scraping the beam in the SPS itself.

Figure 17 shows the result of such a measurement. The bunch length at first increases gradually with intensity and then much more pronouncedly above the threshold. The bunch lengthening below the threshold is due to potential well distortion, a reduction in the effective focusing voltage due to the vacuum chamber impedance. It does not lead to longitudinal emittance growth. Above the threshold, also emittance growth occurs which leads to beam loss later in the acceleration cycle when the available radiofrequency accelerating voltage is no longer sufficient.

4.4.2 Intrabeam scattering.
The first observation of intrabeam scattering was made on the electron storage ring ADA in 1962 [39]. The lifetime of a stored beam was found to depend on the beam intensity and on the energy. The phenomenon was explained as being due to Coulomb collisions between particles resulting in a change in energy sufficiently large for the particles to

be lost from the machine, the so-called Touschek effect. In 1965 it was pointed out that particles could also undergo much less violent multiple collisions resulting in a dilution of the beam density and consequently even reducing the loss rate.

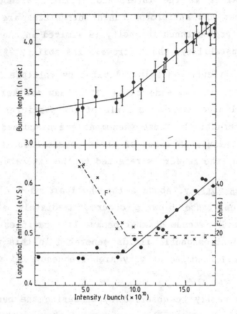

<u>*Fig. 17*</u> *- The microwave instability in the SPS: both the bunch length and the longitudinal emittance increase above a threshold around 9 × 10¹⁰ p/bunch.*

In 1974, A. Piwinski[37] published a theory of intrabeam scattering which predicted the rate of emittance growth in all three phase planes. Essentially the same results were derived in quite a different way in 1982 [40].

The first evidence of intrabeam scattering in proton beams was obtained in the CERN Intersecting Storage Rings in 1975 [41]. It was quickly realised that this effect would be a major limitation to the performance of low energy accumulators like the AA where beam cooling increases the phase-space density by many orders of magnitude until the beam heating due to the intrabeam scattering balances the effect of the cooling. The effect has indeed been observed in the AA and the predictions of the theory verified to a reasonable precision.

Initially, it was thought that intrabeam scattering was only important in low energy machines. This turns out not to be the case for two reasons. Firstly, as the energy increases, the 6-dimensional phase-space density increases due to the adiabatic damping of emittance. In addition, the beam becomes very "cold" in the longitudinal phase plane due to relativistic contraction, resulting in a strong unbalance between the densities in each of the phase-space planes, thus enhancing the rate of intrabeam scattering. It has now been shown that this effect is the most limiting factor for the luminosity lifetime in the SPS due to growth of both horizontal and longitudinal beam dimensions.

The predictions of the Piwinski theory have been verified in some detail also in the SPS[42]. The longitudinal emittance growth of a dense proton bunch can be clearly seen in Fig. 18a which shows the longitudinal bunch profile measured at intervals of 15 min throughout a long store. The bunch soon blows up to completely fill the bucket. In contrast, Fig. 18b shows the same measurements made on an antiproton bunch with more than an order of magnitude less intensity.

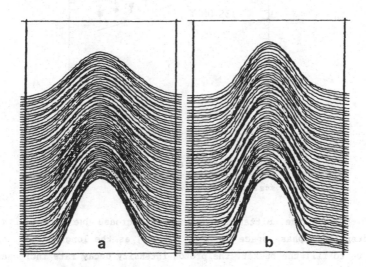

Fig. 18 - Longitudinal bunch profile of a strong proton bunch (a) and a weak antiproton bunch (b); vertical scale adjusted for same graphical height, sampled at 15 min intervals during a store. The vertical lines correspond to the limit of the bucket (5 ns length).

36

Figure 19 shows the horizontal emittance growth rate, measured with a rapid wire scanner, as it varies over a long store. The solid line represents the theoretical growth rate computed from instantaneous measured values of emittance and intensity. The reduction in growth rate as the beam blows up can clearly be seen. Agreement with the experimental points is even better when the small constant emittance growth $\dot{\varepsilon}$ due to multiple scattering on the residual gas is added. At the beginning of the store the emittance growth time is as short as 18 hours, by far the most important factor determining the luminosity lifetime.

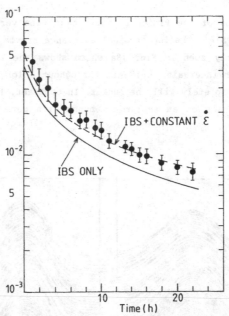

Fig. 19 - *Measured growth rate of transverse emittance and theoretical growth rate for intrabeam scattering (IBS).*

With time, the emittance growth rates decrease due to dilution of the six-dimensional phase space density. However, as the longitudinal emittance blows up to fill the bucket, the proton intensity decay rate increases. Particles near the edges of the bucket are more easily spilled out. The intensity lifetime eventually reaches an equilibrium value of around 50 h, leading to an overall luminosity lifetime of around 30 h. Detailed measurements have shown that for a given longitudinal emittance the proton beam lifetime

is intensity dependent. This effect is almost certainly due to intrabeam
scattering of particles across the bucket separatrix so that they are lost
from the bunch. Intrabeam scattering is therefore the major contributor to
the luminosity decay throughout the duration of a store.

4.4.3 The beam-beam interaction. Intrabeam scattering arises through Coulomb collisions between individual particles in the same bunch. In addition,
each particle experiences the global electromagnetic field of all the other
particles in the same bunch and in the opposing bunch as they collide.

For particles in the same bunch the electric and magnetic effects tend
to cancel: the residual force is proportional to $(1 - \beta^2) = 1/\gamma^2$. However,
at the injection momentum (26 GeV/c) it is not negligible, resulting in an
amplitude-dependent tune spread which must be taken into consideration in
order to avoid crossing low order magnet imperfection resonances. At storage
energy (315 GeV) this effect is no longer important.

For particles in the opposing beam the electric and magnetic forces of
a bunch add, so that a particle gets a transverse kick proportional to
$(1 + \beta^2)$, essentially independent of energy, as it passes through the other
bunch. This kick is highly nonlinear and gives rise to two disturbing
effects.

Firstly, it changes the tune of individual particles according to their
amplitude, resulting in a tune spread. Secondly, due to the periodic nature
of the kicks, nonlinear resonances are excited for certain combinations of
tune values satisfying a relationship of the form

$$n_H Q_H + n_V Q_V = p \qquad (27)$$

where n_H, n_V and p are integers, $n_H + n_V = N$ is the order of the resonance
and $Q_{H,V}$ are the horizontal and vertical betatron tunes.

Nonlinear resonances due to magnetic field imperfections have been the
plague of particle accelerators ever since their conception, but they are
generally restricted to low order, being driven by errors in quadrupoles (N
= 2), sextupoles (N = 3) or octupoles (N = 4), still leaving adequate space
in the tune diagram to accelerate or store beams. It is a characteristic of
the beam-beam force that high order resonances are excited orders of magnitude more strongly than by "naturally" occurring field imperfections. One
may take as a measure of the resonance excitation the linear beam-beam tune

shift ξ which, for a round bunch of N particles with transverse normalised emittance E, is given by

$$\xi = N r_p / \pi E \qquad (28)$$

where r_p is the classical proton radius.

Notice that ξ is independent of energy and of the β^* values at the crossing point. For the SPS the linear tune shift is normally between 3 and 4×10^{-3} per $\bar{p}p$ crossing.

The catastrophic effect of high-order nonlinear resonances on the weak antiproton bunches is now well documented[43]. An example can be seen in Fig. 20 which shows the intensity decay with three strong proton bunches and a single weak antiproton bunch as the machine tune is varied. Figure 20a shows a chart recorder output of the intensity of one of the proton bunches and of the single antiproton bunch on a very sensitive scale. Figure 20b shows the tune diagram between 3rd and 4th order resonances where resonances of order 10, 7 and 11 are indicated. The intensity decay was measured at different tunes indicated by the lines marked 1, 2, etc. The meaning of these lines is the following: the proton bunch (with which the tune is measured) suffers negligible spread from the weak antiproton bunch and sits at the lower point of these lines. The smallest amplitude antiprotons are shifted upwards in both planes, corresponding to the upper point in each line. Larger amplitude antiprotons occupy the space between these two points.

The decay rate of the antiproton intensity is hence extremely sensitive to the tune. As the working point is moved upwards, large amplitude antiprotons straddle the 10th order resonances, resulting in a considerable increase in their decay rate. On the other hand the proton bunch lifetime is practically independent of the tune because the antiproton bunch is too weak to cause resonant excitation. The practical consequence of all this is that the working space is restricted to a very small area between 3rd and 10th order resonances corresponding to the line 1 in Fig. 20b. The total tune spread 2Mξ is of the order of 0.02 for M = 3 bunches per beam and this can just be accommodated. However, if the number of bunches is increased to 6 per beam it will no longer be possible to stay clear of these resonances and a very poor lifetime will result.

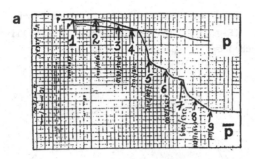

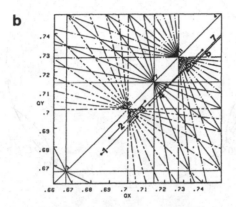

Fig. 20 - *The beam-beam effect in the SPS. a) Intensity decay
(zero suppressed) of a strong proton bunch and a
colliding weak antiproton bunch as the working point
is varied. b) Tune diagram with the working points
at which the intensity decay was measured.*

It is obviously interesting to work with the highest possible beam-beam
tune shift per crossing but with the smallest possible number of crossing
points in order to reduce the total tune spread. With 6 bunches per beam
there are 12 collision points (of which only 2 are of interest for the
physics experiments). Initial tests have shown that the beam lifetime is un-
acceptably low under these conditions.

In order to still obtain a lifetime sufficiently long for physics data
taking with 6 bunches per beam, a scheme has been implemented to separate
the beams horizontally at all collision points except at the two experimental insertions and the one inbetween[44]. Two pairs of electrostatic deflectors produce a global orbit distortion in opposite directions for protons

and antiprotons in such a way that the beams miss one another in 9 of the 10 unwanted crossings (Fig. 21). The improvement in beam lifetime, when the separation is switched on, is spectacular (Fig. 22).

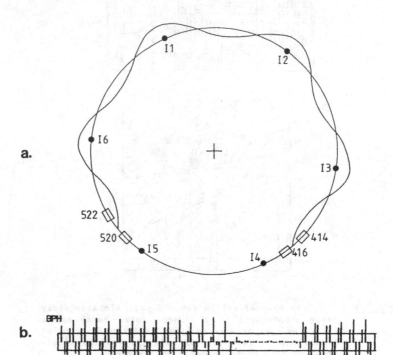

Fig. 21 - a) The bunch separation scheme using electrostatic deflectors. The deflections for protons and antiprotons are in opposite directions.
b) Measured closed orbit distortion when bunch separation is applied.

5. THE FUTURE OF p̄p COLLIDING BEAMS AT CERN

In the light of the success of the p̄p project, an improvement programme was approved in January 1984. Its eventual goal is an order of magnitude increase in the antiproton accumulation rate with a corresponding increase in peak luminosity in the collider, up to around 4×10^{30} cm$^{-2} \cdot$s^{-1}.

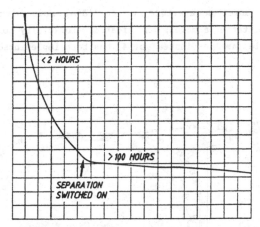

Fig. 22 - Chart recorder output showing the intensity decay
(zero suppressed) of a single antiproton bunch in the
presence of 6 strong proton bunches. Switching the separation scheme on has a pronounced beneficial effect.

The key element is a new ring, the Antiproton Collector (AC) inbetween
the production target and the AA, collecting 10 times as many antiprotons
as the AA did and thus increasing the antiproton accumulation rate by an
order of magnitude. Figure 23 shows the layout of the new facility. The ACOL
project involves also the redesign and reconstruction of the target area and
of the old AA to permit the latter to digest the tenfold increase in antiproton flux. The ACOL project is described in detail in a design study[19]
with updates in many publications.

The increased antiproton flux will allow the number of bunches transferred to the SPS to be increased from 3 to 6 and a bunch intensity of up to
10^{11} $\bar{p}$/bunch. In order to accommodate this in the collider some modifications to the SPS are necessary.

Firstly, as already mentioned, beam separation using electrostatic deflectors will be necessary in order to reduce the effect of the beam-beam
interaction. Secondly, production of 10^{11} $\bar{p}$ in a single bunch in the AA
requires twice the longitudinal emittance. In order to capture this in the
SPS, a new radiofrequency system operating at half the frequency (100 MHz)
of the present radiofrequency system is installed. Finally, in order to

42

provide enough momentum aperture in colliding mode with the increased
longitudinal emittance, a more sophisticated chromaticity correction scheme,
using 5 sextupole families instead of 4, has been implemented.

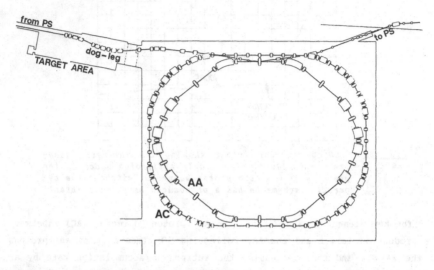

Fig. 23 - *In the course of the ACOL project, a new Antiproton Collector
(AC, outer ring) was built around the existing, but modified,
Antiproton Accumulator (AA). The target area (upper left) was
profoundly reconstructed and a "dog-leg" added to limit the
entry of secondary particles into the hall.*

Running-in of the new Antiproton Accumulator Complex (AAC) together
with the upgraded SPS is scheduled for the last quarter of 1987 and there is
every hope that the peak luminosity will be pushed past the 10^{30} cm$^{-2} \cdot$s^{-1}
mark during 1988.

ACKNOWLEDGEMENTS

We are grateful to S. van der Meer for his encouragement. From
J. Gareyte and L. DiLella we received useful comments and N. Walker helped
us with the references and the editing. Last, but not least, our thanks are
due to Mrs. A. Molat-Berbiers for her patient help with typing and layout.

REFERENCES

1) Budker, G.I., Proceedings, Symposium International sur les Anneaux de Collision, Saclay (1966).

2) Johnsen, K., CERN 84-15 (1984).

3) Budker, G.I., Derbenev, Y.S., Dikansky, N.S., Kudelainen, V.I., Meshkov, I.N., Parkhomchuk, V.V., Pestrikov, D.V., Sukhina, B.N., Skrinsky, A.N., IEEE Trans. Nucl. Sci., NS-22, 2093 (1975).

4) Bramham, P., Carron, G., Hereward, H.G., Hübner, K., Schnell, W. and Thorndahl, L., Nucl. Instrum. and Methods, 125, 201 (1975).

5) Antiprotons for Colliding Beam Facilities, CERN Accelerator School, Geneva 1983, Proceedings: CERN 84-15 (1985).

6) Allaby, J.V., CERN 84-15 (1984).

7) Jones, E., CERN 84-15 (1984).

8) Eichten, T., Hardt, D., Pattison, J.B.M., Venus, W., Wachsmuth, H.W., Worz, O., Jones, T.W., Aubert, B., Chounet, L.M., Heusse, P. and Franzinetti, C., Nucl. Phys. B-44, 333 (1972).

9) Johnson, C.D., CERN AA Long Term Note 26 (1983).

10) van der Meer, S., CERN 62-16 (1962).

11) Bellone, R., del Torre, G., Ross, M. and Sievers, P., CERN/SPS/80-9/ABT (1980).

12) Sacherer, F., CERN/ISR-TH/78-11 (1978).

13) Möhl, D., Petrucci, G., Thorndahl, L. and van der Meer, S., Phys. Rep. 58(2) 73-119 (1980).

14) Möhl, D., CERN 84-15 (1984).

15) Hereward, H.G., CERN 65-20 (1965).

16) Palmer, R.B., private communication (1975).

17) Carron, G. and Thorndahl, L., CERN/ISR-RF/78-12 (1978).

18) Taylor, C.S., CERN 84-15 (1984).

19) Design Study of a Proton-Antiproton Colliding Facility, ed. E.J.N. Wilson, CERN/PS/AA/78-3 (1978).

20) Autin, B., CERN 84-15 (1984).

21) Fiander, D.C., Milner, S., Pearce, P. and Poncet, A., CERN/PS/84-23 (1984).

22) van der Meer, S., CERN/PS/AA/78-22 (1978).

23) Faltin, L., Nucl. Instrum. and Methods 148, 449 (1978).

24) Pedersen, F., Carron, G., Chohan, V., Eaton, T.W., Johnson, C.D., Jones, E., Koziol, H., Martini, M., Maury, S., Metzger, C., Poncet, A., Rinolfi, L., Sherwood, T.R., Taylor, C.S., Thorndahl, L., van der Meer, S. and Wilson, E.J.N.W., Particle Accelerator Conf., Washington DC, 1987. Proceedings to be published in IEEE Trans. Nucl. Sci.

25) Eaton, T.W., Hancock, S., Johnson, C.D., Jones, E., Maury, S., Milner, S., Schnuriger, J.C., Sherwood, T.R., IEEE Trans. Nucl. Sci., Vol. NS-32, No. 5, 3060 (1985).

26) Martini, M., CERN/PS/84-9 (AA) (1984).

27) Jones, E., Pedersen, F., Poncet, A., van der Meer, S., Wilson, E.J.N., IEEE Trans. Nucl. Sci., Vol. NS-32, 2218 (1985).

28) Pedersen, F., Particle Accelerator Conf., Washington DC, 1987. Proceedings to be published in IEEE Trans. Nucl. Sci.

29) Gourber, J.P., Hereward, H.G., and Myers, S., IEEE Trans. Nucl. Sci., Vol. NS-24, 1405 (1979).

30) Pedersen, F., Pirkl, W., Schindl, K., IEEE Trans. Nucl. Sci., Vol. NS-30, 1343 (1983).

31) The 300 GeV Programme, CERN/1050 (1972).

32) Faugeras, P.E., Faugler, A., Hilaire, A., Warmen, A., Proceedings of the 12th Int. Conf. on High-Energy Accelerators, FNAL, Batavia, Illinois, 232 (1983).

33) Boussard, D., Evans, L., Gareyte, J., Linnecar, T., Mills, W., Wilson, E.J.N., IEEE Trans. Nucl. Sci., NS-26, 3484 (1979).

34) Garoby, R., CERN PS/LR/Note 79-16 (1979).

35) Lauckner, R., IEEE Trans. Nucl. Sci., NS-32, 1653 (1985).

36) Boussard, D., CERN-Lab.II-RF/Int/75-2 (1975).

37) Piwinski, A., Proceedings of the 9th Int. Conf. on High-Energy Accelerators, SLAC, Stanford, 405 (1974).

38) Nonlinear Dynamics Aspects of Particle Accelerators, Springer-Verlag, Berlin, eds. Jowett, J.M., Month, M., Turner, S., (1985).

39) Bernardini, C., Corazza, G.F., Di Giugno, G., Ghigo, G., Haissinski, J., Marin, P., Querzoli, R., Touschek, B., Phys. Rev. Letters, 10, 407 (1963).

40) Bjorken, J.D., Mtingwe, S.K., Particle Accelerators, 13, 115 (1983).

41) Hübner, K., IEEE Trans. Nucl. Sci., NS-22, 1416 (1975).

42) Evans, L., Proceedings of the 12th Int. Conf. on High-Energy Accelerators, FNAL, Batavia, Illinois, 229 (1983).

43) Evans, L., CERN 84-15 (1984).

44) Evans, L., Faugier, A., Schmidt, R., IEEE Trans. Nucl. Sci., NS-32, 2209, (1985.

ELASTIC SCATTERING AND TOTAL CROSS-SECTION

A. Martin

CERN, Geneva, Switzerland

G. Matthiae

Physics Department, Second University of Rome,

Rome, Italy

This article is dedicated to the memory of Rodney L. Cool who contributed so much to this field.

1. INTRODUCTION

Proton-proton scattering is probably the oldest collision experiment at accelerators because it was thought to be a way to study nuclear forces. Indeed, in the range of energies up to 300 MeV the description of nucleon-nucleon scattering in term of a potential derived from field theory and dispersion theory, with the addition of a few phenomenological ingredients (as, for example, the Paris potential [1]), has met considerable success. Low-enery proton-antiproton scattering is still being studied in the same range of energies but in this case the short distance part of the potential is ambiguous and there are various candidates with various degrees of success [2]. The very low-energy region is now being studied very carefully by using LEAR, the low-energy antiproton ring of CERN, with which it is possible to go down to laboratory energies as low as 5 MeV and measure scattering lengths. As we shall see later, all this information is relevant even for high-energy scattering because dispersion relations involve all energies, from very low to very high.

In the high-energy region, above 1 GeV, the description of the nucleon-nucleon interaction in terms of a potential certainly breaks down and one must find something else. Up to 1960, a naive and simple description of high-energy nucleon-nucleon scattering was based on the belief that the nucleon has a fixed shape and a fixed size. This explained the apparently constant value of the total cross-section around 40 mb and the fact that the slope of the differential cross-section, b, defined as d[log(dσ/dt)]/dt, is approximately constant with s, the square of the centre-of-mass energy, and t, the square of the momentum transfer. However, this simple minded picture was destroyed by V.N. Gribov [3] who showed that the assumption that the scattering amplitude F(s,t) tends to sf(t) at high-energy is inconsistent with analyticity and unitarity. Specifically, if one continues sf(t) to positive, hence unphysical values of t, f(t) should satisfy the requirements due to analytic continuation of elastic unitarity in the t channel, and this is not the case. The way out is that either the cross-section decreases slowly to zero with increasing energy, or the diffraction peak changes slope and shrinks as the energy increases. It is the second possibility that Nature has chosen, and at the 1962 High-Energy Conference in Geneva it was announced that the diffraction peak was indeed shrinking, i.e. that the slope was increasing with energy in the domain up to 28 GeV in the laboratory system [4]. This was explained by assuming that the scattering amplitude was dominated by a Regge pole, called Pomeron, which behaves like $f(t)\ s^{\alpha(t)}$ with $\alpha(0) = 1$, $\alpha'(0) > 0$. Then the total cross-section is proportional to Im f(0) and the slope at t = 0 is dominated by $2\alpha'(0)\log s$, thus eliminating Gribov's objection. However, other objections arise : besides Regge poles there are now complicated objects like Regge cuts, which might reduce, or even completely remove the shrinking of the diffraction peak. Indeed, in pion-nucleon scattering no shrinking was observed. So far, the dogma of a constant value of the total cross-section was unaffected in the energy range up to 30 GeV, equivalent to an energy of 8 GeV in the centre-of-mass system (c.m.s.).

On the theoretical side, around 1969, Cheng and Wu [5] started developing a toy model using massive QED in which, by assuming a series of dominant diagrams they got an amplitude rising faster than s, i.e. as $s^{1+\varepsilon}$. This would be unacceptable at least for extremely high energies because, as first shown by Froissart, total cross-sections cannot rise faster than $(\log s)^2$ [6,7]. Fortunately, by eikonalising their model, Cheng and Wu ended up with a cross-section which at extremely high energies behaves precisely like $(\log s)^2$ [8]. However it could not be excluded at that time that another acceptable model would produce a cross-section with a slower increase or perhaps approaching a constant.

Despite the predictions of the Cheng-Wu model, it was a surprise to discover at the CERN Intersecting Storage Rings (ISR), in the range of c.m.s. energies from 23 to 62 GeV, that the proton-proton cross-section increased by about 4 mb [9]. From this, using a theorem of Khuri and Kinoshita [10], it could be predicted that the real part of the proton-proton amplitude, known to be negative at low energies, would become positive at ISR energies, if the cross-section continued to increase with energy, and if the difference between the proton-proton and the proton-antiproton cross-sections decreased, a fact consistent with systematic particle-particle and particle-antiparticle cross-section measurements performed at Fermilab [11] at laboratory energies below 400 GeV.

The real part was indeed found to be positive above a c.m.s. energy of 35 GeV and a simultaneous fit of the total cross-section and the real part constrained by dispersion relations led to the prediction that the cross-section would continue to increase at least up to a c.m.s. energy of 500 GeV [12]. A best fit using a simple parametrization gave a cross-section of about 65 mb at the energy of 550 GeV.

Indeed, when proton-antiproton collisions were studied in the CERN Super Proton Synchrotron (SPS) transformed into a Collider, the total cross-section was found to be 62 ± 1.5 mb [13]. Was it just luck ? Other fits [14] gave somewhat higher values (70 mb) for the predicted total cross-section, but this represented anyway a small deviation from the measured value.

The new fact is that we have now a measurement of the real part of the scattering amplitude which gives a ratio ρ of the real to the imaginary part of the forward scattering amplitude $\rho = 0.24 \pm 0.04$ [15], more than two standard deviations above the largest predictable value prediction of about 0.15 [16]. It is essentially impossible to accomodate this result with the usual smooth energy dependence used to fit the total cross-section. Such a value of ρ would indicate that something rather dramatic must happen at higher energies, as discussed later.

Other facts remain to be explained. For example, the elastic cross-section at ISR energies behaves according to geometrical scaling [17], which implies in particular that the ratio of the elastic to the total cross-section is constant. However, between the ISR and the $p\bar{p}$ collider range the ratio σ_{el}/σ_{tot} is found to rise quite noticeably [13].

The existence of a dip at low energies and at the ISR (more marked for pp than for $\bar{p}p$ scattering) and its replacement by a plateau which is more than one order of magnitude higher at the Sp$\bar{p}$S energies [18] is another problem [19].

Unlike the case of hard collisions, where one can use perturbative QCD, including higher order corrections, and where the only unpleasant feature is the uncertainty of the fragmentation mechanism, it is not possible to calculate proton-proton and proton-antiproton scattering directly from the QCD lagrangian. There are models more or less inspired by field theory, Regge calculus, even on a more phenomenological basis, but these are only models and it is probably for this reason that many theoreticians and experimentalists have little interest in this process where theory cannot be proven right or wrong. We do not share this view, for there would be no progress in physics if it were limited to the study of exactly predictable phenomena. There is however an interesting theoretical guide, which is the set of rigorous theorems [20] that one can obtain from the fundamental principles of local field theory. Amazingly, these theorems are very restrictive. Sometimes these theorems, called "asymptotic", are not expected to apply to presently available energies, but in fact some of them do apply already now. Then they are no longer theorems but they give useful trends. We shall try to review the most important of them in the following section.

2. GENERAL RESULTS ON SCATTERING AMPLITUDES

2.1 Introduction

Here, for simplicity, we shall ignore spin most of the time, hoping that this simplification is permitted. Clearly, spin cannot be ignored in low-energy nucleon-nucleon scattering but we hope that this can be done in high-energy scattering. However this remains to be proven. On the experimental side the only indication we have that it might be so, at least in low momentum transfer collisions is the consistency of the results on the total cross-section at the ISR obtained from different experimental information, the extrapolation of the differential cross-section to $t = 0$, the measurement of the luminosity and the measurement of the total interaction rate [21]. From any pair of measurements we can get a value of the total cross-section, but if spin effects were important they would give different results, which is not the case [22]. Small spin effects, unfortunately, are undetectable. Some of the results, like $\sigma_{tot} < (\log s)^2$, have been proven including spin complications, but for others, like the comparison of particle-particle and particle-antiparticle cross-sections, only the spinless case has been treated so far.

We shall also assume that one can define a strong interaction amplitude, while the physical amplitude is a combination of strong and electromagnetic interactions. This familiar separation has been justified to a certain extent but it is only an approximation.

We shall use the following notations. The square of the centre of mass energy is $s = 4(m_p^2+k^2) = 2(m_p E+m_p^2)$, where m_p is the proton mass, k the c.m.s. momentum and E is the laboratory energy. The four-momentum transfer squared is given by $t = -2 k^2 (1-\cos\theta)$ and is negative in the physical region of the s channel. The energy u in the crossed channel is given by $s+t+u = 4m_p^2$.

The scattering amplitude is written as

$$F(s,t) = F(s,\cos\theta) = (\sqrt{s}/2k) \Sigma (2\ell+1) f_\ell(s) P_\ell(\cos\theta) \quad (1).$$

The right hand side represents the partial wave expansion of F in terms of the Legendre polynomials $P_\ell(\cos\theta)$. The partial wave amplitudes $f_\ell(s)$ satisfy the unitarity conditions, $\text{Im} f_\ell(s) = |f_\ell(s)|^2$ in the elastic region defined by $4m_p^2 < s < (2m_p+m_\pi)^2$, where m_π is the pion mass. Otherwise,

$$\text{Im} f_\ell(s) \geq |f_\ell(s)|^2.$$

The differential cross-section of elastic scattering is written as

$$d\sigma/dt = (\pi/k^2)d\sigma/d\Omega = (4\pi/k^2 s)|F|^2 \approx (16\pi/s^2)|F|^2 \quad (2).$$

Except for weight factors due to statistics, the total cross-section is given by the optical theorem

$$\sigma_{tot} = (4\pi/k^2) \Sigma (2\ell+1) \text{Im} f_\ell = (8\pi/k\sqrt{s}) \text{Im} F(s, \cos\theta = 1) \quad (3).$$

2.2 Bounds on the scattering amplitude

The scattering amplitude is the boundary value of an analytic function [23]. For definiteness we shall call s the square of the c.m.s. energy for $p\bar{p} \to p\bar{p}$ and u the square of the c.m.s. energy for $pp \to pp$. For fixed t, where t lies in a region of the complex plane containing the negative real axis, we believe that the scattering amplitude is analytic in a twice cut plane in the variable s (or u). The right hand cut starts at $s = 4m_\pi^2$ but the physical part of the cut starts at $s = 4m_p^2$. For any $s > 4m_p^2$, $\lim_{\varepsilon\to 0}F(s+i\varepsilon,t,u)$ is the physical proton-antiproton amplitude. The left hand cut starts at $u = 4m_p^2$ (or, equivalently, at $s = -t$) and the proton-proton amplitude is given by $\lim_{\varepsilon\to 0} F(s,t,u+i\varepsilon)$. For t real and less

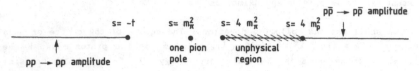

Fig. 1 Analyticity domain of the pp and p̄p amplitudes at fixed momentum transfer.

than zero, F possesses the reality property, i.e. $F(s+i\varepsilon,t) = F^*(s-i\varepsilon,t)$. For $t = 0$ a particularly convenient variable is E, the laboratory energy. Then, if $F(E+i\varepsilon)$ represents the proton-antiproton forward amplitude at energy E, $F(-E-i\varepsilon)$ represents the proton-proton amplitude at the same energy. There are two ways to connect the two amplitudes: to follow a complex path going in between the left and right hand cut or to follow a big half circle in the complex plane and take the complex conjugate of the quantity continued to cross the left hand cut (Fig. 1).

A crucial result [24] is that in this complex domain the scattering amplitude grows less fast than $|s|^N$ for $s \to \infty$. Another important result is the so-called Froissart bound, first derived by Froissart [6] from the Mandelstam representation in 1961 and later proven directly from local field theory and unitarity by Martin [7] (the Mandelstam representation was never proven and might be wrong). The Froissart bound gives primarily a bound on the number of effective partial waves contributing to the scattering:

$$\ell_{max} < C \sqrt{s} \log s \qquad\qquad (4).$$

where C is a constant. This corresponds to a maximum impact parameter b growing like $\log s$. Then from Eq. (1) and the fact that each partial wave amplitude is bounded by unity we get

$$F(s,\cos\theta) < C s (\log s)^2 \text{ and } \sigma_{tot} < C' (\log s)^2 \qquad (5).$$

where C and C' are constants. These bounds can be further refined by noticing that for $0 < t < 4m_p^2$, the amplitude grows at most like s^2 and one then gets [25]

$$\sigma_{tot} < (\pi/m_\pi^2) (\log s)^2 \qquad\qquad (6).$$

Then it is possible to write twice subtracted dispersion relations [26] for $t < 4m_\pi^2$ as

$$F(s,t,u) = A + Bs + \frac{s^2}{\pi} \int_{4m_\pi^2}^{\infty} \frac{Im\, F_{p\bar{p}}(x,t)\, dx}{x^2(x-s)} + \frac{u^2}{\pi} \int_{4m_p^2}^{\infty} \frac{Im\, F_{pp}(x,t)\, dx}{x^2(x-u)}$$

$$+ \text{pole terms .} \tag{7}$$

For $t = 0$, $Im\, F_{p\bar{p}}$ and $Im\, F_{pp}$ are proportional to the total cross-section at least for $s > 4m_p^2$ and $u > 4m_\pi^2$ respectively. However there is an unphysical contribution in the s channel from $s = 4m_\pi^2$ to $4m_p^2$. At high-energy (say above 10 GeV c.m.s. energy) the unphysical cut is negligible. At low energy it is relatively well parametrized in terms of a few poles [27].

We give now a series of constraints following from the Froissart bound. On the integrated elastic cross-section one has,

$$\sigma_{tot}^2/\sigma_{el} < C\,(\log s)^2.$$

About the slope of the diffraction peak, unitarity alone [28] says that

$$d/dt\, \log|Im\, F(s,t)|\big|_{t=0} > \sigma_{tot}^2/\,(36\,\pi\,\sigma_{el}) \tag{8}.$$

If the real part of the amplitude is small or if the ratio Re F/Im F varies slowly with t, this becomes

$$d/dt\,|\log\, d\sigma/dt\,|_{t=0} > \sigma_{tot}^2/\,(18\,\pi\,\sigma_{el}).$$

However from the constraint on ℓ_{max} (see Eq. 4), we get an inequality in the opposite direction

$$d/dt\,|\log(d\sigma/dt)| < C\,(\log s)^2.$$

Notice, however, that for $t < -T < 0$, with $T > 0$, one can prove that if $d\sigma/dt$ is monotonous in some fixed interval in t, then for $t < 0$, i.e. in the physical region,

$$d/dt\,|\log(d\sigma/dt)|_{t<0} < C(T)\,\log s.$$

A very amusing situation is when the Froissart bound is saturated, i.e. $\sigma_{tot} \sim (\log s)^2$. Then the previous inequalities indicate that:

$$d/dt\,(\log\, d\sigma/dt)|_{t=0} \sim (\log s)^2 \text{ and } \sigma_{el} \sim (\log s)^2.$$

A more careful analysis [29] shows that in this case the scattering amplitude and the differential cross-section are given asymptotically by a function of a single variable. One has

$$F(s,t) \simeq F(s,0) \cdot f(\tau) \qquad\qquad (9)$$

where $\tau = t \ (\log s)^2 = C \ t \ \sigma_{tot}$. When the amplitude depends only on the product $t \cdot \sigma_{tot}$ this is called "geometrical scaling", and what happens is that if the Froissart bound is qualitatively saturated, geometrical scaling is reached asymptotically at high energies. The function $f(\tau)$ possesses a "technical property" which is that it is an entire function of order 1/2. An example of a possible function is

$$f(\tau) = (J_1(\sqrt{-\tau})/\sqrt{-\tau})^2 .$$

This example appears explicitly in the model of T.T. Wu and collaborators [8].

Even though a big gap in energy has been recently crossed with the help of the $p\bar{p}$ Colliders at CERN and at Fermilab, it is not clear that we have yet reached the asymptotic regime of $p\bar{p}$ scattering. So the asymptotic theorems that we have discussed should rather be used to suggest trends that any reasonable model should possess.

Now we would like to discuss the comparison of the pp and $p\bar{p}$ amplitudes and cross-sections. Originally, the first comparison theorem was the Pomeranchuk theorem [30]: the difference of the $p\bar{p}$ and pp total cross-sections goes to zero as the energy goes to infinity. This theorem does not go without a certain assumption that the forward amplitude, normalized as in (1) has a modulus bounded by s. Since σ_{tot} is seen to increase, possibly to infinity, this assumption is not necessarily satisfied. It is easy to construct examples in which Re F ~ s log s and the difference $\sigma(p\bar{p})$ - $\sigma(pp)$ approaches a constant different from zero. However, the old Pomeranchuk theorem can be replaced by a new one, first proposed by Eden and Kinoshita and generalized by Grunberg and Truong [31] : if at least one of the two total cross-sections goes to infinity, then their ratio tends to unity. This is proved by making a very clever combination of analyticity and unitarity. We shall illustrate this later by taking the special case of cross-sections behaving like powers of log s.

Another "theorem" of the same category concerns the comparison of differential cross-sections and slopes of the diffraction peak. The theorem states that the ratio [dσ/dt($p\bar{p}$)]/[dσ/dt(pp)] and b($p\bar{p}$)/b(pp) go to unity for s → ∞. In the initial version by Van Hove [32] and by Logunov and collaborators [33], there was an assumption on the structure of the scattering amplitude in the physical region and, in particular, on its phase. Later Cornille and Martin [34] made progress by proving that, if one neglects spin effects, the phase is bounded in the pp and p̄p diffraction peaks, and, in particular, these peaks have equal slopes. It appears that this equality holds pretty well (see Fig. 2) already at ISR energies [35].

Very important consequences of analyticity are the Khuri - Kinoshita theorems on the real part [10]. Dispersion relations allow to calculate the real part of the scattering amplitude from the imaginary part. To discuss this subject, it is useful to introduce the notion of even and odd signature amplitudes. The even signature amplitude is $F^+ = (F_{p\bar{p}}+F_{pp})/2$, while the odd signature amplitude is $F^- = (F_{p\bar{p}}-F_{pp})/2$.

One of the theorems of Khuri and Kinoshita says that if σ_{pp} and $\sigma_{p\bar{p}}$ → ∞ (the Eden - Kinoshita theorem ensures that, if either σ_{pp} or $\sigma_{p\bar{p}}$ → ∞, also the other cross-section has the same behaviour), the real part of the even signature amplitude becomes positive at high energies. If one supplements this with the assumption that the odd signature amplitude is negligible,

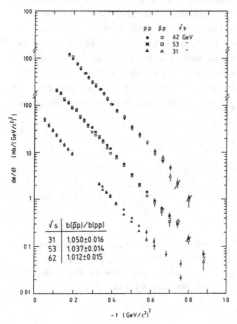

Fig. 2 The shape of the diffraction peak for pp and p̄p scattering at three different ISR energies [35]. The ratio of the slope values is also given.

one predicts that $\text{Re}F_{p\bar{p}}$ and $\text{Re}\ F_{pp}$ become positive at high-energy. It must be understood that with a definite model for the $p\bar{p}$ and pp total cross-sections, this kind of prediction is not an asymptotic theorem. One can in fact predict, with the help of dispersion relations, the energy value at which this happens (remember that $\text{Re}\ F_{pp}$ is negative at low energy).

It is useful here to say a few words about the "standard model" and explain why it is believed, but not proven, that the odd signature amplitude is negligible. Since the original formulation of Regge phenomenology in the 60's, it is believed that the even signature amplitude is dominated by the Pomeron contribution, due to some singularity at $j = 1$ of the complex angular momentum amplitude in the t channel. Only poles at even j materialize as particles in the even signature amplitude. Extra singularities are poles crossing $t = 0$ around $j = 1/2$ and giving contributions to the even amplitude which behave approximately like $s^{1/2}$. It is believed that the odd signature amplitude only possesses that kind of singularity near $j = 1/2$ for $t = 0$. A Regge pole at $j = 1$, $t = 0$ would materialize into a massless particle which does not seem to exist. In case of more complicated singularities we do not really know. To support the belief that the odd signature amplitude is not dominant, we have limited experimental evidence from the comparison of the total cross-section difference $\sigma_{tot}(ab) - \sigma_{tot}(a\bar{b})$. Up to available energies ($\sqrt{s} = 63$ GeV for pp and $p\bar{p}$, less for the other particles), $\Delta\sigma$ is rather well fitted by a $s^{-1/2}$ law. The predictions of the total cross-section at Collider energies from measurements of total cross-sections and real part at the ISR made use of the assumption that the odd signature amplitude is asymptotically negligible as an essential ingredient and turned out to be successful. A compilation of $\Delta\sigma_{tot} = \sigma_{tot}(p\bar{p})-\sigma_{tot}(pp)$, which includes recent ISR data [36,37] is shown in Fig. 3.

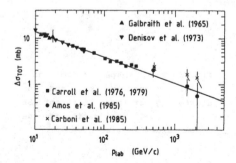

Fig. 3 The total cross-section difference $\Delta\sigma_{tot} = \sigma_{tot}(p\bar{p}) - \sigma_{tot}(pp)$. The line is the result of the fit of Ref. 36.

We now illustrate the Khuri-Kinoshita and Eden-Kinoshita theorems in a simple case where the even and odd signature amplitudes behave like $C_{\pm}E(\log E)^{\alpha\pm}$.

We first consider F^+. We have $F^+(E) = F^+(-E)$ and if

$$F^+(E+i\varepsilon) \sim C_+ E(\log E)^{\alpha^+} \text{ for } E \to +\infty + i\varepsilon,$$

we can continue F^+ by turning in the upper half complex plane and in this way we get

$$F^+(-E+i\varepsilon) \sim -C_+ E (\log E)^{\alpha^+} \text{ and}$$
$$F^+(-E-i\varepsilon) \sim -C_+{}^* E (\log E)^{\alpha^+}.$$

Hence, since $F^+(E) = F^+(-E)$, we have $-C_+{}^* = C_+$, which means that C_+ is pure imaginary. So $F^+ \sim i |C_+| E (\log E)^{\alpha^+}$. However if we are more refined we see that F^+ is not exactly crossing symmetric:

$$F^+(-E+i\varepsilon) \sim i|C_+| (-E) (\log |E|+i\pi)^{\alpha^+} \text{ and}$$
$$F^+(-E-i\varepsilon) \sim i|C_+| (-E) (\log |E|-i\pi)^{\alpha^+}.$$

To get perfect crossing symmetry we should take

$$F^+(E+i\varepsilon) \sim i|C_+| E (\log E-i\pi/2)^{\alpha^+}.$$

Therefore F^+ is dominantly imaginary and

$$\rho_+ = \text{Re } F^+/\text{Im } F^+ = \pi\alpha_+/(2/\log E)$$

which is a special case of the Khuri-Kinoshita theorem. For $\alpha > 0$, $\sigma_{tot}(pp)+\sigma_{tot}(p\bar{p}) \to \infty$ since $\sigma_{tot} \sim (\log E)^{\alpha^+}$ and ρ_+ is positive.

At this point we notice that the substitution $\log|E| \to \log|E|-i\pi/2$ in the expression of the even signature amplitude is equivalent, to the lowest order in $1/\log|E|$, to the lowest order of the "Derivative dispersion relation", because if $F_+ \simeq iCE \, f(\log|E|)$, where f is a smooth function of $\log|E|$, then

$$f(x+\Delta) \simeq f(x) + \Delta \, df(x)/dx$$

and hence we get

$$F_+ \simeq i \, C \, E \, [\, f(\log|E|) - (i\pi/2) \, df/d \, \log|E| \,].$$

However, the substitution $\log|E| \rightarrow \log|E| - i\pi/2$ gives exact crossing while the derivative dispersion relations do not, except for polynomials in $1/\log|E|$. Derivative dispersion relations may be handy, but contain nothing deep and when the imaginary part of the amplitude is not smooth, may be dangerous to use.

We now look at F^-. We have

$$F^- \sim C^- \, E \, (\log E)^{\alpha_-},$$
$$F^-(-E+i\varepsilon) \sim -C_- \, E \, (\log E)^{\alpha_-}$$
and $F^-(-E-i\varepsilon) \sim -C_-^{\;*} \, E \, (\log E)^{\alpha_-}.$

Since $F^-(-E) = -F^-(E)$, C^- is real, but crossing is perfect only by taking $F^-(E) \sim C^- \, E \, (\log E - i\pi/2)^{\alpha_-}$ and F^- is dominantly real and $\text{Im} \, F^-/\text{Re} \, F^- = -\pi\alpha_-/(2 \log E)$. This is an illustration of a general theorem of J. Fischer et al.[38]: if $\sigma^- = \sigma_{tot}(p\bar{p}) - \sigma_{tot}(pp)$ and $\text{Re} \, F^- = \text{Re} \, F(p\bar{p}) - \text{Re} \, F(pp)$ have the same sign, then σ^- goes to zero at high energy. We introduce now the unitarity restriction :

$$|\text{Re} \, F^-|^2 = |\Sigma (2\ell+1) \, \text{Re} \, f_\ell^{p\bar{p}} - \text{Re} \, f_\ell^{pp}|^2 < 2L^2 \, \big[\Sigma (2\ell+1)(\text{Im} \, f_\ell^{p\bar{p}} + \text{Im} \, f_\ell^{pp}) \big] \, ,$$

which gives $2\alpha_- \leq \alpha_+ + 2 \leq 4$. In addition, the condition $|\text{Im} \, F^-| < \text{Im} \, F^+$ gives $\alpha_- \leq \alpha_+ + 1$. This is what defines the "Cornille plot" [39] which is shown in Fig. 4. Now we come to the Eden-Kinoshita theorem :

$$\sigma^-/\sigma^+ = (\log s)^{\alpha_- - 1 - \alpha_+} = (\log s)^{(\alpha_- - 1 - \alpha_+/2) - \alpha_+/2} \, .$$

Therefore, if α_+ is strictly positive, this ratio goes to zero.

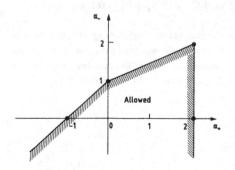

Fig. 4   The 'Cornille plot'.

If we forget all our prejudices on the odd signature amplitude and use only s channel unitarity, as was done by Lukaszuk, Kang and Nicolescu [36] and later by others, we see that there is room for a large odd signature amplitude, the extreme case being $\alpha_- = \alpha_+ = 2$, then

$$F^- \sim C^- E (\log E - i\pi/2)^2$$
$$F^+ \sim C^+ E (\log E - i\ \pi/2)^2.$$

The quantity $|C^-|^2$ is numerically bounded by $|C^*|$ times a constant depending on the coefficient of $\sqrt{s} \log s$ in ℓ_{max}. In this extreme case, the ratio of $\sigma(p\bar{p})$ and $\sigma(pp)$ approaches unity, both cross-sections behave like $(\log s)^2$ and $\sigma(p\bar{p})-\sigma(pp)$ behaves like $\log s$. This is the extreme "odderon" picture [40]. When confronted with experimental data, the problem is to choose between exploiting our ignorance of the even signature cross-section above 550 GeV (soon above 1800 GeV) to explain real part measurements, or to appeal to the "odderon" picture, which so far has not shown up in a clear way in other phenomena.

These general considerations are useful mainly because there is no straightforward way of calculating pp and p$\bar{p}$ scattering from first principles, say from the QCD lagrangian. They give us guidelines for acceptable models.

Though there is no proof that cross-sections go to infinity, most models incorporate the assumption that they do tend to infinity and in fact eventually saturate the Froissart bound. Present models have different origins with different motivations but in fact tend to converge. In these models the basic mechanism tend to produce a cross-section that grows polynomially with energy but iterations and rescattering correct such behaviour.

3. EXPERIMENTAL RESULTS

Measurements of the proton-antiproton total cross-section at $\sqrt{s}$ = 546 GeV were made in the first physics runs of the SPS Collider in 1981/82 with errors at the 10% level by experiments UA1 and UA4. A result with improved accuracy was then obtained by UA4. Later on, UA5 reported a measurement at $\sqrt{s}$ = 900 GeV. The data on elastic scattering come mainly from UA4, a dedicated experiment especially designed for the measurement of the total cross-section, elastic and diffractive scattering. The UA4 experiment measured elastic scattering at $\sqrt{s}$ = 546 GeV in the momentum transfer range $2.10^{-3} < -t < 1.55$ GeV^2 and at $\sqrt{s}$ = 630 GeV in the range $0.7 < -t < 2.2$ GeV^2.

3.1 The experimental technique

Three different methods for measuring the total cross-section at a hadron collider are known. They were developed almost two decades ago at the CERN ISR [9,21].

i) The first method, which is the most direct, requires a measurement of the total number of the elastic (N_{el}) and of the inelastic (N_{inel}) interactions and of the integrated machine luminosity $\mathscr{L}$,

$$N_{tot} = N_{el} + N_{inel} = \mathscr{L} \, \sigma_{tot} \qquad (10)$$

ii) The second method is based on the use of the optical theorem (3). Elastic scattering is measured at low momentum transfer in order to get by a suitable extrapolation, the number of elastic events per unit interval of t at t = 0 which is given by

$$(dN_{el}/dt)_{t\,=\,0} = \mathscr{L}(1+\rho^2)\sigma_{tot}^2/16\pi \qquad (11)$$

where ρ is the ratio of the real to the imaginary part of the forward scattering amplitude. The extrapolation of the measured t-distribution of elastic scattering to the forward direction is to some extent model dependent. The influence of the parameter ρ on the determination of σ_{tot} is on the other hand quite small.

iii) The third method is the combination of i) and ii) and allows a luminosity independent determination of the total cross-section. In fact combining the two expressions (10) and (11) one can eliminate the luminosity and obtain

$$\sigma_{tot}(1+\rho^2) = 16 \ \pi \ (dN_{el}/dt)_{t\,=\,0} \ / \ (N_{el} + N_{inel}) \qquad (12)$$

At the CERN ISR the luminosity was measured with the Van der Meer method [41] to an accuracy of about 1%. Both methods i) and ii) were employed and provided consistent results on σ_{tot} with an error at the 1% level. At the SPS Collider, however, the luminosity could only be inferred from measurements of the beam profile by means of the wire scan technique [42] with an accuracy limited to 10%. As a consequence, in order to get a better precision and avoid the presence of uncontrollable systematic errors, the combined method iii) was adopted by experiment UA4.

Clearly, the measurement of N_{inel} requires detection of events having all possible configurations as can only be achieved in a fully inclusive experiment. In particular, the detector must be able to observe the events of diffractive type, where one of the two colliding hadrons is scattered at small angles and travels undetected inside the vacuum chamber while the other one fragments into a small number of particles which are emitted in a small cone in the opposite hemisphere. These diffractive events are known to contribute as much as 20 % of the inelastic cross-section [43]. Therefore the measurement of the rate of the inelastic interactions requires a large coverage system, able to provide a clear discrimination of beam-beam events against background. The detector used for this purpose by the UA4 group is shown in Fig. 5. It consists of telescopes of wire chambers placed in the long straight section LSS4 of the SPS and arranged in two symmetric arms on both sides of the apparatus of experiment UA2 which surrounds the crossing region. When operated together with the UA2 central detector, this system covers the pseudorapidity interval $-5.6 < \eta < 5.6$, the beam rapidity being 6.3. The discrimination of beam-beam interactions from background is obtained by using the time-of-flight information and by reconstructing the interaction vertex from the observed charged tracks. This system was able to detect inelastic interactions with an efficiency of 99%.

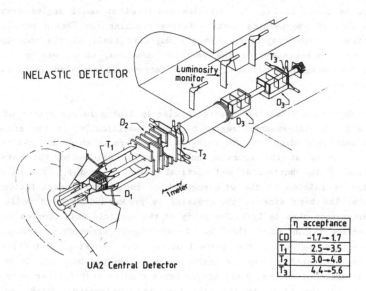

	η acceptance
CD	-1.7 → 1.7
T₁	2.5 → 3.5
T₂	3.0 → 4.8
T₃	4.4 → 5.6

Fig. 5 Layout of one arm of the vertex detector of experiment UA4 (the full detector is symmetric with respect to the crossing point). The central detector of experiment UA2 is also sketched. The η-acceptance of the trigger hodoscopes is indicated.

The study of elastic scattering at the SPS Collider requires the detection of particles emitted at angles of the order of 1 mrad with respect to the circulating beams. This was achieved by exploiting the technique of the "Roman pots" which had been successfully developed at the ISR. The "Roman pots" are movable sections of the vacuum chamber which are connected to the main accelerator pipe by bellows. Each "Roman pot" houses a detector unit consisting of a small high-precision wire chamber, a scintillation counter hodoscope and a trigger counter as shown in Fig. 6. The pots were normally kept in a retracted position, then, once stable beam conditions were reached, they were displaced toward the beam up to a position where background was still tolerable. During data taking the distance between the inner edge of the detectors and the beam was of 1-2 cm. In this way particles travelling close to the beam could be easily detected. In the UA4 experiment the "Roman pots" were arranged in eight telescopes placed symmetrically on the left and on the right side of the crossing region, above and below the machine plane, as sketched in Fig. 7. More technical details about the UA4 experiment are given in Ref. [44].

The selection of elastic events was achieved by requiring the collinearity of the proton and antiproton tracks emerging from the crossing region. As shown in Fig. 7, particles scattered at small angles traverse quadrupoles of the machine lattice before reaching the "Roman pots". The deflection of the trajectories in the magnetic field of the quadrupoles permitted a measurement of the particle momentum, which was useful to reject the background due to low energy secondaries produced in inelastic interactions.

The detection of elastic events is clearly linked to the optics of the machine in the intersection region or, more specifically, to the strength of the machine quadrupoles. The relevant parameters which define the optics of the machine at the crossing point are the values of the betatron functions in the horizontal and vertical plane, β_H and β_V. The betatron function is related to the wavelength λ of the betatron oscillations by $\beta = \lambda/2\pi$. The beam size at the crossing is proportional to $\sqrt{\beta}$ while the beam divergence goes as $1/\sqrt{\beta}$. The study of the particle trajectories in the intersection region shows that, as the quadrupole strength is changed to increase the value of the β-function at the crossing, detection of particles scattered at smaller angles becomes feasible because, in such a case, as sketched in Fig. 7 the trajectories tend to stay further away from the axis of the beam. At the same time the luminosity, which goes as $1/(\sqrt{\beta_H}\sqrt{\beta_V})$ will in fact decrease, but this is not a problem because the differential cross-section of elastic scattering is very large at small

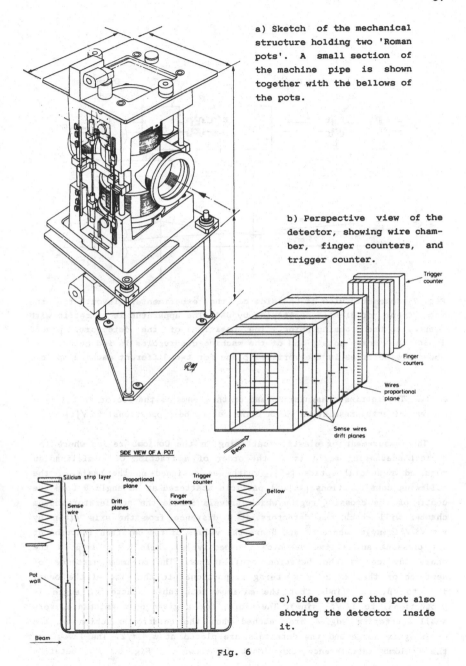

a) Sketch of the mechanical structure holding two 'Roman pots'. A small section of the machine pipe is shown together with the bellows of the pots.

b) Perspective view of the detector, showing wire chamber, finger counters, and trigger counter.

Trigger counter

Finger counters

Wires proportional plane

Sense wires drift planes

Beam

SIDE VIEW OF A POT

Silicium strip layer

Proportional plane

Trigger counter

Drift planes

Finger counters

Bellow

Sense wire

Pot wall

Beam

c) Side view of the pot also showing the detector inside it.

Fig. 6

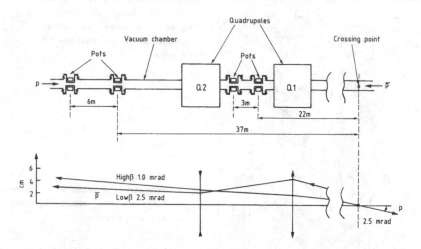

Fig. 7 Schematic view of one side of the experimental layout for the measurement of elastic scattering by UA4. The apparatus is symmetric with respect to the crossing point. The position of the detectors, placed inside the 'Roman pots', and of the machine quadrupoles is indicated. Also shown trajectories in the vertical plane for two different machine optics.

angles. The intrinsic angular spread of the beams is the factor that limits the experimental resolution which turns out to be proportional to $\sqrt{|t|}$.

The measurement of elastic scattering in the Coulomb region where the typical scattering angle is of the order of a fraction of a milliradian required a special optics [45], which was designed on the basis of the following considerations [46]. A particle scattered at an angle θ from the centre of the crossing region which travels inside the accelerator vacuum chamber will reach the detectors at a distance from the axis given by $x = \sqrt{\beta^*}\sqrt{\beta}(\sin\psi)\theta$, where β^* and β are the values of the betatron function at the crossing and at the detector, respectively, while $\psi = \int ds/\beta$ is the phase advance of the betatron oscillations. The minimum distance of approach of the "Roman pots" being proportional to the size of the beam, i.e. to $\sqrt{\beta}$, one finds that the minimum detectable scattering angle is proportional to $1/(\sqrt{\beta^*} \sin\psi)$. Therefore, for a given beam emittance, very small scattering angles are reached when the betatron function at the crossing is large and the detectors are placed at $\psi = \pi/2$. The layout of the Coulomb interference experiment is shown in Fig. 8. The betatron

The high-β insertion for Coulomb scattering at the CERN collider

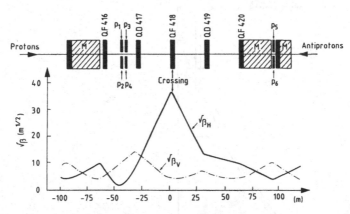

Fig. 8 Sketch of the layout of the Coulomb interference experiment at the CERN SPS Collider. The behaviour of the betatron function is also shown.

function at the crossing in the horizontal plane, where elastic events were measured, was as large as 1100 m.

3.2 Total cross-section

The early Collider measurements of the total cross-section by UA4 [47] and UA1 [48] at $\sqrt{s}$ = 546 GeV, which are plotted in Fig. 9 agreed with the prediction made by Amaldi et al. [12] in 1977 using the ISR data on σ_{tot} and ρ and the constraint provided by the dispersion relations. In Ref. [12] the high-energy behaviour of the total cross-section was described by a leading term of the form $(\log s)^{\gamma}$ and the fit to the data which were available at that time gave $\gamma \approx 2$. These early Collider data gave the first indication that in the energy range from the ISR up to the SPS Collider, the total cross-section continue to rise essentially as was inferred from the ISR data.

Afterwards, a more accurate measurement was performed by UA4 using the combined method discussed in sec. 3.1, inelastic rate and low-t elastic rate, with the result $(1+\rho^2)\sigma_{tot}$ = 63.3 ± 1.5 mb. Taking for the parameter ρ the value ρ = 0.15 as suggested by the dispersion relation fit of Ref. [12], one obtains σ_{tot} = 61.9 ± 1.5 mb. This result is shown in Fig. 10 together with a compilation of lower energy results including the recent p$\bar{\text{p}}$ data from the CERN ISR. In the same experiment, the total

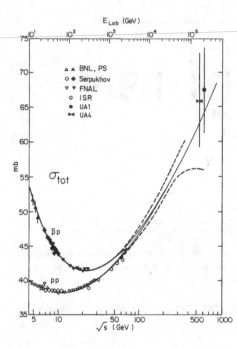

Fig. 9 The early Collider results by UA1 and UA4 on the total cross-section are plotted together with the dispersion relations prediction by Amaldi et al. [12].

cross-section was also determined from the low-t elastic scattering data together with the measurement of the luminosity obtained by the wire scan [42]. The result, $\sigma_{tot} = 61 \pm 3$ mb agrees with the more accurate one obtained using the luminosity independent method.

In a special run of the SPS Collider where the machine energy was cycled between $\sqrt{s} = 200$ GeV and $\sqrt{s} = 900$ GeV, the UA5 experiment [49] measured the ratio of the inelastic cross-sections at these two energies as $\sigma_{inel}(900 \text{ GeV})/\sigma_{inel}(200 \text{ GeV}) = 1.20 \pm 0.02$. From this result the following value of the total cross-section at $\sqrt{s} = 900$ GeV was inferred, $\sigma_{tot} = 65.3 \pm 1.7$ mb, which is also plotted in Fig. 10.

The solid lines in Fig. 10 represent the result of a dispersion relation analysis [36] very similar to that of Ref. [12]. The total cross-section was parametrized as

$$\sigma_{tot} = C_0 + C_1 E^{-\alpha} \pm C_2 E^{-\beta} + C_3 (\log s/s_0)^\gamma \qquad (13)$$

where the positive sign refers to $p\bar{p}$ and the negative sign to pp scattering and the scale factor s_0 is arbitrarily fixed to 1 GeV². The result of the

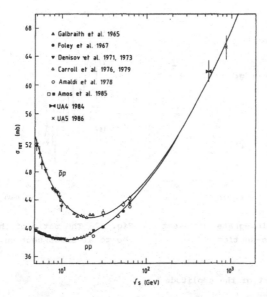

Fig. 10 The total cross-section measurements at the SPS Collider are shown together with a compilation of lower pp and p̄p data. The lines represent the dispersion relations fit of Ref. 36.

fit gave $C_3 = 0.19 \pm 0.01$ mb and $\gamma = 2.02 \pm 0.01$. While the power of the log s term turns out to be close to two, the experimental value of the coefficient C_3 is much smaller than the constant $\pi/m_\pi^2 \approx 60$ mb that appears in the expression of the Froissart bound (Eq. 6). This indicates that, at least within the limits of the simple parametrization (13) the Froissart bound is "qualitatively saturated" at present energies.

The integrated elastic cross-section at $\sqrt{s} = 546$ GeV was obtained by UA4 from the integration of the elastic scattering differential cross-section and found to be $\sigma_{el} = 13.3 \pm 0.6$ mb. This result, presented in Fig. 11 together with lower energy data shows that going from the ISR to the SPS Collider, the elastic cross-section increases faster than the total cross-section. The ratio σ_{el}/σ_{tot} (which is independent of the parameter ρ) is 0.215 ± 0.005 at $\sqrt{s} = 546$ GeV while at the ISR it is about 0.175 as shown in Fig. 12. The observation that the ratio σ_{el}/σ_{tot} is nearly constant in the energy range of the ISR was taken as indication for the onset of "geometrical scaling" [17]. The Collider result shows, however, that "geometrical scaling" has only limited validity in the restricted range of energy which was covered by the ISR.

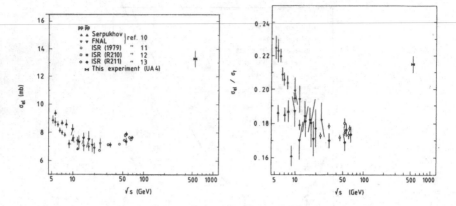

Fig. 11 The integrate d elastic scattering cross-section.

Fig. 12 The ratio of the elastic to the total cross-section.

3.3 The real part of the amplitude

The real part of the hadronic amplitude was measured at $\sqrt{s}$ = 546 GeV in a short dedicated run of the SPS Collider using the special insertion discussed in Sec. 3.1 (see Fig. 8). The measured t-distribution [15] which extends down to a minimum value of $|t|$ of 2.10^{-3} GeV^2 is shown in Fig. 13. The momentum transfer resolution was essentially determined by the angular spread of the beams and given by $\Delta t \approx 0.01\sqrt{|t|}$. The data were analyzed in the conventional way expressing the differential cross-section in the form

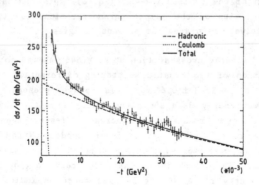

Fig. 13 The differential cross-section in the Coulomb interference region. The result of the fit is shown together with the Coulomb and hadronic contributions.

$$d\sigma/dt = (16\pi/s^2) \mid F_C + F_H \mid^2.$$

The Coulomb amplitude F_C is given by

$$F_C = \alpha \ s \ G^2(t) \ \exp(-i\alpha\phi) \ / \ (2 \mid t \mid)$$

where $G(t)$ is the electromagnetic form factor of the proton. The phase ϕ of the Coulomb amplitude has been discussed by several authors [50]. In the region of momentum transfer which is relevant for the measurement of the parameter ρ, it can be approximated by $\phi = \log(0.08/\mid t \mid) - 0.577$. For the hadronic amplitude, the following conventional and simplified form was used

$$F_H = s \ \sigma_{tot} \ (\rho + i) \ \exp(-^1/_2 b \mid t \mid) \ / \ (16\pi).$$

The result of fitting the measured t-distribution with the constraint $\sigma_{tot}(1+\rho^2) = 63.3 \pm 1.5$ mb, was $\rho = 0.24 \pm 0.04$, where the quoted error includes systematic uncertainties.

This result, plotted in Fig. 14 together with data at lower energies on pp and p$\bar{\text{p}}$ scattering, is appreciably larger than the prediction ($\rho \simeq 0.13$) of the dispersion relation fit of Ref. [36]. If we assume that the measurement of ρ is not affected by a hidden systematic error, various explanations of this discrepancy can be proposed [51].
i) A conventional explanation [52] would be that the total cross-section increases in the few TeV energy range faster than expected. In this case, as can be seen qualitatively from the dispersion relations written in the derivative form

$$\rho \simeq \pi \ (d\sigma_{tot}/d \log s)/(2 \ \sigma_{tot}),$$

the parameter ρ might turn out to be substantially larger than predicted by the standard fit of Ref. [36]. In fact exact dispersion relations show that an anomalous rise of the total cross-section above the energies at which it has been measured can produce an increase of ρ at the presently available energies. A connection between this faster rise of the total cross-section and the rapidly rising "minijet" cross-section observed by UA1 [53] can possibly be established [54].
ii) The determination of ρ certainly depends on the way of parametrizing the hadronic amplitude at low momentum transfer. In some of the current models of high-energy scattering [55,56], the forward diffraction peak has a rather complicated structure with the slope increasing at very low-t and, moreover, with $b_{real} > b_{im}$. If such structure indeed is present, fitting

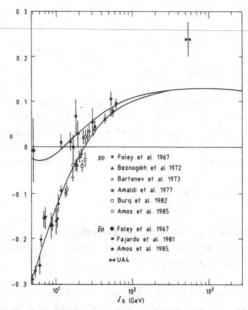

Fig. 14 The Collider result on the parameter ϱ is plotted together with lower-energy pp and p̄p data. The lines are the dispersion relation fit of Ref. 36.

the data with a simple exponential in t, would result in overestimating the actual value of ρ.

iii) A natural explanation of the high value of ρ is provided by the "odderon" picture of Nicolescu et al. [40,57], already discussed in Sec. 2. In this picture the pp total cross-section is predicted to cross the p̄p cross-section at an energy around √s ≃ 100 GeV and rise faster with energy so that asymptotically $\sigma_{tot}(pp)-\sigma_{tot}(p\bar{p}) \sim \log s$. This crossing may seem rather unnatural but cannot be excluded on rigorous grounds. It should also be added that a careful analysis of the odd signature amplitude by Kroll and Igi [58], including the measurements of the p̄p scattering length at LEAR and the UA6 data on the real part [59] disfavours but does not exclude the "odderon" interpretation.

3.4 Elastic scattering

A systematic stydy of elastic scattering at √s = 546 GeV in the range 0.03 ≤ -t ≤ 1.55 GeV² was done by UA4 [18,60], using three different beam optics. For each beam optics a different range of scattering angles was covered. The absolute normalization of the low-t data was obtained from the optical point while the other data samples were normalized by smoothly joining in the overlapping regions.

The measured differential cross-section for $-t \leq 0.5$ GeV^2 is shown in Fig. 15. It was found that for $-t \leq 0.15$ GeV^2 the data can be well described by a single exponential exp(bt). The same is true in the range $-t \geq 0.2$ GeV^2. The slope parameter b has, however, quite different values in the two regions, $b = 15.3 \pm 0.3$ GeV^{-2} for $0.03 \leq -t \leq 0.10$ GeV^2 and $b = 13.4 \pm 0.3$ GeV^{-2} for $0.21 \leq -t \leq 0.5$ GeV^2. A rapid change of the local slope parameter takes place between these two regions of momentum transfer. A fit to the data with two constant slopes leaving also as free parameter the value of momentum transfer t' where the sudden change of slope occurs, gives the result $-t' = 0.14 \pm 0.02$ GeV^2. It should be noted that the exponential form with quadratic t-dependence, exp(bt+ct^2), is not a good parametrization of the data in the range $-t \leq 0.5$ GeV^2.

This behaviour of the Collider data is very similar to that observed earlier at the ISR [61], where a change of slope was found around $-t = 0.1$ GeV^2. As the energy increases from the ISR to the Collider, the

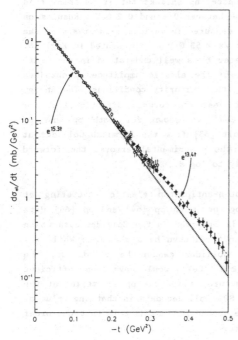

Fig. 15 Differential cross-section of elastic scattering at $\sqrt{s}$ = 546 GeV in the range $0.03 \leq -t \leq 0.5$ GeV^2. Open and full data points refer to two different runs.

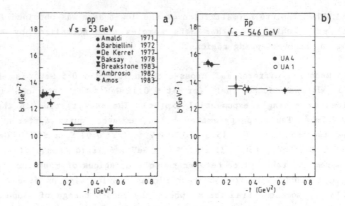

Fig. 16 Dependence of the local slope parameter on t. The horizontal bar indicates the t-interval where the exp(bt) fit was performed. a) Compilation for pp scattering at ∫s = 53 GeV. b) The SPS Collider results.

overall t-distribution of elastic scattering shrinks but it mantains this feature of a rapid variation of slope between 0.1 and 0.2 GeV². Results on the local value of the slope b(t) as measured in various t-intervals at the SPS Collider and in pp scattering at √s = 53 GeV are presented in Fig. 16. The t-dependence of the slope parameter b is well understood in the model of Ref. 62 where the imaginary part of the elastic amplitude is obtained from the production amplitudes using the unitarity conditions. The energy dependence of the slope parameter b near the forward direction,i.e. for $|t| \simeq 0.05$ GeV² and for $|t| \simeq 0.2$ GeV² is shown for both pp and p$\bar{\text{p}}$ scattering in Fig. 17. A recent result [63] from the Fermilab Collider at √s = 1.8 TeV is also shown. Within the experimental errors, the forward slope seems to increase proportionally to log s.

The full t-distribution of proton-antiproton elastic scattering at √s = 546 GeV is shown in Fig. 18 together with pp [64] and p$\bar{\text{p}}$ [65] data from the ISR at √s = 53 GeV. No dip is observed in the Collider data which show a sharp break around -t = 0.9 GeV² followed by a shoulder. While the presence of a narrow dip at the Collider cannot be ruled out, the t-resolution ($\Delta t = 0.06 \sqrt{|t|}$ for -t ≥ 0.5 GeV²) would have been sufficient to observe a prominent dip-bump structure as seen in pp scattering at the ISR. Another striking feature of the SPS Collider data is that the value of the differential cross-section at the second maximum is more that one order of magnitude higher than at ISR energies.

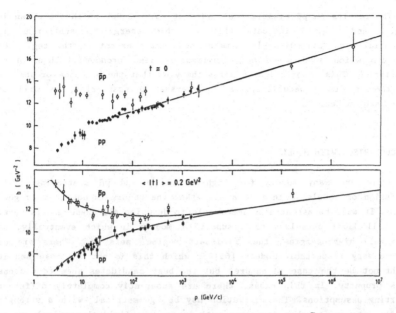

Fig. 17 Compilation of the slope parameters for pp and p$\bar{\text{p}}$ scattering as a function of energy, close to the forward direction and around $|t|$ = 0.2 GeV².

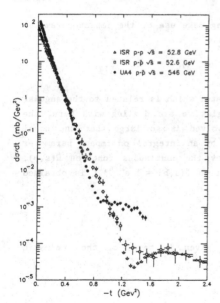

Fig. 18 The differential cross-section of elastic scattering at $\sqrt{s}$ = 546 GeV is shown together with ISR data on pp and p$\bar{\text{p}}$ scattering at $\sqrt{s}$ = 53 GeV.

The results on p$\bar{\text{p}}$ elastic scattering at $\sqrt{s}$ = 53 GeV which are shown in Fig. 2 and Fig. 18 indicate that at this energy p$\bar{\text{p}}$ scattering is essentially undistinguishable from pp scattering except in the region of the dip which appears to be significantly less pronounced than in pp scattering. This observation supports the view that the presence or absence of the dip is a peculiarity of pp versus p$\bar{\text{p}}$ scattering, as will be discussed in Sec. 4.

4. COMPARISON WITH MODELS

There are many models for high-energy pp and p$\bar{\text{p}}$ scattering. This profusion of models is in fact a sign that the theory is not in very good shape. It will be essentially impossible to describe all these models here. We shall limit ourselves to a subset of models in which eventually, at extremely high-energies, the "Froissart regime" sets on. There are of course very respectable models [66] in which this is not the case and it might not be the case in nature, but the best candidates seem to possess this property. In this class, there are apparently completely different starting assumptions. The motivation may be "geometrical" with a strength of the interaction that increases with energy, or field theoretical, or be in the framework of Regge theory with a supercritical Pomeron. The results, however, turn out to be very similar.

Most models of high-energy scattering make use of the impact parameter representation of the scattering amplitude

$$F(s,q) = (is/8\pi) \int e^{iqb} [1 - e^{-\Omega(s,b)}] d^2b \qquad (14)$$

where $q^2 = -t$ and b is the impact parameter which is related to the angular momentum ℓ by $bk = \ell + {}^1/_2$. At the energies we are dealing with here, the number of partial waves which are involved is so large that the usual partial wave expansion can be replaced by an integral on impact parameter while the phase shifts are replaced by the continuous function $\Omega(s,b)$, often called opacity. The profile function $\Gamma(s,b) = 1 - e^{\Omega(s,b)}$ is obtained by inverting Eq.(14)

$$\Gamma(s,b) = (2/i\pi s) \int e^{-iqb} F(s,q) d^2q$$

The inelastic cross-section $\sigma_{in}(s)$ is given in terms of the inelastic overlap function $G_{in}(s,b)$ by

$$\sigma_{in}(s) = \int G_{in}(s,b) d^2b$$

where $G_{in}(s,b) = 1 - |e^{-\Omega(s,b)}|$. The unitarity condition reads

$$G_{in}(s,b) = 2Re\Gamma(s,b) - |\Gamma(s,b)|^2.$$

By assuming, as it is indeed the case, that the scattering amplitude is dominantly imaginary, one may derive the inelastic overlap function from the data. This model independent analysis was performed by Henzi and Valin [67] on the ISR and Collider data. Their result, shown in Fig. 19, has the following intuitive meaning: as the energy increases from the ISR to the Collider, the effective size of the proton becomes larger. In addition the absorption in the central region, i.e. at small values of the impact parameter, approaches its maximum.

These features were qualitatively predicted by the geometrical model of Chou and Yang [68] where the basic idea is that the proton opacity essentially reflects the charge density as given by the electromagnetic form factor. In this model the opacity is assumed to factorize as the product of a function of energy by a function of the impact parameter as $\Omega(s,b) = A(s) K(b)$, where $K(b)$ is given by the transform of the electromagnetic form factor. The function A is determined by the value of the total cross-section. The model makes no prediction on the dependence of σ_{tot} on energy. However, in a later paper [69], Chou and Yang calculated the expected values of σ_{el}/σ_{tot} and of the height of the second maximum as a function of the observed cross-section. Their predictions are qualitatively in agreement with the experimental results at the Collider.

The geometrical picture of Chou and Yang is closely related to the multiple diffraction model of Glauber [70] which has been succesfully

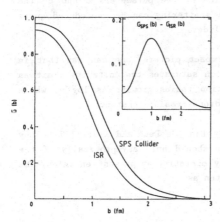

Fig. 19 The inelastic overlap function for $p\bar{p}$ scattering at $\sqrt{s} = 53$ GeV and for pp scattering at $\sqrt{s} = 546$ GeV as obtained by Henzi and Valin [67].

applied to the collisions of hadrons with nuclei. The Glauber model has been also used [55] to describe the SPS Collider data, by assuming that the proton and the antiproton each consist of a cluster of constituent particles ("partons") that interact in pairs as the clusters pass through each other. Proton-antiproton scattering is then calculated in terms of the elementary parton-parton scattering amplitude and of the electromagnetic proton form factor. By studying pp and p$\bar{\text{p}}$ data over a wide range of energy, one may follow the energy dependence of the parton-parton amplitude.

A model which has demonstrated a remarkable predicting power is that one which relies on the investigation by Cheng and Wu [5,8] of the asymptotic behaviour of higher-order diagrams in relativistic quantum field theories. From these studies it was concluded that the total and the elastic cross-sections must increase without bound in the high-energy limit. These theoretical considerations led to the development of a phenomenological model formulated in the impact picture framework which has met with remarkable success. In the later version of this model by Bourrely et al. [71], the opacity function is written in a essentially factorized form as

$$\Omega(s,b) = A(s) \, K(b) + R(s,b)$$

where $R(s,b)$ is a Regge term which is irrelevant at high-energy. The energy dependent factor $A(s)$ is given by

$$A(s) = s^c/(\log s)^{c'} + u^c/(\log u)^{c'}$$

while $K(b)$ is determined from the proton form factor. In this model the Froissart bound is eventually saturated and the ratio σ_{el}/σ_{tot} approaches 1/2 asymptotically. With a proper choice of the parameters c and c' the shape of the t-distribution of elastic scattering can be reproduced quite well both at the ISR and at the SPS Collider.

In the intuitive language of the impact picture, one can say that in this model the inelastic overlap function saturates the unitarity limit as the energy increases, with an interaction radius growing as log s, while the "diffuseness" of the interacting hadrons remains constant.

A different approach was followed by Dias de Deus and Kroll [17]. They assumed that the scaling law (9) which should hold asymptotically if the Froissart bound is saturated, is already operating at present energies. The differential cross-section is then written as

$$d\sigma/dt \ (s,t) = (1/16\pi) \ \sigma_{tot}{}^2(s) \ \{ \ \phi^2(\tau) + \rho^2(s) \ [d/d\tau(\tau\phi)]^2 \ \}.$$

If the values of σ_{tot} and ρ at one energy are known together with the differential cross-section $d\sigma/dt$ as a function of t, then $d\sigma/dt$ at a new energy can be predicted if the values of σ_{tot} and ρ at the new energy are also given. In this model the ratios σ_{el}/σ_{tot} and $b(t = 0)/\sigma_{tot}$ stay constant with energy. If the real part of the amplitude is neglected, this model can be reformulated in the impact parameter representation by saying that the opacity has the special form, $\Omega(s,b) = \Omega[b/R(s)]$ where all the energy dependence of the opacity is included in the scaling parameter R(s). As a consequence one expects $\Omega(b = 0)$ to be energy independent. The geometrical scaling model gives a satisfactory description of pp elastic scattering in the whole energy range of the ISR [17,72]. However, the rise with energy of the ratio σ_{el}/σ_{tot} between ISR and SPS Collider and the level of the differential cross-section at the second maximum cannot be explained. Geometrical scaling is, therefore, a transient phenomena, not to be confused with the asymptotic scaling that we expect at energies which might never be reached.

In the category of Regge inspired models, we have already decided to leave out the "critical Pomeron" [73] which has a very beautiful set of asymptotic predictions, one of them being that the ratio of the elastic to the total cross-section decreases with energy, contrary to the observation. We are left with theories in which the Pomeron is supercritical, i.e. the dressed Pomeron has an intercept $\alpha(0) > 1$. Hopefully such an object might be produced in QCD by the exchange of a chain of gluons pairs, as proposed some years ago by Nussinov [74] and Low [75], so that the basic mechanism might in fact be not terribly different from the one of Cheng and Wu, where one has exchange of "fermion loops" and photon pairs in massive QED.

In these models there are two attitudes : in the first case, one sums the whole series of basic processes and one gets then an amplitude that explicitly does not violate, and in fact saturates the Froissart bound (this is the case of the models of Kaidalov and Ter Martirosyan [76]). The second attitude (taken, for instance, by Capella et al. [77], and more recently by Collins [78] and by Donnachie and Landshoff [79]) consists in noticing that present energies (540 GeV up to 1800 GeV) are very far from the asymptotic regime and that the first term or the first few terms are probably sufficient. Then there is an apparent violation of the Froissart bound, but this violation takes place at energies much beyond the domain where this approximation is valid.

The problem of the dip or shoulder has received much attention and has been discussed by several authors. For example in the model of Islam et al. [80], the scattering in this momentum transfer region is described as due to diffraction and hard scattering which interfere causing the observed structure. In this model the dip is a transient phenomenon, common to both pp and p$\bar{\text{p}}$ scattering, which disappears slowly as the energy increases.

In the model of Donnachie and Landshoff [79] on the other hand, the structure in the -t ~ 1 GeV^2 region is due to a complicated interplay between single-Pomeron, double-Pomeron and three-gluon exchange. The three-gluon exchange amplitude dominates at large momentum transfer and has opposite signs for p$\bar{\text{p}}$ and pp scattering. The interference with this amplitude is responsible for the presence of the dip in pp scattering and its absence in p$\bar{\text{p}}$ scattering at all energies.

5. CONCLUSIONS AND OUTLOOK.

We have seen that the domain of high-energy proton-proton and proton-antiproton scattering is very rich and full of experimental surprises. Some of them are now well accepted like the rise of the total cross-section which is well reproduced by reasonable models. There are other unexpected experimental results which have to be confirmed, like the increase of the ratio σ_{el}/σ_{tot}, the behaviour of the differential cross-section in the region of the dip (or shoulder) which is different for pp and p$\bar{\text{p}}$ and is also strongly energy dependent. The large value of the real part is a puzzle. We think that there is a tremendous need of clarification of the situation both on the experimental and theoretical side. This can be done with existing hadron colliders at CERN and at Fermilab, and possible future colliders as the SSC or the Large Hadron Collider in the LEP tunnel.

We have seen that one major area where lack of experimental information allows all sorts of speculations is that of the particle-particle to particle-antiparticle comparison. This means that experiments like UA6 which indeed make this comparison possible for pp and p$\bar{\text{p}}$ should be vigorously pursued in a wide range of values of t. This also means that an analogous of UA6 should be considered at Fermilab. It would be desirable, of course, to have such a comparison also at the future colliders, but we feel that it would be unrealistic to consider a double machine for pp and p$\bar{\text{p}}$ collisions with elastic scattering and total cross-section as motivations, unless this is at no extra cost. What can be done, however, is to extend the particle-particle to particle-antiparticle comparison to other channels by using fixed target accelerators. Already the Tevatron at Fermilab will allow to explore a domain where nothing is presently known on the comparison of π^+p to π^-p, and K^+p to K^-p, but the machine which could make a big step forward is UNK in the Soviet Union. Then, accurate measurements will make it possible to look for any sign of the "odderon". In addition, measurements of the total cross-section, including those made with the hyperon beams, will make it possible to see whether the additive quark model rules, in particular the famous 3/2 ratio between pp and πp, are satisfied in the region where cross-sections are rising. These rules in fact until now have been seen to work only in an energy domain where the cross-sections are approximately flat.

REFERENCES

1. Lacombe, M. et al., Phys. Rev. C21, 861 (1980).

2. Coté, J. et al., Phys. Rev. Lett. 48, 1319 (1982).
   Lacombe, M. et al., Phys. Rev. C29, 1800 (1984).
   Dover, C.B. and Richard, J.M., Ann. Phys. 121, 47 (1979).

3. Gribov, V.N., Proc. of the 1960 International Conference on High-Energy Physics, Rochester, 1960, p.340.

4. Cocconi, G., Proc. of the 1962 International Conference on High-Energy Physics at CERN, Geneva, 1962, p.883 and Diddens, A.N., ibidem, p.576.

5. Cheng, H. and Wu, T.T., Phys. Rev. Lett.22, 666 (1969) and "Expanding Protons - Scattering at High Energies", M.I.T. Press, 1987.

6. Froissart, M., Phys. Rev. 123, 1053 (1961).

7. Martin, A., Phys. Rev. 129, 1432(1963) and Nuovo Cim. 42, 930 (1966).

8. Cheng, H. and Wu, T.T., Phys. Rev. Lett. 24, 1456 (1970) and Phys. Rev. D1, 2775 (1970).

9. Amaldi, U. et al., Phys. Lett. 44B, 112 (1973).
   Amendolia, S.R. et al., Phys. Lett. 44B, 119 (1973). See also, "A discussion on proton-proton scattering at very high energies", organized by J.M. Cassels and A.M. Wetherell, in Proc. R. Soc. London A335, 409-507 (1973).

10. Khuri, N.N. and Kinoshita, T., Phys. Rev. 137, B720 (1965).

11. Carroll, A.S. et al., Phys. Rev. Lett. 33, 928 (1974), Phys. Lett. 61B, 303 (1976) and 80B, 423 (1979).

12. Amaldi, U. et al., Phys. Lett. 66B, 390 (1977).

13. UA4 Collaboration, Bozzo, M. et al., Phys. Lett. 147B, 392 (1984).

14. Block, M.M. and Cahn, R., Phys. Lett. 120B, 224 (1983) and Rev. Mod. Phys. 57, 563 (1985).

15. UA4 Collaboration, Bernard, D. et al., Phys. Lett. 198B, 583 (1987).

16. Bourrely, C. and Martin, A., Proc. of the ECFA-CERN Workshop on Large Hadron Collider in the LEP Tunnel, Lausanne and Geneva, 1984 (CERN 84-10), vol.1, p.323.
    Block, M.M. and Cahn, R., Phys. Lett. 188B, 143 (1987).

17. Dias de Deus, J. and Kroll, P., Acta Phys. Pol. B9, 157 (1978),
    Kroll, P., 3rd Moriond Workshop, La Plagne, 1983, p.81.

18. UA4 Collaboration, Bozzo, M. et al., Phys. Lett. 155B, 197 (1985) and Phys. Lett. 171B, 142 (1986).

19. For a recent review of high-energy scattering, see :
    Castaldi, R. and Sanguinetti, G., Ann. Rev. Nucl. Part. Sci. 35, 351 (1985) and references therein.

20. Roy, S.M., Phys. Rep. 5C, 125 (1972).
    Fischer, J., Phys. Rep. 76, 157 (1981).

21. CERN-Pisa-Rome-Stony Brook Collaboration, Phys. Lett. 62B (1976) 460. and Nucl. Phys. B145 (1978) 367.

22. Martin, A., Journal de Physique, C2, 727 (1985).

23. Bros, J., Epstein, H. and Glaser, V., Nuovo Cim. 31, 1265 (1965).

24. Epstein, H., Glaser, V. and Martin, A., Commun. Math. Phys. 13, 257 (1969).

25. Lukaszuk, L. and Martin, A., Nuovo Cim. 52A, 122 (1967).

26. Jin, Y.S. and Martin, A., Phys. Rev. 135, B1375 (1364).

27. Söding, P., Phys. Lett. 8, 285 (1964).

28. MacDowell, S.W. and Martin, A., Phys. Rev. 135, B960 (1964).

29. Auberson, G., Kinoshita, T. and Martin, A., Phys. Rev. D3, 3185 (1971).

30. Pomeranchuk, I.Ya., JETP 7, 499 (1958).

31. Eden, R.J., Phys. Rev. Lett. 16, 39 (1966).
    Kinoshita, T., in Perspectives in Modern Physics, New York, 1966,
    p.211.
    Grunberg, G. and Truong, T.N., Phys. Rev. Lett. 31, 63 (1973).

32. Van Hove, L., Phys. Lett. 5, 252 (1963).

33. Logunov, A.A. et al., Phys. Lett. 7, 69 (1963).

34. Cornille, H. and Martin, A., Phys. Lett. 40B, 671 (1972) and Nucl.
    Phys. B48, 104 (1972).

35. Breakstone, A. et al., Nucl. Phys. B248, 253 (1984).

36. Amos, A. et al., Nucl. Phys. B262, 689 (1985).

37. Carboni, G. et al., Nucl. Phys. B254, 697 (1985).

38. Fischer, J. et al., Phys. Rev. D18, 4271 (1978).

39. Cornille, H., Lett. Nuovo Cim. 4, 267 (1970).

40. Łukaszuk, L. and Nicolescu, B., Nuovo Cim. Lett. 8, 405 (1973).
    Kang, K. and Nicolescu, B., Phys. Rev. D11, 2461 (1975).

41. Van der Meer, S., CERN internal report ISR-PO/68-31 (1968).

42. Bosser, J. et al., Nucl. Instrum. Methods A235, 475 (1985).

43. UA4 Collaboration, Bernard, D. et al., Phys. Lett. 186B, 227 (1987).

44. UA4 Collaboration, Battiston, R. et al., Nucl. Instr. Methods A238,
    35 (1985).

45. Faugeras, P.E., CERN Internal note CERN SPS/84-7(ARF), 1984.
    Faugeras, P.E. et al., CERN Internal note SPS/DI-MST/ME/85-03.

46. Haguenauer, M. and Matthiae, G., Proc. of the ECFA-CERN Workshop on
    Large Hadron Collider in the LEP Tunnel, Lausanne and Geneva, 1984
    (CERN 84-10), vol.1, p.303.

47. UA4 Collaboration, Battiston, R. et al., Phys. Lett. 117B, 126
    (1982).

48. UA1 Collaboration, Arnison, G. et al., Phys. Lett. 121B, 77 (1983).

49. UA5 Collaboration, Alner, G.J. et al., Z. Phys. C32, 133 (1986).

50. Buttimore, N.H., Gotsman, E. and Leader, E., Phys. Rev. D18, 694 (1978).
    Cahn, R., Z. Phys. C15, 253 (1982) and references therein.

51. Leader, E., Phys. Rev. Lett. 59, 1525 (1987) and Proc. of the 2nd International Confference on Elastic and Diffractive Scattering, The Rockefeller University, New York, 1987.
    Kluit, P.M. and Timmermans, J., Phys. Lett. 202B, 458 (1988).

52. Martin, A., CERN-TH.4852/87 (1987) and Proc. of the 2nd International Conference on Elastic and Diffractive Scattering, The Rockefeller University, New York, 1987 (K. Goulianos editor).
    Hadjitheodoridis, S. and Kang, K., Phys. Lett. 208B, 135 (1988).

53. UA1 Collaboration, Albajar, C. et al., to appear on Nucl. Phys. B.

54. Innocente, V., Capella, A. and Tran Thanh Van, J., Orsay preprint, LPTHE Orsay 88/30, 1988.
    Margolis, B., University of Winsconsin preprint, MAD/PH/425, 1988.
    See also Durand, L. and Hong Pi, Phys. Rev. Lett. 58, 303 (1987).

55. Glauber, R.J. and Velasco, J., Phys. Lett. 147B, 380 (1984).

56. Bourrely, C., Soffer, J. and Wu, T.T., Phys. Lett. 196, 237 (1987).
    and preprint CPT-87/P.2032 (1987).

57. Bernard, D., Gauron, P. and Nicolescu, B., Phys. Lett. 199B, 125 (1987).
    Nicolescu, B., Proc of the 2nd International Conference on Elastic and Diffractive Scattering, The Rockefeller University, New York, 1987.

58. Igi, K. and Kroll, P., preprint University of Tokyo, UT532 (1988).

59. Breedon, R., Proc. of the XIX International Symposium on Multiparticle Dynamics, Arles, 1988.

60. UA4 Collaboration, Bozzo, M. et al., Phys. Lett. 147B, 385 (1984).

61. Barbiellini, G. et al., Phys. Lett. 39B, 663 (1972).

62. Giffon, M., Nahabetian, R.S. and Predazzi, E., Z. Phys. C36, 67 (1987).

63. Amos, N.A. et al., Phys. Rev. Lett. 61, 525 (1988).

64. Nagy,E. et al., Nucl.Phys. B150,221(1979).

65. Breakstone, A. et al., Phys. Rev. Lett. 54, 2180 (1985).

66. See for instance, Jenkovszky, L.L., Riv. Nuovo Cim. 10, 1 (1987).

67. Henzi, R. and Valin, P., Phys. Lett. 132B, 443 (1983),
    Henzi, R., Proc. of the 4th Topical Workshop on Proton-Antiproton Collider Physics, Bern, 1984, p.314.

68. Chou, T.T. and Yang, C.N., Phys. Rev. Lett. 20, 1213 (1968) and Phys. Rev. 170, 1591 (1968).

69. Chou, T.T. and Yang, C.N., Phys. Rev. D19, 3268 (1979).

70. Glauber, R.J., in High-Energy Physics and Nuclear Structure, New York, 1970, p.207.

71. Bourrely, C., Soffer, J. and Wu, T.T. Nucl. Phys. B247, 15 (1984) and Phys. Rev. Lett. 54, 757 (1985).

72. Amaldi, U. and Schubert, K.R., Nucl. Phys. B166, 301 (1980).

73. Baumel, J., Feingold, M. and Moshe, M., Nucl. Phys. B198, 13 (1982).

74. Nussinov, S., Phys. Rev. Lett. 34, 1286 (1974) and Phys. Rev. D14, 246 (1976).

75. Low, F.E., Phys. Rev. D12, 163 (1975).

76. See for instance, Kaidalov, A.B., Phys. Rep. 50, 157 (1977).

77. Capella, A. and Kaplan, J., Phys. Lett. 52B, 448 (1974).
    Capella, A., Tran Thanh Van, J. and Kaplan, J., Nucl. Phys. B97, 493 (1975).

78. Collins, P.D.B., Proc. of the 1st International Conference on Elastic and Diffractive Scattering, Blois, 1985, p.339.

79. Donnachie, A. and Landshoff, P.V., Phys. Lett. 123B, 345 (1983) and Nucl. Phys. B231, 189 (1983).
Landshoff, P.V., Proc. of the 1st International Conference on Elastic and Diffractive Scattering, Blois, 1985, p.209.

80. Islam, M.M. et al., Nuovo Cim. 81A, 737 (1984).

PROPERTIES OF SOFT
PROTON – ANTIPROTON COLLISIONS.

D.R.Ward

CERN, 1211 Genève 23, Switzerland.
and
Cavendish Laboratory, Madingley Road, Cambridge, England.

INTRODUCTION

The primary reason for building the CERN SPS collider ("Spp̄S") was to reach an energy high enough to produce the W and Z weak bosons. In this the project was triumphantly successful, as described in other contributions to this volume. However, much effort has also been devoted to the study of "soft" processes in which only low p_T particles are produced. Why? The first reason is simply that such events are there. Indeed, more than 99% of all particles produced have $p_T < 2\text{GeV}/c$. Although we have no rigorous theory of such processes (or rather we do not know how to apply QCD in this case) we should still try to understand what is going on, and to systematize the properties of this vast bulk of interactions. Furthermore, the "soft" events constitute a source of background to the other events, which seem, at present, more interesting. It is therefore necessary to have some understanding of the soft events, even if only in an empirical way. Finally we should mention the possibility of exotic effects showing up in the new energy regime opened up at the Spp̄S. Indeed, there have been reports of several curious effects in cosmic ray data at these energies [1], which appear to occur with large cross-sections. It would be foolish not to search for such phenomena, which should show up in "minimum bias" data taken for the study of soft processes.

In subsequent sections we first briefly outline the Collider experiments, stressing those features important for the present topics. The experimental results will then be reviewed, mainly from an empirical point of view, particularly emphasizing the energy dependence of the phenomena studied. Of necessity the choice of data and the discussion will be selective, but we shall try to give full references to the original sources of data, where more details can be found. We shall attempt to cover most aspects of "soft" physics, with the exception of elastic scattering and the total cross-section, which are discussed elsewhere in this volume. A very brief review of the many models which have been proposed to describe various features of these data will then be given.

THE SPS COLLIDER AND THE EXPERIMENTS

The SPS Collider

The data discussed below generally come from periods when the Spp̄S was operating at very low luminosity. These were, for example, the early machine development runs of the machine, without low-β. In the very first runs of the Spp̄S in October–November 1981 [2] luminosities of around $10^{25}\text{cm}^{-2}\text{s}^{-1}$ were achieved, and in 1982 of typically $10^{26}\text{cm}^{-2}\text{s}^{-1}$. Such luminosities were ideally suited for triggering with a "minimum-bias" trigger designed to accept a large fraction of "soft" collisions. Most of the results presented below at $\sqrt{s} = 546\text{GeV}$ come from these early runs. In addition some special runs at different β values were taken to allow UA4 to explore elastic scattering in different t-ranges.

An important run for the study of "soft" physics was the so-called Pulsed Collider run of March–April 1985. In normal Collider running the beam energy was limited to 270GeV (or more recently 315GeV) because of power dissipation in the magnets. In 1982 a method was proposed [3] whereby the beam energy would be cycled repeatedly between 100GeV and 450GeV, enabling one to observe collisions at $\sqrt{s}$ = 900GeV with a duty factor of around 20%. This scheme was successfully implemented and a Pulsed Collider run was carried out in 1985, with luminosities of typically $10^{26}\text{cm}^{-2}\text{s}^{-1}$ and beam lifetimes of a few hours. This run allowed the study of soft $\bar{p}p$ collisions at $\sqrt{s}$ = 900GeV, and incidentally also at $\sqrt{s}$ = 200GeV, giving a useful intermediate point between the old Sp$\bar{p}$S data and data from the CERN ISR.

The experiments

There were two major experiments at the SPS Collider aimed principally at the study of "soft" processes. The UA4 experiment used "Roman pots" [4] set into the beam pipe in the far-forward region to study elastic scattering, and thereby the total cross-section. These topics are discussed elsewhere in this volume. In addition UA4 had some forward drift chambers which, combined with the central detector of UA2 allowed them to study single diffractive dissociation (see section "DIFFRACTION DISSOCIATION" below).

The other detector which has concentrated on soft inelastic processes was UA5 [5,6], shown in Figure 1. The main feature of this detector was a pair of very long streamer chambers, giving acceptance for particles out to pseudorapidity[†] = ±5. A detailed description of this detector and analysis techniques appears in [5]. A large share of the results discussed in this paper come from the UA5 experiment. Incidentally, the UA5 experiment also took data in a test run at the CERN ISR [7,8], meaning that it took data over an unusually wide range of c.m. energy, from 53 to 900GeV.

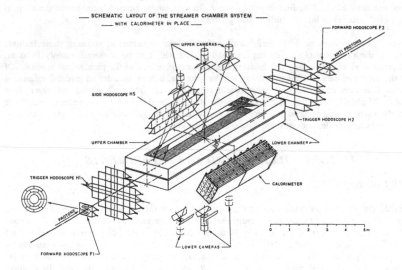

Figure 1:   Schematic layout of the UA5 detector.

[†] $\eta \equiv -\ln\tan\dfrac{\theta}{2}$, where θ is the c.m. polar angle of production)

The UA1 and UA2 experiments were primarily designed for the study of vector bosons and high p_T phenomena. However, especially in the early days of Collider running when the luminosity was low, these experiments made valuable measurements on soft processes. UA1 is a general purpose detector, and thus capable of many studies in the soft physics area, but once the luminosity was high enough their main interests were obviously in other areas. In addition, UA1 took data during the Pulsed Collider run. UA2 is perhaps more specialized towards W, Z and jet physics, but during their first runs they replaced part of their central calorimeter by a "wedge" spectrometer, to observe particle production at 90° in soft collisions.

Triggering

Before discussing the results from the Sp$\bar{p}$S a few points should be made concerning triggering, since the problems are somewhat different from lower energy experiments. The discussion will be centred on the UA5 detector (Figure 1), but the principles are the same for the others. The normal trigger used is called a "two-arm" trigger, requiring hits in the trigger counters at both ends of the detector in coincidence with a beam crossing. UA1 and UA2 used a similar trigger. This is often referred to as a "minimum-bias" trigger, but this term can be misleading. The two-arm trigger accepts most inelastic events, except that it has a low efficiency for single-diffractive (SD) events, which are normally highly asymmetric. The recoil proton or $\bar{p}$ in a diffractive event essentially always passes inside the beam pipe. Therefore to capture SD events a single-arm trigger must be employed, demanding hits in only one set of trigger counters. Such a trigger was very susceptible to background, and was only enabled by UA5 for a few runs, most successfully in the Pulsed Collider run. UA4 were able sometimes to detect the recoil p/$\bar{p}$ in SD events in their pots.

The typical relationship between event classes and triggers is outlined in Figure 2. The precise numerical values vary from experiment to experiment, and from energy to energy. To a first approximation, the two-arm trigger which is normally employed corresponds to non−single-diffractive (NSD) events, and most of the results from the Sp$\bar{p}$S correspond to NSD events. In contrast, the lower energy data with which we wish to compare generally correspond to the full inelastic cross-section. They may in some cases be converted to NSD data using published results on diffraction dissociation. In the Sp$\bar{p}$S data we only have full inelastic (sometimes called "inclusive") data for cases where it was possible to combine single-arm and two-arm triggers. This must be borne in mind when comparing data at different energies.

CHARGED PARTICLE MULTIPLICITIES

Average Multiplicity

There are extensive data at lower energies on charged particle multiplicity distributions and thus on average multiplicities. The bulk of these data come from bubble chamber experiments, where this is one of the most straightforward measurements to make. The experimental problems are somewhat different at a colliding beam machine, where even the best detectors have limited acceptance. At the Sp$\bar{p}$S the UA5 detector has the greatest acceptance for charged tracks. The UA5 correction procedures are described in detail elsewhere [5,9], and the data are given in [5,9,10,11].

The data for $<n_{ch}>$ versus $\sqrt{s}$ in NSD events are shown in Figure 3. Two fits to the data are also shown, a quadratic in $\ln s$ (solid curve):

$$<n_{ch}>_{NSD} = 2.7 - 0.03\ln s + 0.176\ln^2 s \quad : \chi^2/d.f. = 17/9 \tag{1}$$

and a power law (dashed curve):

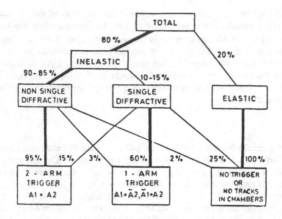

Figure 2:   Relationship between event categories and trigger types. The numerical values given relate to the UA5 experiment at $\sqrt{s} = 900\text{GeV}$.

$$<n_{ch}>_{NSD} = -7.0 + 7.2s^{0.127} \qquad : \chi^2/d.f. = 11/7 \tag{2}$$

The $\ln^2 s$ fit, which gave an entirely adequate description of the data up to ISR energies, is still a reasonable fit, though there is maybe a hint that the data are rising above this fitted curve. Indeed, the power law fit gives a slightly better χ^2, and fits the higher data points better, but the difference is clearly marginal considering the errors on the Sp$\bar{\text{p}}$S data. Similar conclusions follow for the full inelastic $<n_{ch}>$ (not shown). Further discussion of average multiplicities of various particle species appears in [5].

Multiplicity Distributions and KNO scaling

The Sp$\bar{\text{p}}$S data from UA5 and UA1 have revived interest in the subject of KNO scaling. The original KNO hypothesis [12]

$$<n>\sigma_n/\sum_0^\infty \sigma_n \xrightarrow[s\to\infty]{} \psi(z) \ ; \ where \ z = n/<n> \tag{3}$$

was based on Feynman scaling [13], which has long been known to be violated in the central region [14]. Nonetheless, KNO scaling, which implies that the *shape* of multiplicity distributions should *asymptotically* become energy independent, has proved a useful framework for the discussion of the energy dependence of multiplicity distributions. Data up to ISR energies had shown that the full inelastic distributions did not scale [14]. One reason for this might be the mixing of SD and NSD processes, with different energy dependences, and indeed the NSD data up to $\sqrt{s} = 63\text{GeV}$ were compatible with KNO scaling [15]. However, it should be pointed out that there are fewer data available for NSD than for full inelastic, and that the different experiments use different techniques for the separation of the SD events.

The early UA5 data from the Sp$\bar{\text{p}}$S [16] showed that KNO scaling between FNAL and Sp$\bar{\text{p}}$S energies was excluded for NSD events. Later data from UA5, by which time ISR data were available

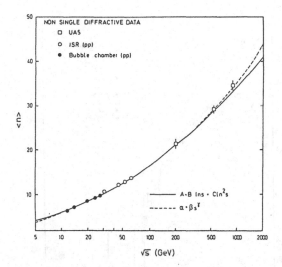

Figure 3:   Average charged multiplicity in NSD events vs. $\sqrt{s}$. The curves are described in the text.

also, confirmed the effect [17]. This is shown by the data of Figure 4, which compares the UA5 546GeV data in KNO-scaled form with lower energy data. The UA5 data clearly follow a broader distribution.

The effect is confirmed by the 200 and 900GeV NSD data [10]. For example, Figure 5 shows the C-moments[†] as a function of $\sqrt{s}$ up to 900GeV, and also the fractions of events having $z > 1.5, 2$ or 2.5. The C-moments and the fractions of high multiplicity events are rising throughout the Sp$\bar{p}$S energy range, and the effect is most pronounced for the highest multiplicities.

For the full inelastic data, KNO scaling was already ruled out by the ISR data [14]. For example Figure 6 shows a "Wróblewski" plot, of D_2,[††] against $<n>$. The data up to ISR energies lie nicely on straight lines. However, unlike the NSD case, the inelastic data cross the $<n>$-axis not at the origin but at $<n> = \alpha$, where α is close to 1. This corresponds to KNO scaling in the variable $(n - \alpha)$ rather than n, and may be regarded as "asymptotic KNO scaling" as n becomes large [18,5]. However, the Sp$\bar{p}$S data from UA5 show clear deviations from even this behaviour, and thus seem to show no sign of asymptotic scaling. Another extension of KNO scaling, "KNO-G scaling" [19], is close to this $(n - \alpha)$ scaling behaviour, and is equally incompatible with the Sp$\bar{p}$S data. A recent analysis, in the context of a statistical model, has shown that the data can only be accomodated in such a modified KNO framework if the parameter α is allowed to vary with $\sqrt{s}$ [20].

[†] The C-moments, defined by $C_q = <n^q> / <n>^q$, should be independent of c.m. energy in the case of ideal KNO scaling.

[††] The D-moments are defined by $D_q = <(n - <n>)^q>^{1/q}$, and D_q should be proportional to $<n>$ in the case of perfect KNO scaling. Only D_2 is shown in Figure 6, but the higher moments show the same behaviour.

90

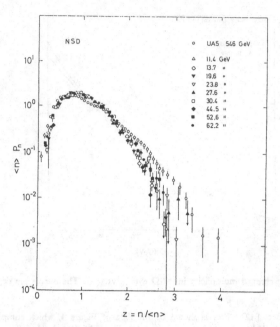

Figure 4: Test of KNO-scaling for NSD multiplicity distributions. Note that the UA5 546GeV distribution is broader.

However, there were indications from the early UA1 data [21,22] that KNO scaling held reasonably well in the central region, $|\eta| < 1.5$, and this appeared to be confirmed by the UA5 data [17], though there were some doubts whether the trigger conditions were strictly comparable between the ISR and Sp$\bar{\text{p}}$S data. Recently, UA5 has made a systematic study of multiplicity distributions at $\sqrt{s}$ = 200, 546 and 900GeV, for various rapidity regions, $|\eta| < \eta_c$ [23,24,11]. In Figure 7 the energy dependence of some C_2 for various values of η_c is shown. We see that the value of C_2 grows with energy for all values of η_c, except for $\eta_c \lesssim 0.5$, where C_2 is rather constant, and maybe even starts to fall with $\sqrt{s}$. The higher C-moments show similar behaviour, though possibly flattening out at slightly lower values of η_c. These data have passed through the same correction procedures at each energy, and are therefore free of many of the uncertainties which affected the earlier comparisons. This suggests that the earlier observation of approximate scaling for $\eta_c = 1.5$ may have been somewhat fortuitous.

Negative Binomial Description of Multiplicity distributions

Recently UA5 have analyzed their multiplicity data in a different way, which gives an alternative view to KNO scaling. They have observed [25,10] that the NSD multiplicity distributions at all energies may be very well described by the negative binomial distribution:

$$P(n; <n>, k) = \begin{bmatrix} n+k-1 \\ k-1 \end{bmatrix} \frac{(<n>/k)^n}{(1+<n>/k)^{n+k}} \tag{4}$$

This distribution had previously been proposed for such an application by several authors (e.g. refs. [26,27,28]). The UA5 fits are very good, though of course one cannot exclude the possibility that the

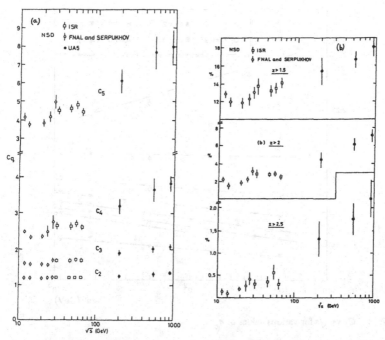

Figure 5: (a) C-moments for NSD data vs. $\sqrt{s}$. (b) fraction of events having multiplicity greater than some multiple of $<n>$, vs. $\sqrt{s}$.

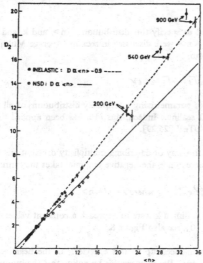

Figure 6: Plots of D_2 versus $<n>$ for inelastic and NSD data.

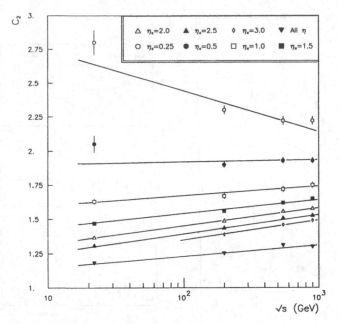

Figure 7: C_2 vs. $\sqrt{s}$ for various values of η_c

fits are only approximate; indeed, UA5 suggest that a negative binomial distribution for underlying clusters might fit equally well given the statistical errors. There is also some hint in the latest UA5 data that the quality of the fit may be less good for the full pseudorapidity range at the highest Sp$\bar{p}$S energy, $\sqrt{s} = 900$GeV [11].

Two parameters are required to specify this distribution, $<n>$ and k, and it turns out that both vary smoothly with $\sqrt{s}$. Fits to $<n>$ were discussed in section "Average Multiplicity" above, whilst k may be very well fitted by the form

$$k^{-1} = -0.104 + 0.058\ln\sqrt{s} \quad ; \sqrt{s} \; in \; GeV \tag{5}$$

This leads one to a simple way of parametrizing multiplicity distributions at all energies, using the fits to $<n>$ and k instead of KNO scaling. Indeed, this idea has been applied to "predict" multiplicity distributions at the SSC, $\sqrt{s} = 40$TeV [25,29].

It is of interest to compare this way of describing multiplicity distributions with the KNO scaling scheme. First we note that for large $<n>$ the negative binomial takes the form:

$$<n> P(n;k) \approx [k^k/\Gamma(k)]z^k e^{-kz} \quad ; \quad where \; z = n/<n> \tag{6}$$

This would correspond to KNO scaling if k were to approach a constant value. However up to $\sqrt{s} = 900$GeV there is no sign of this [10], see also Figure 8.

What then of the KNO scaling seen up to ISR energies? In the negative binomial picture this scaling would appear to be fortuitous. We may see this by noting that the moment γ_2 is given by:

$$\gamma_2 \equiv C_2 - 1 = 1/<n> + 1/k. \tag{7}$$

The data of Figure 8 show that $1/<n>$ falls with $\sqrt{s}$, while $1/k$ rises, and over the FNAL/ISR energy region they add to give a broad minimum. Similar formulae apply to the higher C-moments, and thus approximate KNO scaling ensues. Since the parameters $<n>$ and k have smooth and simple energy dependence one might infer that the KNO scaling up to ISR energies was accidental. Certainly the data of Figure 5 do not conclusively establish the existence of scaling; arguably the FNAL/ISR data have neither a long enough lever arm in energy nor sufficient precision.

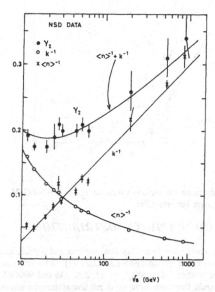

Figure 8:   The smooth variation of $<n>^{-1}$ and k^{-1} with $\sqrt{s}$. Thus approximate KNO scaling appears for $10 \leq \sqrt{s} \leq 100\text{GeV}$.

Another point of view is that the KNO scaling up to $\sqrt{s} = 63\text{GeV}$ is significant, and that the scaling violations at the Sp$\bar{p}$S reflect the onset of new physics. The likeliest candidate for such a new component is a rising hard or semi-hard scattering component. For example the authors of ref [30] have argued that this is a consistent view of the data. Indeed, there is little doubt that such processes are becoming more important, since UA1 have evidence for such a contribution (see section "Mini-jets" below). This topic will be discussed further under "Microscopic models" below. It is not clear, however, how to reconcile such a picture with the smooth behaviour of the negative binomial parameters from FNAL/ISR energies (where jets are unimportant) to the highest Sp$\bar{p}$S energy.

We should also note that the negative binomial distribution gives excellent fits to multiplicity distributions in limited rapidity intervals [31,11], see Figure 9. The parameter k rises approximately linearly with the size of the rapidity interval, approaching a value ≈ 1.5 for very small intervals. As $\sqrt{s}$ increases, then k rises for all rapidity intervals (in contrast to C_2, for which the trend is reversed in small intervals — this is because C_2 includes dependence on $<n>$ as well as k). Note that a Poisson distribution is a special case of a negative binomial with $k = \infty$. Thus the multiplicity distributions are relatively broader, and deviate increasingly from a Poisson form, as the rapidity interval is made smaller. Further discussion of the negative binomial distribution in the context of models appears in "Statistical models" below.

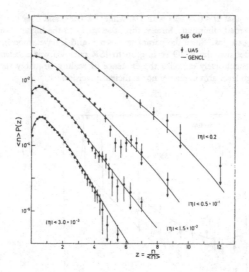

Figure 9:   Multiplicity distributions for various values of η_c at $\sqrt{s} = 546\text{GeV}$. The curves show the quality of the negative binomial fits.

PSEUDORAPIDITY DISTRIBUTIONS

Measurements of charged particle pseudorapidity distributions and particle multiplicities near $\eta = 0$ were among the first observations made at the Sp$\bar{\text{p}}$S [32,33,21,22]. The latest and most precise data come from UA5 [34], who have analyzed data at $\sqrt{s} = 53, 200, 546$ and 900GeV with essentially the same techniques. These data should therefore have small relative systematic errors, and be particularly suitable for study of energy dependences. Both NSD and inelastic data have been presented, and the general conclusions drawn from both are similar.

In Figure 10 the UA5 NSD data are shown. We see that both the height and width of the pseudorapidity distributions increase with $\sqrt{s}$. The charged particle density at $\eta = 0$, $\rho(0)$, is plotted against $\sqrt{s}$ in Figure 11. The growth of $\rho(0)$ with $\sqrt{s}$ confirms the violation of Feynman scaling seen at the ISR [14]. Good fits to the energy dependence of $\rho(0)$ (for the inelastic data) are given by the forms:

$$\rho(0) = 0.01 + 0.22\ln s \quad ; \quad \chi^2/d.f. = 6.0/7 \tag{8}$$

or

$$\rho(0) = 0.74 s^{0.105} \quad ; \quad \chi^2/d.f. = 4.8/7 \tag{9}$$

The χ^2-values clearly do not allow one to choose between these forms.

The question of Feynman scaling in the beam fragmentation region is however not settled. In an attempt to look at this question the UA5 inelastic data for $\rho(\eta)$ have been plotted (Figure 12) against the variable η_{beam}

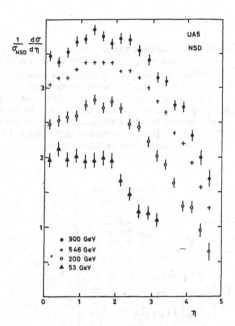

Figure 10: Charged particle NSD η distributions from UA5.

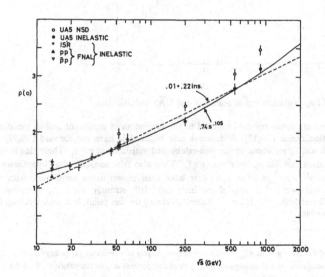

Figure 11: Energy dependence of $\rho(0)$ for both NSD and inelastic data. The fits are described in the text.

$$\eta_{beam} \equiv \eta - Y_{beam} \; ; \quad where \, Y_{beam} = \ln(\sqrt{s}/m_p) \qquad\qquad (10)$$

which approximates to the particle rapidity in the beam's rest frame. We see that the data from $\sqrt{s} =$ 53 to 900GeV roughly scale (to $\pm$ 10%) in the region $|\eta_{beam}| < 2.5$.[†] One simple way to describe the data in the beam fragmentation frame would be by a rise along an energy-independent curve until the (energy-dependent) plateau level $\rho(0)$ is reached, where the distribution flattens out. On closer inspection, however, there is perhaps a hint that the higher energy data fall off slightly more steeply than the lower energy points, suggesting some scaling violations. The data do not seem to be quite precise enough, or the rapidity range not quite wide enough, to resolve this question. Maybe the UA7 experiment, who have just presented their first results on photon production in the region $5 < |\eta| < 7$ [35] using very forward calorimetry will be able to shed further light on this question.

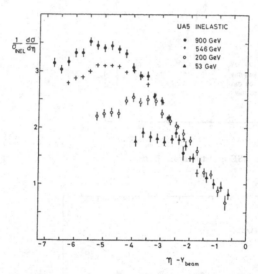

Figure 12: Test of fragmentation region scaling in the UA5 inelastic data.

Some interpretations of cosmic ray data have led to the suggestion of significant scaling violations in the beam fragmentation region [36,37]. Wdowczyk and Wolfendale have put forward [38,39] an elegant way to link such scaling violations to the well-established violations at $\eta = 0$. Their idea is simply to replace the Feynman variable x_F by $x' = x_F(s/s_0)^{\tau}$. They also take account of a (hypothesized) decrease in the inelasticity[††] with $\sqrt{s}$. The UA5 data have been shown to be compatible with the Wdowczyk-Wolfendale framework, but only if the inelasticity falls strongly with $\sqrt{s}$, dropping by around 50% between 53 and 900GeV. There is no direct evidence on this point, but such a strong fall seems somewhat implausible.

[†] For a typical particle p_T of 0.4GeV/c a value $\eta_{beam} > -2.5$ corresponds roughly to Feynman-x ($x_F = 2p_{||}/\sqrt{s}) > 0.05$. One should note that the UA5 acceptance, $|\eta| < 5$, corresponds roughly to $|x_F| < 0.3(0.065)$ at $\sqrt{s} = 200(900)$GeV. Thus the UA5 data do not really probe the high-$|x_F|$ part of the distribution.

[††] The inelasticity, k, is the fraction of the c.m. energy *not* carried off by leading particles.

UA5 have also presented data on charged particle pseudorapidity distributions as a function of charged multiplicity. The data for NSD events at $\sqrt{s}$ = 200 and 900GeV are shown in Figure 13. We note that the growth of particle density with multiplicity is greatest in the central region. At fixed multiplicity, as $\sqrt{s}$ increases, the distribution becomes broader and the plateau lower. Both these effects may be expected to have a mainly kinematic origin [54]. In Figure 14 we show the dependence of $\rho_n(0)$ on n_{ch}. The rise of $\rho_n(0)$ with n is faster than linear, reflecting the narrowing of the distributions with n. The rise is greater at lower $\sqrt{s}$, but the ratio between the values of $\rho_n(0)$ at two different energies for the same value of n seems to be independent of n. If the central density scaled to the overall average $(\rho_n(0)/\rho(0))$ is plotted against the KNO variable $(z=n/<n>)$ a striking energy independent scaling behaviour is seen (Figure 14).

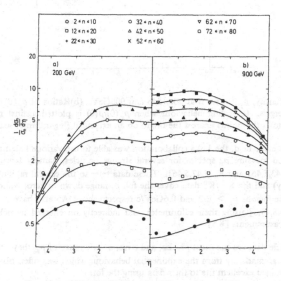

Figure 13:   Pseudorapidity distributions in different ranges of n_{ch}

TRANSVERSE MOMENTUM

Single particle distributions and $<p_T>$

In order to put together a complete picture of this topic at the SppS we need to use data from several of the experiments:

- UA1 have made measurements of charged particle p_T spectra using their central detector, both at 546GeV [40] (in the region $|y| < 2.5$, and 0.3GeV/c $< p_T <$ 10GeV/c) and in the Pulsed Collider run from 200 to 900GeV [41] (reaching lower values of p_T, down to 0.15GeV/c).

- UA2 used their wedge spectrometer to measure charged particle spectra near $\eta = 0$ in the range 0.4GeV/c $< p_T <$ 2GeV/c [42]. They were able to use time-of-flight measurements to separate pions, kaons and protons over various smaller ranges of p_T.

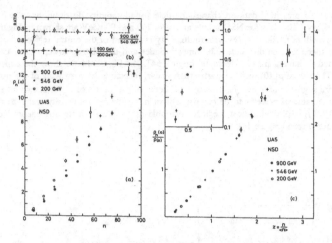

Figure 14:   (a)Central density, $\rho_n(0)$, vs. n at 200, 546 and 900GeV. (b)Ratios of $\rho_n(0)$ between pairs of energies. (c)Scaling behaviour when $\rho_n(0)/\rho(0)$ is plotted against n/<n>. The inset shows the points nearest the origin on an expanded (logarithmic) scale.

• Despite having no magnetic field, the UA5 collaboration was able to use various kinematic constraints to identify and measure p_T spectra for several strange particles, namely $K^{\pm}$ and K^0_s [43,44,45,5,46], Λ^0 [43,5,46] and Ξ^- [47,5,46]. These data refer to the central rapidity region (|y| < 2 to 3.5 typically) and the $K^{\pm}/K^0$ data cover the full p_T-range down to zero, while the Λ^0 and Ξ^- data cover the regions p_T > 0.2 and 0.6GeV/c respectively. UA5 also have some data on photon spectra [48,49,5] using their calorimeter, and indirectly on charged particles using multiple scattering measurements [8,5].

The UA1 charged particle data are the most precise, and cover a wide range of p_T; they are shown in Figure 15. The data diverge markedly from the exponential behaviour which was often observed at low energies. UA1 have obtained excellent fits to their data using the form:

$$E\frac{d^3\sigma}{dp^3} = \frac{Ap_0^n}{(p_0+p_T)^n} \tag{11}$$

and have used such fits to extrapolate to $p_T=0$, and thus estimate $<p_T>$. It has however been suggested [50] that near $p_T=0$ an exponential form like

$$E\frac{d^3\sigma}{dp^3} = Be^{-bm_T} \quad ; \text{ where } m_T=\sqrt{(p_T^2+m^2)} \tag{12}$$

should be expected. The UA5 data on kaons (shown in Figure 16, and compared with the UA2 data, which agree well) have therefore been fitted with a hybrid form, a power law at high p_T and an exponential in m_T at low p_T[†]

[†] The two functions and their first derivatives were constrained to be continuous at the crossover point, chosen to be $p_T=0.4$GeV/c.

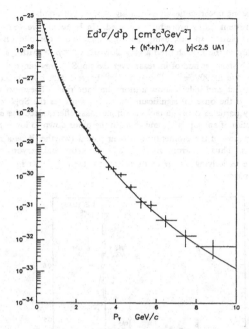

Figure 15: p_T-distribution for charged particles from UA1 at 546GeV.

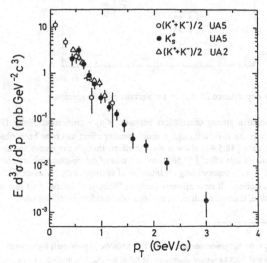

Figure 16: p_T-distribution for kaons from UA5 and UA2 at 546GeV.

100

There are therefore some uncertainties in estimating $<p_T>$, but trends with $\sqrt{s}$ should be correctly measured where a consistent fitting procedure was used. In Figure 17 we show the dependence of $<p_T>$ on $\sqrt{s}$ for various particle species, using the Spp̄S data and a compilation of data up to ISR energies [51]. The data up to the ISR showed a very slow rise of $<p_T>$ with $\sqrt{s}$. The UA1 charged particle data show a much stronger rate of increase over the Spp̄S energy range, from ≈ 0.39GeV/c at $\sqrt{s} = 200$GeV to ≈ 0.44GeV/c at 900GeV. The kaon and baryon data have much larger errors, but they clearly confirm the effect, and indeed show a more dramatic rise.[†] The most natural explanation for this increased growth is the onset of significant jet production over the Spp̄S energy region. The greater increase for heavy particles is presumably a simple mass effect, since the effects of resonance decays (and jet fragmentation if appropriate) tend to push the pions down to lower p_T. The heavy particles carry off a larger fraction of the momentum of their parent (whether it be a resonance or a parton) for kinematic reasons. Thus the strong rise in $<p_T>$ for kaons and baryons may be expected to reflect the changes in the underlying dynamics more directly than the data for all charged particles, which are dominated by pions.

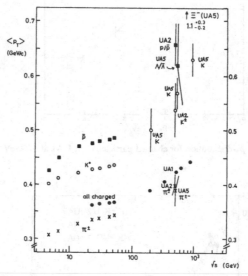

Figure 17:   Energy dependence of $<p_T>$ for various particle species.

UA1 have also noted a strong correlation between $<p_T>$ and multiplicity (Figure 18). This effect had not been seen at the ISR, although a much weaker effect has now been observed [52]. The UA5 kaon and photon data [48,5,44] show a similar effect, though the errors are much greater. One of the first interpretations of this effect [53,50] was in terms of quark-gluon plasma formation, the flattening off in the rise of $<p_T>$ representing an increase of entropy with no associated increase of temperature, i.e. a phase transition. It now appears (section "Mini-jets" below) that the correlation is connected with jet production, though a full quantitative understanding is still awaited.

[†] The UA5 estimate of $<p_T>$ for Ξ^- production at 546GeV, 1.1±0.3GeV/c, is particularly high, though the statistical error is large; one should also note that UA5 have poor acceptance for Ξ^- at low p_T. The preliminary data at 200 and 900GeV [46] already hint that this was an upward fluctuation.

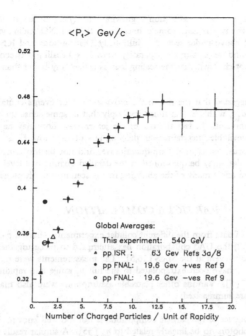

Figure 18:   Correlation of $<p_T>$ with dn/dη for UA1 data at 546GeV.

"Mini-jets"

A recent study which may be of importance in accounting for some of the features of "soft" collisions has been UA1's observations [41] of copious jet production in "minimum-bias" events, so-called "*mini-jets*". In essence UA1 took their standard jet-finding algorithm which they use for high-p_T events and ran it on "minimum-bias" triggers. They were able to convince themselves on a number of grounds that the "jets" found by this procedure were reasonably reliable down to $E_T \approx 5$GeV. They find that a large fraction of "minimum-bias" events contain such "jets", rising from ≈6% at $\sqrt{s} = 200$GeV to ≈17% at $\sqrt{s} = 900$GeV. The reliability of this jet identification at such low energies is not entirely accepted yet,[†] but even if only a fraction of these events really come from "semi-hard" scattering their influence on what we normally refer to as "soft" physics may be important.

For example, the abrupt increase in $<p_T>$ with $\sqrt{s}$ (Figure 17) could well be accounted for by the onset of significant jet production above ISR energies. Furthermore, UA1 have separated their events into those with and without jets. The jet events have average multiplicity roughly double that of no-jet events. Thus the average p_T of high multiplicity events should be greater than low multiplicity events, as observed (Figure 18). Indeed, UA1 see that $<p_T>$ for jet events is roughly constant at ≈0.5GeV/c, while the non-jet events still show some correlation between $<p_T>$ and multiplicity (maybe owing to non-identified jets or jets below 5GeV).

[†] For example, the UA5 collaboration have found [54] that their cluster Monte Carlo produces "mini-jets" at something like half to two-thirds of the rate seen by UA1, and with rather similar properties. In essence a cluster of $E_T = 5$GeV looks very much like a jet of the same E_T. This does not by any means imply that the UA1 interpretation is incorrect, but it does indicate that other interpretations might be possible, and thus that caution is called for.

Since the jet component of the cross-section is growing with $\sqrt{s}$, and the jet events are of higher than average multiplicity, we may envisage some connection with the KNO scaling violations observed at the Sp$\bar{p}$S, as mentioned above under section "Multiplicity Distributions and KNO scaling". UA1 have suggested that the jet and no-jet samples separately show KNO scaling between 200 and 900GeV (in $|\eta| < 2.5$), though it is not clear whether the scaling seen is in fact any better than that for the combined data sample.

Finally, UA1 have suggested that the rise in the cross-section for events containing jets roughly equals the absolute rise in σ_{tot} with $\sqrt{s}$, and thus they imply that in some sense the rise of σ_{tot} is due to the hard-scattering component. In fact the rise of the jet cross-section may be rather too sharp. However, in view of the inevitable uncertainties in defining jets of such low E_T as 5GeV, we should probably regard all this discussion as only semi-quantitative, and not be too concerned by this. In summary, even if the UA1 data may be questioned at the detailed quantitative level, they clearly indicate a mechanism which can affect many of the changing trends seen in the soft physics data above.

PARTICLE COMPOSITION

By using many pieces of data from the different collider experiments (charged particles, K, Λ, Ξ, γ from UA5; K, p from UA2) the UA5 collaboration has attempted to piece together a picture of the composition of a typical "soft" event at the Sp$\bar{p}$S [16,5]. The measurements were made in various different kinematic regions, and have been extrapolated to the full p_T range and rapidity range for comparison, as described in ref [5]. Various other plausible assumptions were also made, as detailed in Table 1, where the results are summarized.

The multiplicity dependence of the particle composition is not too well known. The UA5 photon data show the number of photons to be linearly related to n_{ch} [55]. A similar result seems to hold for the kaons [43,44], though the statistics are meagre. Also UA5 have observed that the average multiplicity of events containing a Ξ^- is as expected assuming a multiplicity-independent Ξ/charged ratio. We can thus conclude that there is no evidence requiring any variation of the particle composition with multiplicity.

We note that the data show a substantial excess of photons compared to $\pi^+ + \pi^-$ (though the errors are large; data at 200 and 900GeV will be of interest, but are not available at the time of writing). Naïvely one would expect that the yields of π^+, π^- and π^0 would be equal (from isospin symmetry), and it has been suggested therefore that substantial production of η-mesons could be indicated [55]. An η/π^0 ratio of 30±10% would account for the observations.

In Figure 19 we see the variation of the $K^{\pm}/\pi^{\pm}$ ratio with $\sqrt{s}$; the rise above kaon production threshold is flattening out, though the ratio seems to be still rising. Also in Figure 19 we show the K/π ratio as a function of p_T for data at $\sqrt{s} = 53$GeV and 546GeV. The same rise with p_T is seen at both energies, suggesting a connection between the rise of $<p_T>$ with $\sqrt{s}$ and the rise of the K/π ratio [44]. If we take η production to account for the photon excess we deduce that of the "stable" particles (i.e. after strong decays) produced at $\sqrt{s} = 546$GeV $\approx 10\%$ are kaons, and $\approx 8\%$ are baryons or antibaryons. For comparison, a similar analysis applied to ISR data yields 9% and 5% respectively [16]. Thus baryon production in particular seems to have increased with $\sqrt{s}$. This appears to be confirmed internally by the Sp$\bar{p}$S data; between 200 and 900GeV baryon production seems to have increased by a factor ~ 2 [46,49], whilst $<n_{ch}>$ has risen by only around 50%.

The K/π ratio has been used to derive the "strangeness suppression factor", λ, which is the ratio of the number of $s\bar{s}$ quark pairs produced to the number of $u\bar{u}$ (or $d\bar{d}$) pairs. This factor is used, for example, in fragmentation models such as the Lund string model. It has recently been important in

Table 1: Particle composition of a typical event at $\sqrt{s} = 546\text{GeV}$.

Particle Type	$<n>$	Source of estimate.
All charged	29.4±0.3	UA5 [5]
$K^0 + \bar{K}^0$	2.24±0.16	{UA5 [44]
$K^+ + K^-$	2.24±0.16	{charged + neutral assumed equal
$p + \bar{p}^{(\dagger)}$	1.45±0.15	UA2 [42], exapolated by UA5[5]
$n + \bar{n}^{(\dagger)}$	1.45±0.15	assumed equal to $p + \bar{p}$ $^{(\dagger\dagger)}$
$\Lambda + \bar{\Lambda} + \Sigma^0 + \bar{\Sigma}^0$	0.53±0.11	UA5 [5]
$\Sigma^+ + \Sigma^- + \bar{\Sigma}^+ + \bar{\Sigma}^-$	0.27±0.06	assume $\Lambda = (\Sigma^+ + \Sigma^0 + \Sigma^-)$; $\Sigma^+ = \Sigma^0 = \Sigma^-$
$\Xi^- + \bar{\Xi}^+$	0.10±0.03	UA5 [47]
$\Xi^0 + \bar{\Xi}^0$	0.10±0.03	assumed equal to $\Xi^- + \bar{\Xi}^+$
γ	33±3	UA5 [55]
$e^+ + e^-$	0.41±0.04	Dalitz pairs, based on γ
$\pi^+ + \pi^-$	23.9±0.4	Subtract other sources from n_{ch}

(†) *Note that leading baryons, one baryon and one antibaryon per event, are not included here. We assume one leading charged baryon in computing the $\pi^+ + \pi^-$ yield.*

(††) *There are estimates of the $\bar{n}$ yield from the UA5 calorimeter[48,49] which are consistent with this assumption.*

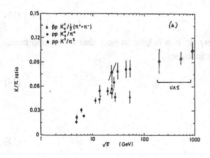
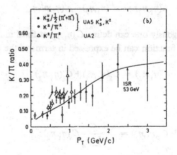

Figure 19: (a)$K^{\pm}/\pi^{\pm}$ ratio vs $\sqrt{s}$. (b)$K^{\pm}/\pi^{\pm}$ ratio vs p_T

interpreting UA1's results on $B^0 - \bar{B}^0$ mixing, since it determines the relative numbers of B^0_s and B^0_d mesons produced. However, the derivation of the λ parameter from the measured K/π ratio is not unambiguous, since one needs to exclude the $q\bar{q}$ pairs formed in resonance decay from the calculation. Analyses of data below Sp$\bar{p}$S energies gave conflicting conclusions: an analysis by Malhotra and Orava [56], based on a statistical quark model of Anisovich and Kobrinski [57], has suggested an energy independent value for λ around 0.29, while the 'quark counting' method proposed by Wróblewski [58] favours a gentle rise with c.m. energy. At Sp$\bar{p}$S energies the two methods give consistent results within errors [44,67], and are compatible with either a constant or a slowly rising value of λ. The values estimated at the Sp$\bar{p}$S are 0.29±0.05, 0.30±0.03 and 0.33±0.03 at $\sqrt{s} = 200$, 546 and 900GeV respectively [44,45].

The Λ/p ratio seems to be roughly 1/3 right up to $\sqrt{s} = 900$GeV [46], about the same as at the ISR and lower energies. In contrast the Ξ/Λ ratio has grown strongly, judging from the UA5 546GeV data, though the errors are large (and asymmetric) [47]. It has recently been suggested that copious Ξ production would be a good signature for quark-gluon plasma production [59]. However, in view of the limited statistics on which the UA5 Ξ data are based it might be prudent to await the equivalent data at $\sqrt{s} = 200$ and 900GeV before drawing conclusions. The first preliminary indications at 900GeV [46] indeed hint that the 546GeV measurement could be an upward fluctuation, though the Ξ/Λ ratio is probably still rising.

CORRELATIONS

In the present short review it is impossible to give a comprehensive review of this complicated subject. The latest UA5 data are well summarized in a recent paper [60], to which the reader is referred for details. Earlier results were presented in refs [61,62]. There are three main methods which have been used to investigate correlations: two-particle correlations, forward-backward multiplicity correlations and rapidity gap correlations. We have to try to make a coherent interpretation of all these data. It turns out that a cluster model together with the observed behaviour of multiplicity distributions is able to give a reasonable description of the data.

Two-particle correlations

These studies have followed generally similar lines to ISR and bubble chamber experiments. Starting from the single- and two-particle densities in rapidity, $\rho^I(\eta)$ and $\rho^{II}(\eta)$, one defines the two particle correlation function:

$$C(\eta_1, \eta_2) = \rho^{II}(\eta_1, \eta_2) - \rho^I(\eta_1)\rho^I(\eta_2). \tag{13}$$

Analogously one can define $C_n(\eta_1, \eta_2)$ at fixed multiplicity n. It is helpful to note that the overall correlation function can be expressed in terms of the C_n's as follows:

$$C(\eta_1, \eta_2) = C_S(\eta_1, \eta_2) + C_L(\eta_1, \eta_2), \tag{14}$$

where

$$C_S(\eta_1, \eta_2) = \sum_n \frac{\sigma_n}{\sum \sigma_n} C_n(\eta_1, \eta_2); \tag{15}$$

$$C_L(\eta_1, \eta_2) = \sum_n \frac{\sigma_n}{\sum \sigma_n} (\rho^I(\eta_1) - \rho^I_n(\eta_1))(\rho^I(\eta_2) - \rho^I_n(\eta_2)). \tag{16}$$

These correlation functions are shown for the UA5 data in Figure 20. The C_S term has a narrow peak around $\eta_1 = \eta_2$, and reflects short-range correlations at fixed multiplicity, n. The C_L term arises because of the mixing of events of different n, which have different single particle densities, ρ_n^I. This term has a broad distribution in $\eta_1 - \eta_2$, and is thus often referred to as a "long-range" component, though the expression "mixing term" might be more appropriate. Clearly C_S is the interesting term which tells us about the effect of correlations on the two-particle densities ρ_n^{II}. One sees from Figure 20 that the change with $\sqrt{s}$ is mainly in the mixing term C_L, while the short-range term changes rather little.

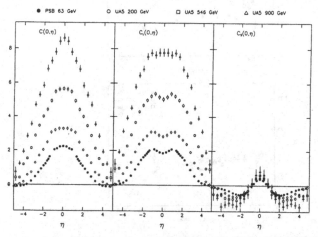

Figure 20: Correlation functions C, C_L and C_S for UA5 data at $\sqrt{s} = 200$, 546 and 900GeV.

UA5 have extracted the C_n correlation functions from data, and interpreted them in a cluster model. In such a model the correlation function should be fitted by a Gaussian superimposed on some (predicted) background. The width of the Gaussian peak gives the decay width, δ_n, of the clusters at multiplicity n, while the height of the Gaussian may be related to the cluster decay multiplicity, specifically to $<\kappa(\kappa-1)>/<\kappa>$, where κ is the number of charged particles into which a cluster decays. In Figure 21 we show the values of δ_n and $<\kappa(\kappa-1)>/<\kappa>$ versus $z(=n/<n>)$ for the UA5 data at 200, 546 and 900GeV, with ISR data [63,64] for comparison. We see that the cluster width δ_n shows little multiplicity dependence, maybe a gentle fall, and also little dependence on $\sqrt{s}$, maybe a slight rise. The decay multiplicity moment $<\kappa(\kappa-1)>/<\kappa>$ again varies only weakly with multiplicity (gently rising in the UA5 data), but its energy dependence is far from clear. The UA5 data, all analysed identically, show essentially no dependence on $\sqrt{s}$, but lie above the ISR data. However, the two ISR experiments clearly disagree with each other, so maybe the internal agreement of the SppS data is the significant observation here. The UA5 data are shown [60] to be consistent with production of clusters having the same properties at all values of $\sqrt{s}$ and n. The average cluster multiplicity, $<\kappa>_0$,[†] is approximately 1.6 for a Poisson decay multiplicity distribution. The distribution of κ is of course unknown, but a Poisson form is not unreasonable, though a different assumption would

---

[†] The angle brackets with subscript $_0$ ($<\kappa>_0$) signify that clusters decaying into only neutral particles are included in the mean. If the subscript $_0$ is omitted then $\kappa=0$ is excluded from the moments. For example a Poisson distribution with $<\kappa>_0 \approx 1.6$ will have $<\kappa> \approx 2.0$.

106

clearly lead to a different estimate of $<\kappa>$. The distribution of κ will influence the form of the $<\kappa(\kappa-1)>/<\kappa>$ vs. z curve (Figure 21). For example, in ref [60] it is shown that a Poisson form fits well (under the assumptions that the overall multiplicity distribution is negative binomial, and that size of clusters is independent of the number of clusters formed), though other possibilities cannot be excluded.

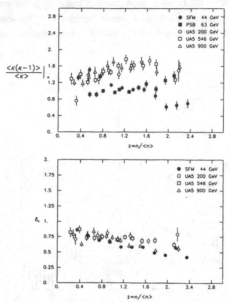

Figure 21: (a) $<\kappa(\kappa-1)>/<\kappa>$ vs z. (b) δ_n vs z

Forward-backward multiplicity correlations

The idea here (following ref [65]) is to study correlations between the charged multiplicities in two symmetric forward and backward rapidity intervals. If the multiplicities are n_F and n_B respectively then the correlation coefficient, b, is defined by:

$$b = \frac{cov(n_F, n_B)}{\sqrt{var(n_f) \cdot var(n_b)}} \qquad (17)$$

In fact it turns out that $<n_B>$ is linearly related to n_F, i.e. $<n_B> = a + b \cdot n_F$, and it can easily be shown that the estimate of b from a least squares fit is the same as that defined from the covariance (in the case of symmetric intervals of rapidity).

In Figure 22 the UA5 data [60,61] are compared with the ISR data [65] in the same three rapidity intervals, $0<|\eta|<4$, $0<|\eta|<1$ and $1<|\eta|<4$. In all cases the correlation increases with $\sqrt{s}$. Most interest has centred on the case $1<|\eta|<4$, since there is a gap of two units of rapidity between the two regions, and it is therefore unlikely that clusters of the type discussed under short-range correlations could give particles in both intervals. This correlation was small at ISR, but is now significant at the Sp$\bar{p}$S. This might be thought to be evidence for a growing long range correlation.

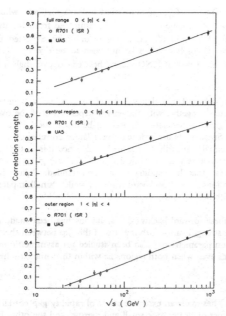

Figure 22: Variation of the forward-backward correlation parameter, b, with $\sqrt{s}$

To understand these data it is helpful to recast the formula for b. The following identity may be shown to hold:

$$b \equiv \frac{D_S^2/<n_S> - 4<d_S^2(n_F)>/<n_S>}{D_S^2/<n_S> + 4<d_S^2(n_F)>/<n_S>} \tag{18}$$

where $n_S = n_F + n_B$. D_S^2 is the variance of the n_S distribution, and $d_S^2(n_F)$ is the variance of the n_F distribution at fixed n_S. The angle brackets denote averages over n_S. The first term in both numerator and denominator is determined merely by the overall multiplicity distribution. The second term reflects forward-backward fluctuations at fixed multiplicity. In a very naïve cluster model (clusters of fixed multiplicity and zero decay width randomly emitted into the forward and backward regions) this second term would just equal the cluster decay multiplicity κ. In a more realistic model one needs to allow for the width of the κ distributions, and for leakage effects which mean that not all decay products of a cluster fall into the selected rapidity region. It turns out that these two corrections act in opposite directions, and therefore roughly cancel. A more detailed discussion may be found in [60].

One can therefore recast the formula for b as:

$$b = \frac{(D_S^2/<n_S> - \kappa_{eff} f_{leak})}{(D_S^2/<n_S> + \kappa_{eff} f_{leak})} \tag{19}$$

where κ_{eff} ($= 1 + <\kappa(\kappa-1)>/<\kappa>$) takes account of the finite spread of cluster multiplicities. Note that κ_{eff} involves the same combination of moments of the κ distribution as was inferred from the short range correlations. The term ζ_{leak} takes account of leakage effects, while the product $\kappa_{eff}\zeta_{leak}$ is of the same order as $<\kappa>$. The value of κ_{eff} may be estimated from the data, and seems to be roughly independent of multiplicity and of $\sqrt{s}$, as we would have expected from the analysis of short-range correlations. The rise of b with $\sqrt{s}$ is thus seen to be due to the increase in the term $D_{s}^{2}/<n_s>$, which would rise like $<n_s>$ if KNO scaling held exactly, or slightly faster in view of the observed scaling violations.

The observed forward-backward correlations therefore can be accounted for in terms of short-range correlations (clusters), together with the energy dependence of the overall multiplicity distribution. In one sense the growing correlation is seen to be a statistical effect; as $<n>$ grows the fractional forward-backward fluctuations are reduced and a larger correlation is seen. Of course, the real question is why the multiplicity distribution is so broad, since this is the source of the correlations.[†] An impact parameter picture (see "Statistical models" below) gives a natural explanation for this, as a consequence of the fact that the colliding objects are extended. A large impact parameter will give low multiplicity (in both forward and backward regions), while a small impact parameter will yield high multiplicity (again in both hemispheres).

A systematic study of these forward-backward correlations has been made in [60], for example studying how the correlation strength varies with the size of the gap between the "forward" and "backward" regions. The correlation parameter has also been studied for asymmetric regions; for example a substantial correlation remains even when both regions lie within the same c.m. hemisphere.

Rapidity gap correlations

A recent ISR paper [66] has made an extensive study of rapidity gap correlations. Their conclusion was that two different types of cluster, one small and narrow, and the other large and broad, were required to describe the data. The UA5 Sp$\bar{p}$S data have now been analyzed in a similar way [60], and have been shown to lead to a different conclusion; they are compatible with the UA5 cluster Monte Carlo program, which employs only one type of cluster, whose parameters are chosen to fit the short-range and forward-backward correlation data above.

The basic idea is to study the distribution, $F_k(G)$, of pseudorapidity gaps (G) between pairs of particles having exactly $k-1$ charged particles lying between them in pseudorapidity. A normalised correlation function is defined by

$$R_k(G) = F_k(G)/\tilde{F}_k(G) - 1, \tag{20}$$

where $\tilde{F}_k(G)$ is the distribution expected in the absence of correlations, which can simply be computed in terms of the single-particle rapidity distributions. The correlation function R_k is computed as a function of charged multiplicity, m, and then averaged over m to improve statistics. Some of these correlation functions are shown in Figure 23 for the 546GeV data. The curves show that the UA5 cluster model gives an excellent account of the data for all values of k. A similar conclusion holds at $\sqrt{s} = 200$ and 900GeV, with the same cluster parameters.

[†] For example random particle emission would give $\kappa_{eff} = 1$, $\zeta_{leak} = 1$ and $D_s^2/<n_s> = 1$ for a Poisson multiplicity distribution, and hence b = 0. Indeed, in e⁺e⁻ collisions the multiplicity distribution is not far from a Poisson, and the forward-backward correlations are small.

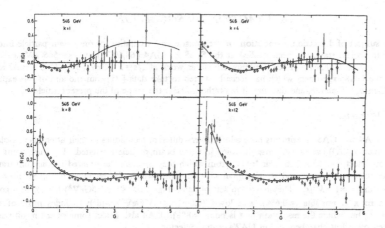

Figure 23:   Rapidity gap correlations at $\sqrt{s} = 546$GeV, for several values of k.

Summary of correlations

Our main conclusion is that a common approach, in terms of a cluster model with $<\kappa> \approx 2$, gives a satisfactory description of the correlation data at the Sp$\bar{p}$S. The cluster parameters (multiplicity, decay width) do not seem to vary significantly over the Sp$\bar{p}$S range. There may be some differences from ISR data, but it is hard to assert this with confidence given the differences between the experiments and their analysis techniques. The other necessary input in interpreting the data is the overall multiplicity distribution, and this is still not understood in any quantitative way. However, despite its success, a cluster model is not the only approach which can deal with these correlation effects, and some other models are discussed below.

One can also speculate on the nature of the "clusters". All we really learn from the data is their decay width ($\delta \approx 0.75$ units of pseudorapidity) and decay multiplicity ($<\kappa> \approx 2$, if all-neutral decays are excluded), and the latter depends on the (unknown) decay multiplicity distribution. Clearly conventional resonances contribute to the clusters, but it is not clear that this is sufficient. An estimate of $<\kappa>$ from resonance decays, based on extrapolating known yields of low mass resonances (i.e. 0^-, 1^-, 2^+ mesons and $\frac{1}{2}^+, \frac{3}{2}^+$ baryons) from lower energies and some plausible assumptions, arrives at ≈ 1.4 [60,67], and also too low a value of $<\kappa(\kappa-1)>/<\kappa>$. Maybe there are many higher mass resonances whose production we have no reliable means to estimate. Or maybe local quantum number conservation, as would occur naturally in a picture like the Lund string model, is responsible. Or maybe we need to allow for the emission of many soft gluons (like mini-jets, or of still lower p_T), which could well be approximated as clusters in a model such as described here.

DIFFRACTION DISSOCIATION

The subject of diffraction dissociation, or the quasi-elastic excitation of one beam particle into a relatively low mass state, has been studied at the Sp$\bar{\text{p}}$S by UA4 [68,69,70] at $\sqrt{s}=546$GeV, and by UA5 (at 200 and 900GeV [71,72], building on earlier work at $\sqrt{s}=546$GeV [5]). One should also mention the UA8 experiment which has recently started to take data [73], but the aim of this experiment is high-p_T diffraction, and therefore it lies rather outside the scope of the present article.

Single Diffraction

The UA4 and UA5 experiments have adopted very different techniques in their attempts to isolate single-diffractive (SD) events, i.e. those in which just one beam particle is excited. UA4 use events in which they detect the recoil $\bar{\text{p}}$ in their "pots", from which they measure the squared momentum transfer to the recoil particle, t, and its Feynman-x value, from which the recoiling mass, M, may be inferred (using $x_F \approx 1 - M^2/s$). They have two sets of data: high-t ($0.5 \leq |t| \leq 1.5 \text{GeV}^2$) for which a good resolution in x is possible ($\approx 0.6\%$), and low-t ($0.04 \leq |t| \leq 0.35 \text{GeV}^2$) which includes most of the cross-section, but where the resolution in x is poor ($\approx 8\%$). UA4 also detect some of the recoil particles in their own drift chambers and in UA2's central detector.

UA5 never detect the recoil particle, and so cannot measure M or t. Instead they exploit the different triggering characteristics of SD and NSD events (Figure 2). By using observed trigger rates for single- and two-arm triggers and using Monte Carlo estimates of trigger efficiencies they infer the cross-sections for SD and NSD events. Clearly the largest uncertainty in this method is in the models used for the Monte Carlo work.

The conventional picture of diffraction dissociation is that the cross-section should peak at small masses, given by the coherence condition $M^2/s \leq (m_\pi R)^{-1}$, where R is the interaction radius. Thus $M^2/s \leq 0.15$, though in practice the diffractive peak is usually clearly seen for $M^2/s \leq 0.05$, and this cut will be generally used in the data discussed below. Based on Mueller-Regge theory the dominant form of the mass distribution should be $d\sigma/d(M^2/s) \sim (M^2/s)^{-1}$. The pre-Sp$\bar{\text{p}}$S data and many of the theoretical aspects are reviewed in ref [74].

Some of the UA4 high-t data [68] are shown in Figure 24. We note a number of important points. Firstly a clear peak is seen at small masses, which we associate with diffraction dissociation. The data fall steeply with $|t|$. The form of the mass spectrum is consistent with $1/M^2$, apart from low masses where the experimental resolution smears the data out. At fixed t the invariant cross section, $\dfrac{s}{\pi} \dfrac{d^2\sigma}{dt dM^2}$, scales with M^2/s from ISR to Sp$\bar{\text{p}}$S energies. All these observations are in accord with expectations from Mueller-Regge theory.

UA4 have also studied the fragmentation or decay properties of the excited system [69] in their high-t data. They find that the pseudorapidity distributions of the recoil particles broaden with M in the way that would be expected for p_T-limited phase space fragmentation, but in conflict with a picture in which the particles are emitted isotropically. Furthermore, it appears that the average charged multiplicity depends on M^2 in the same way that $<n>$ for NSD events depends on s. UA5 reach the same conclusion, but for the full t range. Instead of looking at different M-values they study different ranges of multiplicity, and compare with Monte Carlo calculations for p_T-limited and isotropic "decay" [5,71], as shown in Figure 25. Indeed, UA5 have been able to use the shapes of the pseudorapidity distributions to make a crude estimate of $<p_T>$, quoting a value in the range 0.35 to 0.55GeV/c, thus very compatible with NSD events. Note that it is for the high masses (or multiplicities) that the conclusion about p_T-limited decays is clear, and the Sp$\bar{\text{p}}$S has given us the opportunity to excite large masses diffractively (e.g. $M^2/s = 0.05$ corresponds to M$\approx$200GeV/c^2 at $\sqrt{s}=900$GeV).

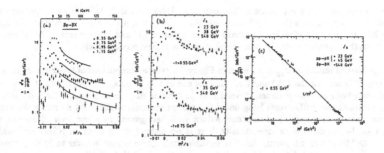

Figure 24: UA4 high-t data. (a)Mass distributions, showing clear diffractive peaks. (b)Test of scaling in M^2/s at fixed t. (c)Showing the $1/M^2$ dependence of the cross-section.

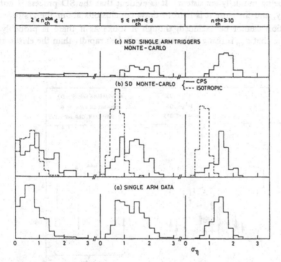

Figure 25: Histograms of $\sigma_\eta = \sqrt{<(\eta - <\eta>)^2>}$. (a)Single arm triggers in 546GeV UA5 data. (b)SD models with p_T-limited (solid) or isotropic (dashed) decay. (c)NSD background in the single-arm sample.

The estimation of the full cross-section for single-diffraction presents both experiments with difficulties, though. The UA5 method is simple in principle, but is dominated by systematic errors in the trigger efficiencies, owing to uncertainties in the models. There appear to be two main uncertainties. The first is the shape of the Feynman-x distribution for leading baryons close to $|x| = 1$ in NSD events, since an NSD event having a proton with $|x| \geq 0.95$ will kinematically fake an SD event, and thus needs to be subtracted. UA5 favour a $1 - |x|$ form, but there are no direct data on this point, and a flat distribution has also been considered. The second uncertainty is the M^2/s distribution for SD events at very low masses, since such events are seldom detected. UA5 assume a $(M^2/s)^{-1}$ form right down to threshold. UA4's problem is that they have to use their low-t data, which have a very poor resolution on x (or M), and they therefore need to make cuts to separate SD and NSD events. They choose to make cuts on the rapidity gap between the recoil $\bar{p}$ and the closest recoil particle.

The published results on σ_{SD} are shown in Figure 26, with a compilation of lower energy data for comparison. We see that the UA4 and UA5 data are not in very good agreement, the UA4 point lying higher, though both have large errors. Several reasons for this may be suggested. The UA5 method explicitly subtracts any NSD events lying under the diffractive peak, whilst UA4 do not make such a subtraction, though they imply that their cuts reduce such background to a low level. However, this must be model-dependent, since kinematically there is no distinction between an NSD event with a proton at $|x| \geq 0.95$ and an SD event. Another possible explanation is that in order to fit their rapidity gap distribution UA4 needed to double the SD cross-section for $M^2 < 4 \text{GeV}$ compared with a $1/M^2$ form. This corresponds to an extra ~2mb of cross-section in a region where UA5 have low detection efficiency, and may go some way to reconciling the data. The best we can state at present is that the SD cross-section in the Sp$\bar{p}$S energy region is ~5–9mb, and thus not much higher than at the ISR. We should also note, however, that the lower energy data, which again use a variety of different techniques, are far from being mutually consistent. It is evident that the SD process is not at all easy to identify experimentally, and therefore it is hard to draw conclusions about the $\sqrt{s}$ dependence of σ_{SD}. A constant cross-section cannot be excluded, though it looks as if there is probably a gentle rise. However, it seems clear that σ_{SD} is rising with energy much less rapidly than the elastic cross-section.

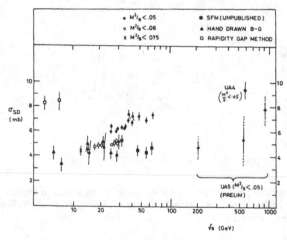

Figure 26: σ_{SD} vs $\sqrt{s}$. For sources of data see ref [71,70]

Double Diffraction

The process of double-diffraction (DD), in which both beam particles are excited into low mass states, is even harder to identify convincingly, but UA5 have made an attempt to estimate the cross-section [71,72]. The method is a simple extension of their SD procedure, by sub-dividing the two-arm trigger sample by some cut designed to enhance double-diffraction (e.g. requiring a large rapidity gap in the central region). They then use Monte Carlo simulations for SD, DD and non-diffractive (ND) events to deduce the proportions of each in the data sample. Clearly the results are dominated by systematic model uncertainties. The problem is that most DD events (with p_T-limited fragmentation) look very much like ND events, and therefore any cut designed to separate them has a low efficiency (or else a low purity). The results quoted by UA5 are $\sigma_{DD} = 4.0 \pm 2.5$mb at $\sqrt{s} = 900$GeV, and 3.5 ± 2.2mb at $\sqrt{s} = 200$GeV.

The DD cross-section may be predicted in terms of SD and elastic scattering if factorization of the diffractive vertices be assumed to hold. The following relations may be expected to hold [74]:

$$ \sigma_{DD} = \frac{\sigma_{SD}^2}{4\sigma_{el}} \cdot \frac{b_{SD}^2}{b_{el}b_{DD}} \quad , \text{with} \quad b_{DD} = 2b_{SD} - b_{el}, \tag{21} $$

where the b's are the slopes of the t-distributions, assumed approximately exponential. If we take $\sigma_{el} = 13.6 \pm 0.6$mb, $b_{el} = 15$GeV^{-2}, $b_{SD} = 8 \pm 1$GeV^{-2} and $\sigma_{SD} = 7 \pm 1$mb we find that σ_{DD} should be in the region of $2-4$mb, not inconsistent with the UA5 estimate. So it appears that double diffraction is probably not a large part of the inelastic cross-section. Furthermore it is not clearly distinguishable from the non-diffractive events, which probably makes its neglect elsewhere in SppS analyses justifiable.

EXOTIC PHENOMENA

Centauro Events

Before the advent of the SppS our only information about particle interactions in this energy region came from cosmic ray experiments. A number of unusual phenomena have been reported in high energy cosmic ray data, of which the most spectacular and inexplicable are the "Centauro" events [1], observed in the Mount Chacaltaya emulsion chamber experiment, and recently by the joint Chacaltaya-Pamir group [75]. Among the striking features of these events are high hadronic multiplicity[†] and high p_T (around 1.7 ± 0.7GeV/c is estimated). However, the most curious property of the Centauro events is the apparent absence of produced photons (and thus π^0). This has led to speculation that the produced particles might be baryons and antibaryons.

[†] The Centauro events are estimated to have a c.m. energy of around 1 to 2TeV. The detection threshold of the experiment is around 1TeV in the laboratory system, which means that the observed particles come from only a part of the forward c.m. hemisphere. The total hadronic multiplicity therefore depends on an extrapolation to the unseen part of phase space. The Chacaltaya group favour a "fireball" type model, and on this basis estimate a hadronic multiplicity around 100 [1], but an equally valid interpretation of the events is through a p_T-limited phase space model, which would lead to a much higher hadronic multiplicity, around 200 [78]. Another uncertainty in interpreting the Centauro events arises because only the electromagnetic part of a hadronic shower is detected. The fraction of a hadron's energy which goes into electromagnetic particles, k_γ, depends on the particle type, being ~0.33 for pions, and ~0.2 for baryons. This leads to uncertainties in estimating the incident energy, and the $<p_T>$ of the produced particles.

There are a number of other exotic types of event which have been claimed in these cosmic ray data, e.g. "mini-Centauros" (like Centauros, but of lower multiplicity), "Chirons" (high p_T, very narrow jets including hadrons) and "Geminions". There was therefore the hope, before the Spp̄S started operation, that some very spectacular phenomena might be seen. It should be emphasized that the cosmic ray experiments are dealing with small numbers of events (up to a few hundreds) at these high energies, so that the exotic events should show up with large cross-sections if they exist, and therefore "minimum-bias" data is the appropriate place to look for them.

In the early 546GeV data both UA5 [55,67] and UA1 [76] made searches for Centauro events, with negative results. The UA5 method was to identify discrete photons via their conversions in a steel beam pipe. They then looked for events with high charged multiplicity and low photon multiplicity. In a sample of 3600 NSD events no events having the properties expected for Centauros were found. The UA1 study used a calorimetric technique. For each event they compared the energy in the first 4 radiation lengths of their calorimeters (which should be predominantly associated with electromagnetic particles) with the energy beyond 12 radiation lengths (mainly hadronic). Again the distributions looked normal, with no unexpected population of events in the region expected for Centauros, in a sample of 48000 events.

However, the Centauros seen in cosmic ray events were all at c.m. energy above 546GeV. Therefore this first negative result at the Spp̄S might indicate a threshold just above. Indeed, a study of the $\sqrt{s}$ dependence of the various unusual events in the cosmic ray data [77] gave some indication of such a threshold, and suggested that at least some types of these exotic events should be produced at the $5-10\%$ level if the Spp̄S energy could be raised to $\sqrt{s} = 900$GeV. This was in fact one of the main motivations for the Pulsed Collider run of 1985.

The UA5 detector was enhanced in two ways for the Centauro search at 900GeV. A "photon converter", shaped so as to give roughly uniform photon detection as a function of pseudorapidity, was placed between the beam pipe and the streamer chambers for certain runs, and a 90° calorimeter was installed to give some energy measurements, and in particular to detect neutrons or antineutrons which might be created in Centauro events. The UA5 900GeV analysis is presented in refs [78,49], using all the characteristic features of Centauro events: high p_T, high multiplicity and photon depletion. For example, Figure 27 shows plots of the number of observed photons against observed charged multiplicity for several event samples — for data at 900GeV, "normal" Monte Carlo and for a Monte Carlo representing the properties of Centauro events. Clearly the data look normal, and there is no population of events in the Centauro region; Figure 27 also shows a sample of events with a cut demanding high E_T in the UA5 calorimeter, which should enhance the proportion of Centauro events, but still there are no candidates. From these results upper limits on Centauro production were deduced, depending on the assumed $<p_T>$, typically of the order of $1-4‰$ of inelastic events at the 95% confidence level. In Figure 28 we show these limits. UA5 also searched in their single-arm triggers for possible diffractively excited Centauro events, but again with negative results.

How then are we to interpret the non-observation of Centauros at the Spp̄S? Clearly there are several possibilities. We may simply not have reached sufficiently high c.m. energies at the Spp̄S and have to await results from the Tevatron. Maybe the cosmic ray data or their interpretation are just wrong for some reason. Or maybe the cosmic ray events are not the result of hadron-hadron collisions. We know for example that the target is a nucleus, and so may be the projectile, though nuclei are not likely to penetrate so deep in the atmosphere, and also in this case we would expect to see some nuclear fragments, which would be of low p_T. Possibly the explanation lies with some exotic component in the cosmic ray spectrum, like globs of quark matter with nucleons accreted around them, as suggested in ref [79]. Another possibility is an anomalous interaction of multi-TeV photons [80], which could help to explain both Centauro events and some other anomalous effects in cosmic ray data.

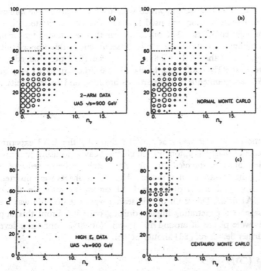

Figure 27: Plots of observed photon multiplicity against charged multiplicity. (a) UA5 data at √s = 900GeV. (b) Normal Monte Carlo. (c) Centauro Monte Carlo. (d) Data having high calorimeter energy.

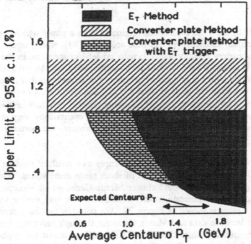

Figure 28: Upper limits (at 95% c.l.) on Centauro production at √s = 900GeV.

Free Quarks

Although present ideas based on QCD do not expect free quarks to be produced, it is clearly appropriate to search for such objects when a new accelerator opens up a new energy regime. At the Sp$\bar{\mathrm{p}}$S the UA2 collaboration has made such a search at √s = 546GeV (see ref [81], updated with greater statistics in ref [82]). Their 90° spectrometer, installed for the early Collider runs, was equipped with scintillation counter hodoscopes which were used to measure the ionization of produced particles. A maximum likelihood method was used to make the best estimate of the ionization based on the signals in seven or eight counters. The outcome was that no candidates having ionization less than 0.7 times that for a minimum ionizing particle were seen. From this upper limits on the production of free quarks were placed at the level of 2.8 10^{-6} and 5.6 10^{-5} per charged particle for charges $\pm 1/3$ and $\pm 2/3$ respectively (at 90% confidence level). These limits apply to very light quarks; since the study was made at 90° the momenta of the particles are low, and so the limits rise quite rapidly with mass.

Monopoles

A search for magnetic monopoles was made at the Sp$\bar{p}$S by the UA3 experiment [83]. Their technique was to use sheets of plastic (kapton) in which the heavily ionising monopoles should leave tracks which were then developed by chemical etching. After etching the presence of small holes in the plastic sheets would signify the passage of monopoles. The plastic sheets were inserted both inside and outside the UA1 beam pipe, and also around the UA1 central detector, in the hope of tracking monopoles through the UA1 field. The result of this search was that no hole was found which could be attributed to the passage of a penetrating highly ionizing particle. From this upper limits on the production of monopoles were placed at around 10^{-32} cm² (at 90% confidence level). These limits are for low masses; for further details ref [83] should be consulted.

Rapidity Fluctuations ("Spikes")

It is not clear that this topic should be classified as "exotic". However, the results were thought by some to be surprising when they were first presented, and it was suggested that they might indicate the formation of "hot spots" of quark-gluon plasma. The explanation now appears to be more mundane.

The idea for this study by the UA5 experiment originated with a paper which suggested the possible emission of "gluon-Cerenkov radiation" in hadronic collisions [84], which would lead to the emission of many particles in a narrow rapidity interval, with no jet-like structure in the azimuthal angle. The experimental method is to take a fixed window in pseudorapidity, say $\Delta \eta = 0.5$, and for each event to find the position in rapidity where this window contains the largest number of particles, which is denoted n_{max}. In this way one finds the greatest local rapidity density of each event. In the UA5 data [85,86,5] events were seen in which up to 15 particles were emitted in a half unit of pseudorapidity, i.e. $dn/d\eta \approx 30$ or ten times the overall average. Some of these events look very striking; Figure 29 shows the rapidity distributions for a few such events. Furthermore, the particles show no jet-like clustering in azimuth, ϕ.

The question is then whether such events can arise simply as a result of random fluctuations, given the observed rapidity distributions and knowledge of short range correlations. The answer appears to be yes, based on a comparison with the UA5 cluster Monte Carlo, which incorporates these effects [54,29]. The Monte Carlo seems to generate "spike" events at a rate very similar to that seen in the data. A Monte Carlo spike event is shown in Figure 29. We could in fact have anticipated this from the agreement noted earlier between data and Monte Carlo for the high order rapidity gap correlations (section "Rapidity gap correlations"), since these correlations tell us about the probability of finding large numbers of particles in various small rapidity intervals. We therefore conclude that the spike events principally arise from the random superposition of particles and clusters.

The $\sqrt{s}$ dependence of these events has also been studied by UA5, and an interesting scaling behaviour was noted. In Figure 30 we show a plot of $< n_{max} >$ against n_{ch}, for data at $\sqrt{s} = 200$, 546 and 900GeV. All the data show the same behaviour. Furthermore, if a small subset of the data having high E_T in the UA5 calorimeter is used, the points still lie on the same curve. The ability of the UA5 Monte Carlo to reproduce these data is also shown in Figure 30.

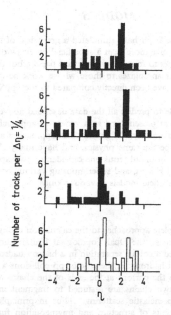

Figure 29: Rapidity distribution of typical "spike" events. The solid histograms are data, and the open ones Monte Carlo, both at $\sqrt{s} = 546\,\text{GeV}$.

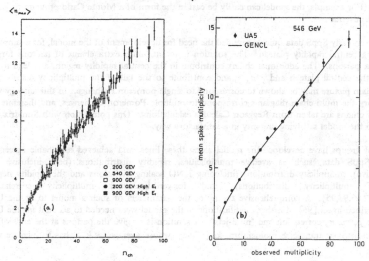

Figure 30: (a) Average "spike" multiplicity $<n_{max}>$ vs. n_{ch}. (b) 546GeV data compared with the UA5 Monte Carlo.

MODELS

The "soft-physics" data from the Spp̄S have stimulated a great deal of theoretical activity, and in the present review it is impossible to give more than a superficial overview of the main types of model used. An attempt will be made to refer to many of the relevant papers, but the bibliography is certainly far from complete, and the author apologizes to those whose work he cannot include. Emphasis will be given to those models which have been directly compared to the Spp̄S data.

In principle QCD should be able to predict all the data described above. In practice we have no precise way to make the calculations at present, and so a variety of approaches have been followed in an attempt to describe the data approximately. Of course these approaches are not orthogonal to each other — they are all trying to describe the same physics, and have many elements in common. We outline some of the main ideas here. In broad terms one can identify two approaches — a microscopic approach starting from quarks and gluons, and superimposing various configurations of partons to make up a hadronic collision, and statistical models of various kinds.

Microscopic models

One of the older and most complete approaches to the calculation of high energy soft processes is the **Dual Parton Model**, also often called the Dual Topological Unitarization (DTU) scheme. Early versions of this scheme [87] envisaged a colour separation in a hadron-hadron collision, leading to the formation of two chains or strings of hadrons: in the case of pp̄ collisions a long $qq - \overline{qq}$ chain and a short $q - \overline{q}$ chain.[†] The momenta of the partons at the ends of the chains are determined from measured structure functions, and the two chains are assumed to fragment independently, in the way observed in e^+e^- collisions or deep inelastic scattering. Thus in principle the model had no free parameters, just requiring measurements of structure and fragmentation functions. Furthermore, in principle the Dual Parton Model gives a complete description of soft hadronic processes, since any measurable quantity can be predicted given the parton distributions and a model for the chain fragmentation. For example, the model can easily be cast in the form of a Monte Carlo program to generate complete events.

However, early Spp̄S data quickly revealed the need for development of the model, for example to fit the rising central rapidity plateau. The solution is to introduce extra chains of hadrons formed between sea partons. The additional chains contribute in the central rapidity region, and lead to the growth of the central plateau and $<n_{ch}>$ and contribute to the tail of the multiplicity distribution. The two-chain picture may be shown to correspond to single pomeron exchange. In this analogy with Regge theory the multi-chain diagrams correspond to multiple Pomeron exchanges, and therefore the numbers of chains are taken from Reggeon Calculus calculations. This connection with S-matrix theory leads to the model satisfying unitarity in a satisfactory way.

Several groups have developed the model along these lines, and achieved reasonable agreement with the Spp̄S data, such as average multiplicities, rapidity distributions (both inclusive and semi-inclusive), multiplicity distributions (including KNO scaling violations and the rapidity dependence of multiplicity distributions), and forward-backward multiplicity correlations [89,90,91,92,93,94,95]. A comprehensive tuning of the parameters of such a model to fit the UA5 data is described in ref [96]. Further modifications to the model were needed to account for the UA1 data on the p_T vs. n_{ch} correlation and "mini-jets". A solution is to give the partons at the ends of the chains some intrinsic transverse momentum, k_T [97,98], whose average, $<k_T>$ has to increase with

---

[†] One prediction of this picture was for differences between pp and pp̄ collisions, since in the former case two overlapping $q - qq$ chains would be formed. The early version of this model was shown to disagree with data [7], but the subsequent introduction of multiple chains led to a better description [88].

$\sqrt{s}$, to more than 1GeV at the Sp$\bar{p}$S. The origin of this k_T could be either semi-hard scattering [100,99] which can be explicitly grafted onto the model or multiple gluon emission, which seems to give very similar results [100].

Other models have been proposed which try to extend perturbative QCD methods into the soft physics region. For example, various soft-gluon bremsstrahlung or QCD branching models have been found to give a good description of the UA5 multiplicity distribution [101,102,103]. A considerable growth of activity in this field has, however, been stimulated by the UA1 mini-jet results, which naturally suggest that one could try to extend the successful QCD description of high E_T jets at the Sp$\bar{p}$S collider down to lower transverse energies. The idea that a growth in parton-parton scattering could be linked to the growth of the total cross-section is not new [104]. The basic idea is that one could represent the total cross-section as:

$$\sigma_{tot} = \sigma_{soft} + \sigma_{jet}(p_T^{min}) \tag{22}$$

where σ_{soft} is assumed constant with energy. Because of the large density of partons near $x = 0$ the fraction that will scatter with $p_T > p_t^{min}$ will grow with $\sqrt{s}$. A choice of p_T^{min} in the region $2 - 4$ GeV/c seems to give a reasonable growth of σ_{tot}.

A number of recent papers have tried to estimate the effect of such "semi-hard" scattering. Qualitatively we can see that the jet events will have higher multiplicity and higher p_T and that the growth of such events could account for KNO scaling violations, and the correlation of $<p_T>$ with multiplicity (e.g. refs [105,106]). A detailed study of charged multiplicity distributions in this picture has been given in refs [107,108,30]. For example, the idea in refs [108,30] is that the KNO scaling observed up to ISR energies is exact, and represents the behaviour of the soft component. One can then extract the properties of the jet events from the observed violations of KNO scaling at the Sp$\bar{p}$S, and finds reasonable agreement with expectations. However, it should be pointed out that UA2 have studied in detail the changeover between hard and soft events in their data [109], and it is not clear that the data (on shape variables for example) in the intermediate region can be adequately described by a simple superposition of two independent models, of hard and soft events.

Perhaps the most complete approach of this kind, insofar as it is embodied in a complete Monte Carlo model which can predict any observed distribution, comes from the Lund group [110,111]. This is an extension of the Lund PYTHIA string model for high p_T processes. The basic idea is to calculate hard scattering above some cutoff p_T^{min}, which is chosen to be around 1.6GeV/c. With such a low cutoff the probability of multiple interactions becomes significant, especially at the highest Sp$\bar{p}$S energies, and this is claimed to be an important factor in fitting the high multiplicity tail of the multiplicity distribution. The remainder of the total cross-section is made up of "soft" events, which are assumed to consist of two strings, rather as in the original version of the Dual Parton Model. The strings in the hard events are stretched in such a configuration that the two string picture is approached in the low p_T limit. The strings fragment according to the standard Lund scheme. A further significant feature of this model is that it takes into account the important impact parameter structure of hadron-hadron interactions. Since hadrons are extended objects, we may suppose that the probability for hard scattering(s) is much greater for small impact parameters than large. In this model a double-gaussian form of the impact parameter distribution is chosen to give the best fit to the data. In addition, diffractive events are incorporated, and found to be necessary in order to account for the observed yield of the low multiplicity events. After including all these (plausible) refinements reasonable descriptions of the data are found, including multiplicity distributions, forward-backward correlations and "spike events".

A different model of soft processes has also come from the Lund group, in the form of a program called FRITIOF [112,113]. Some years ago a version of the Lund model was proposed [114] consisting of one string stretched between the colliding hadrons. This model worked well in the projectile fragmentation regions, but yielded too few particles in the central region. In the new model a hadron is treated like a vortex line in a colour superconducting medium. In a soft collision the hadrons overlap, and the net effect is to give two excited stringlike fields extended along the collision axis. These may then emit soft gluons, and fragment according to the standard Lund prescription. It is found that an adequate description of data can be found with just two strings; the additional strings which are required in the Dual Parton Model being effectively replaced by soft gluon emission. At present hard scattering is not included in this model, and this will clearly be needed to describe fully some features of the Sp$\bar{p}$S data.

Statistical models

A wide range of geometrical and/or statistical approaches to multiparticle production in the Sp$\bar{p}$S data can be gathered together under this heading. Several authors have emphasized the importance of the finite size of the hadrons, and thus of accounting for variations of collision properties with impact parameter (e.g. refs [115,116]). The general idea is that small impact parameters (central collisions) lead to violent collisions and high multiplicities, while large impact parameters (peripheral collisions) lead to small multiplicities. In contrast with pointlike e^+e^- collision process, the wide range of impact parameters in hadronic interactions leads to a much broader multiplicity distribution than in e^+e^-. Furthermore a peripheral collision will tend to give a low multiplicity in both forward and backward directions, while a central collision will give both high multiplicities. In this way strong forward-backward multiplicity correlations emerge naturally [116,117]. Other features of the data discussed in the context of this geometrical picture have been semi-inclusive rapidity distributions [118] and correlations between $<p_T>$ and multiplicity [119] (on the supposition that higher hadronic temperatures are achieved in central collisions).

Another model which has been extensively compared with the Sp$\bar{p}$S data is the "three-fireball" model of the Berlin group. Their basic idea is that in a hadronic collision some fraction of the energy of the incoming particles is effectively stopped, and forms a central fireball, while two more fireballs are formed by the excited projectile and target. In the original form of their model the three fireballs were taken to have equal energies, and the multiplicity distribution of their decay was computed by treating the excitation as two uncoupled harmonic oscillators [120,121]. This model gave quite a simple explanation of the multiplicity distributions up to ISR energies, both in the full phase space, where all three fireballs contribute, and in the central region, where only the central fireball gives decay products. In order to accomodate the UA5 observations of KNO scaling violations into this picture it appears necessary to assume that the fraction of energy going into the central fireball rises rather rapidly with $\sqrt{s}$ [122,123]. A number of other observations appear to have a statistical origin in this scheme, for example semi-inclusive pseudorapidity distributions [123] and the multiplicity distributions in limited rapidity regions can be well described [124,125], so can forward-backward multiplicity correlations [125], and even the difference between the multiplicity distributions in "jet" and "no-jet" UA1 data [126].

The UA5 results on fitting the Negative Binomial distribution to multiplicity distributions have stimulated a good deal of theoretical interest. This Negative Binomial distribution occurs in Quantum Optics, representing the distribution of photons from k independent sources (k is one of the parameters of the Negative Binomial (equ. (4)). Indeed, the distribution had been proposed for particle physics applications some years ago, based on analogy with these ideas, [27,26,127], and has recently been revived (e.g. the stochastic cell model of [28], which imagines particle emission from the surface of a small chaotic source). In this picture, then, the parameter k represents the number of sources (or phase space cells). In effect the Negative Binomial arises from the convolution of k independent Poisson distributions. It therefore seems reasonable that k should fall as the rapidity window is reduced, in accordance with observation. However, as $\sqrt{s}$ increases, the rapidity range grows, so one would naïvely expect the number of cells to increase, whereas in fact k falls with $\sqrt{s}$. Also in this picture it is not obvious how to interpret non-integer values of k.

Other authors have also pursued the analogy with Quantum Optics, and have particularly emphasized the rôle played by the inelasticity (or the leading particle effect) in hadronic collisions [128]. Their physical picture is that hadronic interactions are dominantly interactions of sea partons, while the valence quarks pass through unaffected and form the leading particles. In a statistical picture the average charged multiplicity should rise like $s^{1/4}$, much faster than observed. This can be accounted for if the inelasticity falls (and its distribution narrows) with increasing energy.[†] Of course the spread of inelasticities may just be another way of considering the spread of impact parameters discussed above. It is important to allow for the spread of inelasticities, which means that different events have different effective energies. In this picture, a reasonable description of the moments of the multiplicity distribution can be obtained with a single source [129]. A recent analysis along these lines of multiplicity distributions in different rapidity ranges has claimed to find evidence for two sources, one of which has bremsstrahlung-like behaviour, while the other has the properties expected for a quark-gluon plasma [130].

The Negative Binomial distribution is not the most general form of multiplicity distribution which could arise in these statistical approaches. A number of authors have suggested the use of the "Generalised Glauber-Lachs" (GGL) distribution (also known as the Perina-McGill distribution) [131], which takes account of both a coherent and a chaotic component in the particle emission. The Negative Binomial form is recovered in the case where the coherent component tends to zero. Recently an attempt has been made to fit the GGL form to the UA5 data [132]. The conclusion seems to be that the GGL formula does not give better fits than the Negative Binomial, which means that these data are not able to give meaningful information about the parameters of the GGL distribution, and specifically about the degree of chaoticity. Another extension of the Negative Binomial, the "Perina-Horák" distribution, has also been applied to the Sp$\bar{\text{p}}$S data [133].

Another quantum statistical interpretation of the Negative Binomial distribution is in terms of partial stimulated emission [134]. In this picture particles may be emitted either independently or the emission may be enhanced by Bose-Einstein interference with particles already present. This model is shown to lead to a negative binomial multiplicity distribution, where the parameter k^{-1} is interpreted as the average fraction of the particles already present which stimulate the emission of an additional particle. This model has no problem with fractional values of k. Since the rapidity range of Bose-Einstein interference is presumably short, we may expect k^{-1} to be larger for small rapidity windows, and as the particle density rises with $\sqrt{s}$ then k^{-1} should rise with $\sqrt{s}$. Both of these expectations are in accord with observation. However, some more recent data give problems for the Stimulated Emission model. In a bubble chamber experiment at $\sqrt{s} = 22\text{GeV}$ it is found that k^{-1} is smaller by about a factor of 2 for the multiplicity distribution of negative particles than for all particles [135], whereas in the stimulated emission model k^{-1} should be greater since a larger fraction of the particles are able to cause stimulated emission in the negatives-only case. Further, it appears that when a cut on the azimuthal angle is made in the Sp$\bar{\text{p}}$S data k^{-1} is essentially unchanged, [23,24], which is again hard to accomodate in the stimulated emission model.

A quite different source for a Negative Binomial distribution is also discussed in [134]. This is a sort of cluster model. The idea is that a Negative Binomial distribution can be formed by the convolution of a Poisson with a Logarithmic distribution. Clusters (more recently named "clans") are taken to be independently emitted with a Poissonian distribution, while the number of particles into which each cluster decays is assumed to follow a Logarithmic distribution (inspired by a cascade picture). This picture has no difficulty in accomodating the dependence of k^{-1} on rapidity and azimuthal cuts, and the behaviour for negatives-only. The parameters of the constituent Poisson and Logarithmic distributions may be inferred from data, and the surprising result is that the average number of clusters

---

[†] It may be interesting to note that a similar fall of inelasticity was required by the Wdowczyk-Wolfendale scaling idea alluded to under "PSEUDORAPIDITY DISTRIBUTIONS" above.

remains roughly constant (at about 8) from the ISR to the top of the Sp$\bar{p}$S energy range. The size of the clusters (in both multiplicity and in rapidity spread) has to grow with $\sqrt{s}$. However, there seems no obvious reason for the number of clusters to be constant. It should also be noted that the convolution of Poisson and Logarithmic distributions is by no means the unique way to arrive at a Negative Binomial [136].

Finally it may be worth mentioning the UA5 cluster model **GENCL** [54,29,132]. It is perhaps presumptuous to call this a model, since its original purpose was to give an accurate simulation of the UA5 data in order to compute corrections, and therefore some experimental results were directly fed in as inputs; in some sense it is a "no-physics" model. However, since GENCL turns out to fit several features of the data which were by no means input there may be something to be learnt from it. It is basically a longitudinal phase space multi-cluster model. Data on the overall charged multiplicity distribution, p_T-distribution and rapidity distribution are essentially put in by hand, or parameters in the program are tuned to fit them. Particles are generated in clusters, decaying isotropically into around 2 charges particles, and leading particle effects are introduced. Energy and momentum are conserved. As seen above, this model gives a satisfactorily consistent description of the Sp$\bar{p}$S correlation data. In addition it fits well the rapidity distributions at fixed multiplicity, the multiplicity distributions in different rapidity windows, the observed yield of "spike" events, and even to a surprising extent the rate of "mini-jet" events and most of their properties. What we should probably learn from this is that many of the pieces of data from the Sp$\bar{p}$S are interrelated. Any model which fits the multiplicity distribution and has appropriate short-range correlations will probably also reproduce the long-range correlation data too, or if the multiplicity and rapidity distributions are well fitted then spike events will probably appear. The key feature of GENCL in fitting the multiplicity distributions seems to be cluster formation (a point also stressed in the "minimal model" of ref [137]), whilst the key to fitting the semi-inclusive rapidity distributions is mainly energy-momentum conservation. Therefore the success of this very simple ad hoc model in fitting the data (probably better than any of the "physics" models to date) should counsel us to take care in assessing which predictions of the models are really significant.

SUMMARY

In this short review we have endeavoured to give a general overview of the "soft physics" data from the CERN SPS collider, and a very brief sketch of some of the many models which have been put forward to try to account for these data.

In the early days, before the Sp$\bar{p}$S started operation, there was much excitement and anticipation of exotic phenomena in this new energy regime, speculation fuelled by the Centauro and other bizarre events seen in cosmic ray data. Therefore the first results from the Sp$\bar{p}$S looked disappointingly normal, lying on simple extrapolations from lower energy data. However, more careful study of the data revealed a number of interesting and unexpected features. Among these we may list the following:

• The violation of KNO scaling, particularly for the full phase space, but now established in other kinematic regions too.

• The strong rise of $<p_T>$ with $\sqrt{s}$ above ISR energies, especially for heavy particles, kaons and baryons.

• The marked correlation between $<p_T>$ and multiplicity.

• The observations of "minijets" by UA1.

- Strong growth of "long range" forward-backward multiplicity correlations (though in a cluster model this seems less surprising, being a statistical effect linked to the development of multiplicity distributions).

- The impressive success of the negative binomial distribution in fitting all charged multiplicity distributions.

- The apparently small value of the single-diffractive cross-section, though this has proved a difficult measurement, and the experiments do not agree too well.

In addition, there is still no evidence for any of the exotic event classes claimed by cosmic ray experiments.

These new experimental results have encouraged an extensive revival of interest in theoretical models of soft processes. The experiments have now almost finished their work (for soft physics). Indeed the Sp$\bar{p}$S experiments may be the last generation of experiments to study complete events in this way, since the acceptance of detectors at the SSC (or LHC) will certainly be confined to the central region, and the same is largely true for the Tevatron Collider. However, an understanding of soft physics is of importance for these very high energy colliders, since multiple soft events could be an important source of background. Therefore for this reason if no other theoretical work is likely to continue for some time.

We now have several complete QCD-inspired models (complete in the sense that they are cast in a Monte Carlo framework and can predict any observed distribution) which give a reasonable description of most features of the data; notably the various versions of the Dual Parton Model model, and two derivatives of the Lund string model. These have many features in common, needing two long strings or chains, together with some additional source of central production, either additional chains or soft bremsstrahlung of gluons. All these models indicate the importance of "semi-hard" scattering at Sp$\bar{p}$S energies, which helps to account for the first four of our "unexpected" results above. In addition there are many interesting ideas in the statistical approaches, and again many recurring themes, such as the need to allow for clustering, and the importance of taking into account leading particle effects (e.g. via the impact parameter dependence, or the spread of inelasticities). We may hope that these data from the Sp$\bar{p}$S will constitute a useful basis for theoretical work for some years yet.

ACKNOWLEDGEMENTS

The author would particularly like to thank the many physicists in the UA5 experiment with whom he has worked for the last few years, and from whom he has learned much about the physics discussed here.

REFERENCES

(1)   Lattes,C.M.G. et al., Phys.Rep. C65 151 (1980).

(2)   The Staff of the CERN p$\bar{p}$ Project, Phys.Lett. 107B 306 (1982).

(3)   Rushbrooke,J.G., CERN/EP 82−6 (1982).

(4)   UA4 Collab., Battiston,R. et al., Nucl.Inst.Meth. A238 35 (1985).

(5)   UA5 Collab., Alner,G.J. et al., Phys.Rep.C, to be published.

124

(6)   UA5 Collab., Phys.Scr. $\underline{23}$ 642 (1981).

(7)   UA5 Collab., Alpgård,K. et al., Phys.Lett. $\underline{112B}$ 183 (1982).

(8)   Weidberg,A.R., Ph.D. Thesis, University of Cambridge (1982).

(9)   Åsman,B., *From tracks on film to corrected multiplicity distributions* Univ. of Stockholm Report, USIP 85−17 (1985).

(10)   UA5 Collab., Alner,G.J. et al., Phys.Lett. $\underline{167B}$ 476 (1986).

(11)   Åsman,B., Presentation at the EPS Conf. on HEP, Uppsala (1987).
        UA5 Collab., paper in preparation.

(12)   Koba,Z.,Nielsen,H.B.,Olesen,P., Nucl.Phys. $\underline{B40}$ 317 (1972).

(13)   Feynman,R.P., Phys.Rev.Lett. $\underline{23}$ 1415 (1969).

(14)   Thomé,W. et al., Nucl.Phys. $\underline{B129}$ 365 (1977).

(15)   Breakstone,A. et al., Phys.Rev. $\underline{D30}$ 528 (1984).

(16)   UA5 Collab., Alpgård,K. et al., Phys.Lett. $\underline{121B}$ 209 (1983).

(17)   UA5 Collab., Alner,G.J. et al., Phys.Lett. $\underline{138B}$ 304 (1984).

(18)   Wróblewski,A., Acta Phys.Pol. $\underline{B4}$ 857 (1973).

(19)   Szwed,R. and Wrochna,G., Zeit.Phys. $\underline{C29}$ 255 (1985).

(20)   Blazek,M., Zeit.Phys. $\underline{C32}$ 309 (1986).

(21)   UA1 Collab., Arnison,G. et al., Phys.Lett. $\underline{107B}$ 320 (1981)

(22)   UA1 Collab., Arnison,G. et al., Phys.Lett. $\underline{123B}$ 108 (1983)

(23)   Fuglesang,C., Proc.XVII Int.Symp. on Multiparticle Dynamics, Seewinkel (1986).

(24)   Carlson,P., Proc.XXIII Int.Conf. on HEP, Berkeley (1986).

(25)   UA5 Collab., Alner,G.J. et al., Phys.Lett. $\underline{160B}$ 199 (1985).

(26)   Giovannini,A., Nuov.Cim. $\underline{15A}$ 543 (1973).

(27)   Knox,W., Phys.Rev. $\underline{D10}$ 65 (1974).

(28)   Carruthers,P. and Shih,C.C., Phys.Lett. $\underline{127B}$ 242 (1983).

(29)   Ward,D.R., *Low p_T physics at the $Sp\bar{p}S$ Collider* Proc. SSC Workshop on Physics Simulations at High Energy, Madison p.208 (1986).

(30)   Martin,A.D. and Maxwell,C.J., Phys.Lett. $\underline{172B}$ 248 (1986).

(31)   UA5 Collab., Alner,G.J. et al., Phys.Lett. 160B 193 (1985).

(32)   UA5 Collab., Alpgård,K. et al., Phys.Lett. 107B 310 (1981).

(33)   UA5 Collab., Alpgård,K. et al., Phys.Lett. 107B 315 (1981).

(34)   UA5 Collab., Alner,G.J. et al., Zeit.Phys. C33 1 (1986).

(35)   UA7 Collab., Paré,E., Presentation to EPS Conf. on HEP, Uppsala (1987).

(36)   Gaisser,T.K., Phys.Lett. 100B 425 (1981).

(37)   Wdowczyk,J. and Wolfendale,A.W., Nature 306 347 (1983), and references therein.

(38)   Wdowczyk,J. and Wolfendale,A.W., Nuov.Cim. 54A 433 (1979).

(39)   Wdowczyk,J. and Wolfendale,A.W., J.Phys. G10 257 (1983).

(40)   UA1 Collab., Arnison,G. et al., Phys.Lett. 118B 167 (1982).

(41)   UA1 Collab., Cerardini,F., Proc.Int. Europhysics Conf. on HEP, Bari, 363 (1985).

(42)   UA2 Collab., Banner,M. et al., Phys.Lett. 122B 322 (1983).

(43)   UA5 Collab., Alpgård,K. et al., Phys.Lett. 115B 65 (1982).

(44)   UA5 Collab., Alner,G.J. et al., Nucl.Phys. B258 505 (1985).

(45)   UA5 Collab., Ansorge,R.E. et al., *Kaon Production at 200 and 900GeV c.m. energy* submitted
        to Zeit.Phys.C;
        UA5 Collab., W.Pelzer, Proc. VI Int.Workshop on p̄p Physics, Aachen (1986);
        Ovens,J.E.V., Ph.D. Thesis, RAL-T-026, University of Cambridge (1986).

(46)   Jon-And.K., Presentation at EPS Conf. on HEP, Uppsala (1987).

(47)   UA5 Collab., Alner,G.J. et al., Phys.Lett. 151B 309 (1985).

(48)   French,K.A., Ph.D. Thesis, University of Cambridge (1984).

(49)   Dewolf,R.S., Ph.D. Thesis, University of Cambridge (1986).

(50)   Hagedorn,R., Riv.Nuov.Cim. 6 no.10 (1983).

(51)   Rossi,A.M. et al., Nucl.Phys. B84 269 (1975); and references therein.

(52)   Breakstone,A. et al., Phys.Lett. 183B 227 (1987).

(53)   van Hove,L., Phys.Lett. 118B 138 (1982).

(54)   UA5 Collab., Alner,G.J. et al., Nucl.Phys. B291 445 (1987).

(55)   UA5 Collab., Alpgård,G.J. et al., Phys.Lett. 115B 71 (1982).

(56)   Malhotra,P.K. and Orava,R., Zeit.Phys. C17 85 (1983).

(57)   Anisovich,V.V. and Kobrinski,M.N., Phys.Lett. 52B 217 (1974).

(58)   Wróblewski,A., Acta Phys.Pol. B16 217 (1985).

(59)   Jacob,M. and Rafelski,J., Phys.Lett. 190B 173 (1987).

(60)   UA5 Collab., Ansorge,R.E. et al., *Charged Particle Correlations in p̄p collisions at c.m. energies of 200, 546 and 900 GeV* Submitted to Zeit.Phys.C.

(61)   UA5 Collab., Alpgård,K. et al., Phys.Lett. 123B 361 (1983).

(62)   Böckmann,K. and Eckart,B.I., Proc. XV Symp. on Multiparticle Dynamics, p.155, Lund (1984).

(63)   Amendolia,S.R. et al., Nuov.Cim. 31A 19 (1976).

(64)   Bell,W. et al., Zeit.Phys. C22 109 (1984).

(65)   Uhlig,S. et al., Nucl.Phys. B132 15 (1978).

(66)   Bell,W. et al., Zeit.Phys. C32 335 (1986).

(67)   Müller,Th., Ph.D. Thesis, Univ. of Bonn, BONN IR − 83 − 21 (1983).

(68)   UA4 Collab., Bozzo,M. et al., Phys.Lett. 136B 217 (1984).

(69)   UA4 Collab., Bernard,D. et al., Phys.Lett. 166B 459 (1986).

(70)   UA4 Collab., Bernard,D. et al., Phys.Lett. 186B 227 (1987).

(71)   UA5 Collab., Ansorge,R.E. et al., Zeit.Phys. C33 175 (1986).

(72)   Schmickler,H., Ph.D. Thesis, Univ. of Bonn, BONN − IR − 86/34 (1986).

(73)   Schlein,P., *Recent Developments in Inelastic Diffraction Scattering* Proc. XXIII Int.Conf on HEP, Berkeley, p.1331 (1986).

(74)   Goulianos,K., Phys.Rep. 101 169 (1983).

(75)   Borisov,A.K. et al., Phys.Lett. 190B 226 (1987).

(76)   UA1 Collab., Arnison,G. et al., Phys.Lett. 122B 189 (1983).

(77)   Rushbrooke,J.G., Proc.XXI Int.Conf. on HEP, C3 − 177, Paris (1982).

(78)   UA5 Collab., Alner,G.J. et al., Phys.Lett. 180B 415 (1986).

(79)   Bjorken,J.D. and McLerran,L., Phys.Rev. D20 2353 (1979).

(80)   Halzen,F., Hoyer,P. and Yamdagni,N., Phys.Lett. 190B 211 (1987).

(81)   UA2 Collab., Banner,M. et al., Phys.Lett. 121B 187 (1983).

(82)   UA2 Collab., Banner,M. et al., Phys.Lett. 156B 129 (1983).

(83)   Aubert.B. et al., Phys.Lett. 120B 465 (1983).

(84)   Dremin,I.M., Sov.J.Nucl.Phys. 33 726 (1981).

(85)   Carlson,P., Proc. IV Workshop on p$\bar{\text{p}}$ Physics, Bern, p286 (1984).

(86)   Rushbrooke,J.G., Proc.Workshop on p$\bar{\text{p}}$ Options for the Supercollider, Chicago (1984).

(87)   Capella,A. et al., Phys.Lett. 81B 68 (1979).
         Capella,A. et al., Zeit.Phys. C3 329 (1980).

(88)   Capella,A. et al., Phys.Lett. 114B 450 (1982).

(89)   Kaidalov,A.B., Phys.Lett. 116B 459 (1982).
         Kaidalov,A.B. and Ter-Martirosyan,K.A., Phys.Lett. 117B 247 (1982).

(90)   Aurenche,P.,Bopp,F.W. and Ranft.J., Phys.Lett. 114B 363 (1982).

(91)   Capella,A. and Tran Thanh Van,J., Zeit.Phys. C23 165 (1984).

(92)   Aurenche,P.,Bopp,F.W. and Ranft.J., Zeit.Phys. C23 67 (1984).

(93)   Capella,A. et al., Phys.Rev. D32 2933 (1985).

(94)   Kanki,T., Nucl.Phys. B243 44 (1984).

(95)   Aurenche,P.,Bopp,F.W. and Ranft.J., Zeit.Phys. C26 279 (1984).

(96)   Holl,B., Diploma Thesis, Univ. of Bonn, BONN − IR − 86/15 (1986).

(97)   Capella,A. and Krzywicki,A., Phys.Rev. D29 1007 (1984).

(98)   Aurenche,P.,Bopp,F.W. and Ranft.J., Phys.Lett. 147B 212 (1984).

(99)   Capella,A. et al., Phys.Rev.Lett. 58 2015 (1987).

(100)   Aurenche,P.,Bopp,F.W. and Ranft.J., Phys.Rev. D33 1867 (1986).

(101)   Pancheri,G. et al., Phys.Lett. 151B 453 (1985).

(102)   Hayot,F. and Navelet,H., Phys.Rev. D30 2322 (1984).

(103)   Durand,B. and Sarcevic,I., Phys.Lett. 172B 104 (1986).

(104)   Cline,D. et al., Phys.Rev.Lett. 31 491 (1973).

(105)   Pancheri,G. and Rubbia,C., Nucl.Phys. A418 117c (1984).

(106)   Gaisser,T.K. and Halzen,F., Phys.Rev.Lett. 54 1754 (1985).

128

(107)   Pancheri,G. and Srivastava,Y., Phys.Lett. 159B 69 (1985).

(108)   Gaisser,T.K. et al., Phys.Lett. 166B 219 (1986).

(109)   UA2 Collab., Preprint CERN EP/87−79; submitted to Zeit.Phys.C.

(110)   Sjöstrand,T., Preprint FERMILAB-Pub-85/119-T

(111)   Sjöstrand,T. and Van Zyl,M., Phys.Lett. 188B 149 (1987).

(112)   Andersson,B. et al., Nucl.Phys. B281 289 (1987).

(113)   Nilsson-Almqvist,B., Comp.Phys.Comm. 43 387 (1987)

(114)   Andersson,B. et al., Phys.Lett. 69B 221 (1977).
          Andersson,B. et al., Phys.Lett. 71B 337 (1977).

(115)   Barshay,S., Phys.Lett. 116B 193 (1982).

(116)   Chou,T.T. and Yang,C.N., Phys.Lett. 116B 301 (1982).

(117)   Chou,T.T. and Yang,C.N., Phys.Lett. 135B 175 (1984).

(118)   Chou,T.T., Yang,C.N. and Yen,E., Phys.Rev.Lett. 54 510 (1985).

(119)   Barshay,S., Phys.Lett. 127B 129 (1983).

(120)   Liu Lian-Sou and Meng Ta-Chung, Phys.Rev. D27 2640 (1983).

(121)   Chou Kuong-Chau, Liu Lian-Sou and Meng Ta-Chung, Phys.Rev. D28 1080 (1983).

(122)   Cai Xu and Liu Lian-Sou, Nuov.Cim.Lett. 37 495 (1983).

(123)   Cai Xu, Liu Lian-Sou and Meng Ta-Chung, Phys.Rev. D29 869 (1984).

(124)   Chao Wei-Qin et al., Phys.Rev. D35 152 (1987).

(125)   Cai Xu et al., Phys.Rev. D33 1287 (1986).

(126)   Chao Wei-Qin et al., Phys.Rev.Lett. 58 1399 (1987).

(127)   Giovannini,A., Nuov.Cim. 24A 421 (1974).

(128)   Fowler,G.N. et al., Phys.Lett. 145B 407 (1984).
          Fowler,G.N. et al., Phys.Rev. D35 870 (1987).

(129)   Fowler,G.N. et al., Phys.Rev.Lett. 56 14 (1986).

(130)   Fowler,G.N. et al., Phys.Rev.Lett. 57 2119 (1986).

(131)   Biyajima,M., Prog.Theor.Phys. 69 966 (1983).
          Biyajima,M., Phys.Lett. 137B 225 (1984).
          Biyajima,M., Phys.Lett. 140B 435 (1984).

(132)   Fuglesang,C., Ph.D. Thesis, Univ of Stockholm (1987).

(133)   Blazek,M., Zeit.Phys. $\underline{C26}$ 455 (1984).

(134)   Giovannini.A. and van Hove,L., Zeit.Phys. $\underline{C30}$ 391 (1986).

(135)   Adamus,M. et al., Phys.Lett. $\underline{177B}$ 239 (1986).

(136)   Shih,C.C. et al., Phys.Rev. $\underline{D34}$ 2710 (1986).

(137)   Fialkowski,K., Phys.Lett. $\underline{B173}$ 197 (1986).

The Physics of Hadronic Jets

R. K. Ellis,
Fermi National Accelerator Laboratory, USA.

W. G. Scott,
University of Liverpool, UK.

Contents

1   The Emergence of Jets

The first results on jets from the CERN SPS $p\bar{p}$ collider [1,2] had a very significant impact on the field of High Energy Physics. Although jets were already a well established phenomenon in e^+e^- annihilation, and (to a lesser extent) in deep inelastic lepton scattering, the observation of clear, uncontroversial jets in hadron-hadron collisions had been long awaited. The first results on jets from the collider represented the successful culmination of years of experimental effort, carried over from the CERN ISR and elsewhere, on a difficult and surprisingly subtle experimental problem. This section reviews these early results, beginning with a brief sketch of the historical context in which they were originally presented.

1.1   Historical Perspective

The motivation for looking for jets in hadron-hadron collisions had been clear since the early seventies when parton model ideas first became widely known. It was foreseen that the constituent partons within the colliding hadrons, whatever their exact nature, would themselves collide and scatter to produce jets. It was hoped that by studying the jets, one could learn about the properties and interactions of the partons. The rapid developments in perturbative QCD theory in the late seventies sharpened the older parton model expectations into quantitative QCD tests. In particular the theory quite definitely predicted jets in hadron-hadron collisions, with properties and production rates which could be reliably calculated on the basis of measurements in e^+e^- annihilation and deep inelastic lepton scattering.

The earliest experimental investigations, carried out at the CERN ISR, revealed copious production of high transverse momentum particles [3] in qualitative agreement with parton model expectations. Following this discovery, for almost a decade, experimenters working at the CERN ISR and elsewhere, struggled with the problem of isolating the jets and measuring their properties. In spite of the evident simplicity of the basic theoretical idea, the problems encountered in interpreting the results of these experimental investigations were far from simple. At the root of it all, was the lack of any real theoretical understanding of the soft physics processes which dominate the total cross-section in hadron-hadron collisions at collider energies and below. The behaviour of the soft physics background and the influence of the event selection called into question the significance of the effects which were observed. In retrospect, there is every reason to suppose that these experiments were indeed, for the most part, studying fundamental parton-parton interactions [4]. Nevertheless at the time of the start-up of the SPS collider, the consensus of opinion was that the existence of jets in hadron-hadron collisions had not been adequately demonstrated.

It was the SPS collider which provided the first clear uncontroversial evidence for jets in hadron-hadron collisions [1,2,5]. The existence of jets at the SPS collider was established convincingly very soon after the initial running period with comparative ease. The initial centre of mass energy of the SPS collider was almost ten times the highest energy available at the ISR and considerably more than ten times the highest energy available in comparable fixed target experiments. With increased centre of mass energy came an improved signal to noise and a quantitative if not qualitative reduction in the problem of the soft physics background. The experimenters who worked at the collider benefited from the experience gained at the ISR and the various fixed target experiments. In particular the need for large solid-angle calorimetry and the importance of the so-called transverse-energy trigger [6,7] were by then fully appreciated.

1.2   The Transverse Energy Trigger

Even at the energies of the SPS collider the majority of interactions which occur do not give rise to high tranverse momentum jets. The dominant event configuration is one in which the available energy is shared between particles having a relatively low transverse momentum (< 1 GeV) with respect to the beam direction, a feature of high energy hadron-hadron interactions which is true over a wide range of energies. Although phenomenologically familiar, the production of particles at low transverse momenta is poorly understood theoretically, because it involves soft physics mechanisms beyond the scope of perturbative QCD.

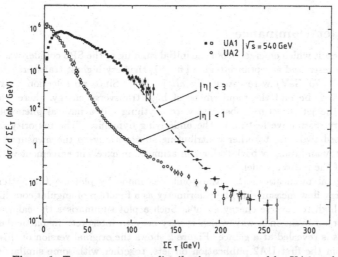

Figure 1: Transverse energy distributions measured by UA1 and UA2.

It was this preponderance of soft physics events which had made it so difficult to establish the significance of the jets in the earlier, lower energy experiments. There was always the possibility that fluctuations in the soft physics background would mimic the production of jets. The observation of jets in event samples obtained using triggers which clearly favoured jet-like topologies could not be considered proof of the existence of jets. Futhermore the use of such triggers invalidated attempts to assess the significance of the jets by studying the transition between non-jet-like and jet-like topologies, since to a large extent only the jet-like topologies survived the trigger. It had been argued for some time, however, that a threshold requirement on the total scalar sum of the transverse momenta of all the secondaries in an event would discriminate effectively against events with only low transverse momentum secondaries, while at the same time doing nothing to bias the data in favour of a specifically jet-like topology. This was the basic motivation underlying the implementation of the so-called transverse-energy trigger, which played a vital role in establishing the credibility of the jets at the SPS collider.

In practice the transverse energy ($\sum E_T$) of an event is defined by the scalar sum of the transverse energies of all the calorimeter hits in the event. Transverse energy distributions ($d\sigma/d\sum E_T$) for $\bar{p}p$ collisions at $\sqrt{s} = 540$ GeV based on data from UA1 [8,9] and UA2 [10] are shown in Figure 1. Clearly the transverse energy measured in a practical calorimeter with a limited angular acceptance falls short of that which would be measured in an idealised calorimeter with full angular coverage. The transverse energy distribution measured by UA1 and plotted for $|\eta| \leq 3$ is close to ideal. The UA2 data are restricted to the region $|\eta| \leq 1$. Note that the absolute cross section $d\sigma/d\sum E_T$ depends quite strongly on the angular acceptance for intermediate values of the transverse energy $\sum E_T \simeq 50 - 150$ GeV.

1.3 Two-Jet Dominance

The startling result which emerged from the initial running at the SPS collider was that, at least for the restricted acceptance region $|\eta| \leq 1$, the very highest transverse energy events ($\sum E_T \geq 70$ GeV) were essentially *all* jet-like. Since no selection whatever had been applied, beyond the requirement of high transverse energy, there was no question that the jet structures observed were anything but a natural phenomenon, reflecting a very distinctive feature of the underlying dynamics. The majority of the events comprised two jets, together contributing substantially to the measured total transverse energy and roughly back-to-back in azimuth, in excellent agreement with the expectations of the parton-model.

In the original publications [1,2] the point was made by plotting the pattern of transverse energy flow measured in the calorimetry as a function of angular coordinates for individual high transverse energy events. Such a plot summarises the calorimeter information for an event in a way which is very easy to interpret: in particular the presence of jets is revealed at a glance. Figure 2 shows the original version of this plot reproduced from the first UA2 publication on jets, together with some similar, more recent plots from the analysis by UA1 [9] based on the full acceptance region. Such plots have since become an essential tool in the analysis of all aspects of collider data.

In an updated version [11,12] of their original analysis the UA2 collaboration has plotted the mean centre of mass sphericity (suitably defined) as a function of $\sum E_T$ for the central acceptance region $|\eta| < 1$ for $\sqrt{s} = 630$ GeV. The result is shown in Figure 3a. The abrupt fall in the mean sphericity signals the onset of two-jet dominance at $\sum E_T \simeq 70$ GeV, in excellent agreement with the result of the original analysis. The analysis by UA1 [9] shows that the onset of two-jet dominance occurs less abruptly and at higher values of $\sum E_T$ ($\sum E_T \simeq 200$ GeV) if the full acceptance region is used, as shown in Figure 3b. Figure 3b shows the fraction (H_2) of the total transverse energy

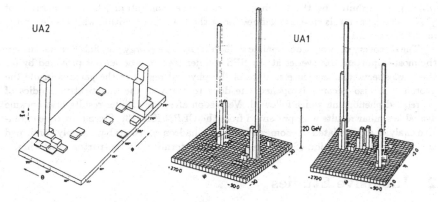

Figure 2: Transverse energy flow plots for high transverse energy events from UA1 and UA2.

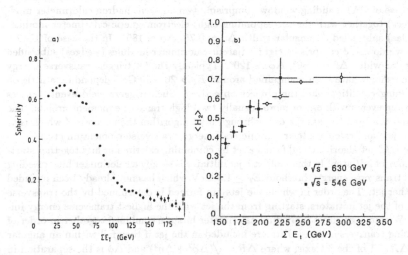

Figure 3: a) The mean sphericity as function of $\sum E_T$ measured by UA2. b) The fractional contribution to the total $\sum E_T$ from the two highest transverse momentum jets measured by UA1.

which is contributed by the two highest transverse momentum jets as a function of $\sum E_T$. The jets, in this case, are defined using the UA1 jet algorithm which is described in the Section 2.1.

The discovery of two-jet dominance at high transverse energy justifiably ranks amongst the most important discoveries at the SPS collider, not only because it provided by far the most spectacular evidence to date of the physical reality of the partons inside the proton, but also because it opened the door to many of the quantitative studies of jet related phenomena which followed. Very soon after the collider results first became available similar results were presented from the CERN ISR [13]. Since that time extensive analysis of jet related phenomena in hadron-hadron collisions have amply confirmed the original interpretation of the two-jet events in terms of parton-parton scattering.

2   Inclusive Studies

2.1   Jet Triggers and Jet Algorithms

With the significance of the jets established, both UA1 and UA2 implemented hardware jet triggers [14,15] requiring one or more localised transverse energy depositions in the calorimetry exceeding specified transverse energy thresholds. The degree of localisation demanded was crudely matched to the width of the jets which were actually observed. In the case of UA1 a sliding window comprising two adjacent hadron calorimeter modules (cees) together with the corresponding eight electromagnetic calorimeter modules (gondolas) subtended an angular width $\Delta\eta \simeq 0.75$, $\Delta\phi \simeq 180°$. In the case of UA2 a sliding window in ϕ comprising eight adjacent calorimeter modules (wedges) subtended an angular width $\Delta\theta \simeq 80°$, $\Delta\phi \simeq 120°$. Typically the jet trigger transverse energy (E_T) thresholds have been maintained around $E_T > 20 - 30$ GeV depending a little on the running conditions and the physics objectives. Such triggers yield event samples comprising events with one or more (usually two) high transverse momentum jets. The next step has been to pass the data through a jet algorithm the purpose of which is to find the jets and calculate their parameters (energy, transverse momentum etc.).

The UA1 jet algorithm [14] forms a jet by combining calorimeter hits together starting from a jet initiator. In this context a jet initiator is simply a calorimeter hit exceeding a given transverse energy threshold $E_T \geq 1.5$ GeV which has not already been included in another jet. The order in which the jets are formed is determined by the transverse energy of the jet initiators, starting from the jet with the highest transverse energy initiator. Within a given jet individual calorimeter hits are combined together in order of decreasing transverse energy. Hits are included in the jet if they fall within an angular range $\Delta R \leq 1$ of the jet axis, where $\Delta R = \sqrt{(\Delta\phi^2 + \Delta\eta^2)}$ and $\Delta\phi$ is the separation in azimuth between the hit and the jet axis and $\Delta\eta$ is the separation in pseudorapidity. The algorithm is iterative to the extent that the jet axis is redetermined at each stage using the hits that have been included in the jet so far. Finally the energy and momentum of each jet is computed by taking respectively the scalar and vector sum over the associated calorimeter hits.

The UA2 algorithm [1,10] combines calorimeter hits occuring in adjacent calorimeter cells to form clusters. In the UA2 apparatus individual calorimeter cells subtend angles

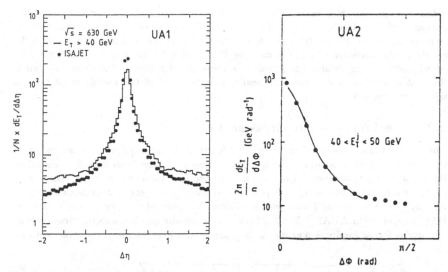

Figure 4: Jet transverse energy profiles plotted as a function of pseudorapidity and azimuth.

$\Delta\theta = 10°$ and $\Delta\phi = 15°$. Hits will be combined in the same primary cluster if they occur in adjacent calorimeter cells and satisfy a transverse energy threshold requirement $E_T \geq 0.4$ GeV. Primary clusters may be split into secondary clusters in a second pass [16], but in practice most jets give rise to only one cluster. The energy of the jet is taken to be the energy of the cluster and the direction of the jet is determined by the centroid of the energy deposition on the front face of the calorimeter.

Figure 4 shows the average transverse energy profile for jets with $E_T > 30$ GeV as a function of pseudorapidity η measured in the UA1 experiment [14]. For this plot the pseudorapidity is referred to the beam axis but measured from the axis of the jet and the transverse energy is integrated over the half cylinder in ϕ containing the jet axis. Also shown in Figure 4 is the transverse energy profile as a function of azimuth ϕ measured in the UA2 experiment [17]. For this plot the azimuth is measured relative to the axis of the higher E_T jet. On the basis of Figure 4 it is concluded that, although the jets have well defined narrow core, the total energy of the jet is distributed over a rather broad angular range. It is also apparent that there is a non-negligible background contribution to the transverse energy flow uncorrelated with the jets themselves.

Average corrections are usually applied [18,15] to account for energy belonging to the jet which is not included in the jet by the jet algorithm, and for the effect of energy 'splashing' randomly into the jet from the rest of the event. Such corrections depend on the details of the fragmentation process, on the behaviour of the hadrons in the calorimeter, and on the jet algorithm itself, and are generally evaluated using detailed monte-carlo simulations. Such effects lead to corrections to the jet energy which are typically at the level of 10% (or less) but which are clearly experiment dependent.

The use of a jet algorithm simplifies comparison of experiment with theory to the extent that jet data may then be compared directly with theoretical predictions at the parton level, largely independent of the details of the fragmentation process. Clearly some arbitrariness inevitably enters when one jet algorithm is chosen in preference to another. The algorithms actually used by UA1 and UA2 differ quite significantly in detail and both algorithms have evolved somewhat with time. The extent to which different algorithms give consistent results is a measure of the usefulness of the jet algorithm approach.

2.2 The Inclusive Jet Cross section

The one jet trigger discussed in Section 2.1 lends itself naturally to a measurement of the inclusive jet cross-section. The inclusive jet cross section is measured by counting the number of jets observed in a given acceptance interval and dividing by the corresponding integrated luminosity. Thus events containing two jets contribute twice to the inclusive jet cross section if both jets fall within the specified acceptance window. Both UA1 [2,14,19,20] and UA2 [1,23,10,15] have given results on the inclusive cross section $d^2\sigma/d\eta dp_T$ for jet production at $\eta = 0$. In practice of course the measured cross sections

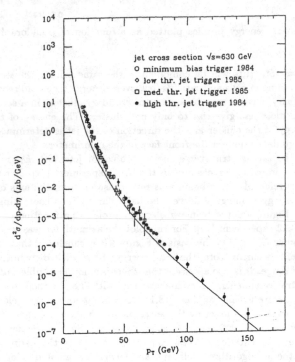

Figure 5: The inclusive jet cross-section at $\sqrt{s} = 630$ GeV measured by UA1. The solid curve is the QCD prediction based on Duke and Owens structure functions assuming $\Lambda_{QCD} = 0.2$ GeV.

are always averaged over a finite acceptance interval, typically over an angular range $|\eta| \le 1$.

Our knowledge of the inclusive jet cross section at collider energies has advanced considerably since the time of the first publications on jets. A selection of UA1 data on the inclusive jet cross section at $\sqrt{s} = 630$ GeV is shown in Figure 5 as a function of the transverse momentum p_T of the jet. The data shown are based on four statistically independent data sets. Data from ref.[19] in the range $p_T = 30 - 50$ GeV are based on a two jet trigger and have been discarded. Data for which $p_T < 15$ GeV overlap with the so-called minijet region which is discussed in Section 6 and are not included in this plot.

The theory of jet production in hadron-hadron collisions is discussed in Section 3. In Figure 5 the solid curve represents the leading order QCD prediction based on Duke and Owens structure functions [21] assuming $\Lambda_{QCD} = 0.2$ GeV. The prediction is absolute and no normalisation adjustment has been made to fit the data. The data and the theoretical curve track each other closely over many orders of magnitude in rate, making this comparison among the more spectacular and convincing of QCD tests. In particular, since gluon-gluon interactions dominate the theoretical prediction for the cross section at low p_T (see Section 3), this data constitutes rather direct proof of the existence of the gluon self-coupling in QCD [22]. More critically, it is clear that over most of the p_T range the UA1 data tend to lie above the theoretical curve by about a factor of two. It is possible that higher order QCD corrections may modify the predictions for the inclusive cross-section. However, at least in the case of the UA1 data, systematic errors in the jet energy scale, due to uncertainties in the calorimeter calibrations and in the jet energy corrections, lead to a systematic error in the jet cross section which leaves the normalisation of the cross section uncertain at the level of a factor of two. In the case of the UA2 data [15] the systematic error on the jet cross section is somewhat smaller ($\simeq \pm 50\%$).

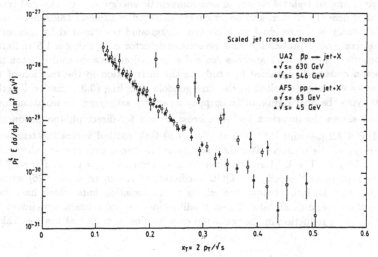

Figure 6: Scaled inclusive jet cross sections as functions of $x_T = 2p_T/\sqrt{s}$.

Finally in Figure 6 we show the scaled cross section $p_T^4 E d\sigma/dp^3$ measured by the UA2 collaboration [15] as a function of the scaling variable $x_T = 2p_T/\sqrt{s}$. If scaling held exactly then the scaled cross section would be a function of x_T only, independent of $\sqrt{s}$. The collider data are shown for $\sqrt{s} = 546$ GeV and for $\sqrt{s} = 630$ GeV, and within themselves are consistent with exact x_T scaling. Also shown are data for pp collisions from the AFS collaboration at the ISR for $\sqrt{s} = 45$ GeV and $\sqrt{s} = 63$ GeV. Relative to the ISR data, the collider data show a very substantial deviation which is largely attributable to QCD scale breaking effects.

2.3 Direct Photons

High transverse momentum direct photon production is a process closely related to the production of high transverse momentum jets. From an experimental point of view, the study of direct photon production has several advantages with respect to the study of jets: the energy resolution of the electromagnetic calorimeter is generally better for photons than it is for hadrons, and systematic uncertainties on the photon energy scale are smaller. Furthermore, since photons do not fragment the direction and energy of photons is straightforwardly measured in the calorimeter without the need of a jet algorithm which is required to reconstruct a jet. Only the relatively low rate for the production of direct photons and the non-negligible background contribution from jet production processes have limited the usefulness of the direct photons for making quantitative QCD tests.

In the collider experiments the events with direct photons with $E_T > 10 - 15$ GeV are selected by the hardware trigger in connection with the study of W-production and decay through the electron channel (W$\rightarrow e\nu$). The dominant background in the photon sample is from jets containing one or more π^0s (or η^0s) which decay to photons, which are not individually resolved in the calorimeter. This background may be reduced substantially by requiring an *isolated* cluster of electromagnetic energy with no charged tracks pointing at or near the cluster, and no jet energy detected in adjacent calorimeter cells. In the UA2 analysis the residual multiphoton background is estimated by measuring the average conversion probability in a preshower detector comprising a 1.5 radiation length convertor. In the UA1 analysis some discrimination between multiphoton and single photon clusters is obtained by studying the distribution in the fraction of the cluster energy which is deposited in the first gondola sampling (3.3 radiation lengths) and by studying the distribution of accompanying energy satisfying the isolation cut.

Figure 7 shows the invariant inclusive cross section for direct photon production measured by UA2 [24] and UA1 [25] at $\sqrt{s} = 630$ GeV plotted versus the transverse momentum p_T of the photon. The UA2 results are averaged over the pseudorapidity interval $| \eta | < 0.8$. The UA1 results have been corrected to $\eta = 0$. In Figure 7 the solid curve represents the order α_s^2 QCD prediction of Aurenche et al. [26] without modification to take into account the effects of the isolation cuts which have been applied to the experimental data. These modifications are experiment dependent but typically lead to a reduction in the predicted cross section of $\simeq 30\%$ at low p_T. Taking

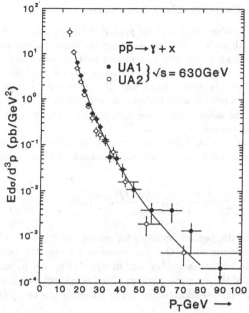

Figure 7: The inclusive cross section for direct photon production at $\sqrt{s} = 630$ GeV measured by UA2 and UA1.

into account systematic errors on the the measured cross sections of $\simeq 20 - 25\%$, it is concluded that the data are in acceptable agreement with the QCD prediction in both shape and normalisation, over the full p_T -range explored.

A discussion of the theory of direct photon production in QCD with emphasis on the role of next-to-leading order corrections is given in Section 3.4

3 Two Jet Physics

The study of two-jet production offers the most promising approach to the development of quantitative QCD tests using hadron beams. The dominant source of two-jet events at the collider is parton-parton elastic scattering proceeding by purely strong interactions. On the face of it, the values of the space-like momentum transfer (Q^2) acessible in two jet events at the collider are far beyond those acessible in present day deep-inelastic lepton scattering experiments. In this section we review some of the main results obtained in two-jet studies at the collider. We begin with a brief summary of the theory of two-jet production in hadron-hadron collisions in leading order perturbative QCD.

3.1 Theory of the Two-Jet Cross Section

In QCD two-jet events result when an incoming parton from one hadron (for example the antiproton) interacts with an incoming parton from the other hadron (for example the proton) to produce two high transverse momentum partons which are observed as jets. From momentum conservation the two final state partons are produced with equal and opposite momenta in the subprocess cms frame. If only two partons are produced therefore, and the relatively small intrinsic transverse momentum of the incoming partons is neglected, then in the laboratory frame the two jets will be back-to-back in azimuth and balanced in transverse momentum.

The two-jet cross section may be written as a sum of terms each representing the contribution to the cross section due to a particular combination of incoming (i, j) and outgoing (k, l) partons:

$$\frac{d^3\sigma}{dy_1 dy_2 dp_T^2} = \frac{1}{16\pi s^2} \sum_{i,j} \sum_{k,l} (\frac{f_i(x_1, Q)}{x_1})(\frac{f_j(x_2, Q)}{x_2}) \overline{\sum} \mid M(ij \to kl) \mid^2 \frac{1}{1 + \delta_{kl}} \quad (1)$$

where the $f_i(x, Q)$ represent the parton densities for partons of type i ($i = u, \bar{u}, d, \bar{d}, g$ ... etc.), evaluated at momentum scale Q. For massless partons the rapidities and pseudorapidities may be used interchangeably and in eqn. (1) y_1 and y_2 represent the laboratory pseudorapidities of the outgoing partons. The laboratory rapidity (y_0) of the two-parton system and the cms pseudorapidity (y) of either of the two jets are given by:

$$y_0 = (y_1 + y_2)/2 \quad (2)$$

$$y = (y_1 - y_2)/2 \quad (3)$$

Note that the subprocess cms scattering angle θ depends only on the difference of the pseudorapidities of the two jets.

$$cos\theta = tanh(y) \quad (4)$$

The longitudinal momentum fractions of the incoming partons x_1 and x_2 are then given by:

$$x_1 = x_T e^{y_0} cosh(y), \quad x_2 = x_T e^{-y_0} cosh(y) \quad (5)$$

where $x_T = 2p_T/\sqrt{s}$. Expressions for the leading order matrix element squared $\overline{\sum} \mid M \mid^2$ averaged and summed over initial and final state spins and colors are given in Table 1 [27]. The $\overline{\sum} \mid M \mid^2$ will depend on $cos\theta$ and, through the factor $g^4 = (4\pi\alpha_s)^2$, on the momentum scale Q.

Given a knowledge of the parton densities from deep inelastic scattering experiments, eqn. (1-5) may be used to make leading order QCD predictions for jet production in hadron-hadron collisions. For example, the inclusive jet cross section may be obtained by integrating eqn. (1) over the rapidity of one of the jets. As discussed in Section 3.4, the exact definition of the momentum scale Q, appearing in eqn. (1), remains somewhat arbitrary so long as only leading order formulae are used. In the case of the inclusive jet cross section, reasonable choices for Q (eg. $Q = p_T$, $Q = p_T/4$ ...etc.) give predictions which differ from each other at the level of a factor two or so, making the inherent theoretical error comparable to the systematic error in the measurement (see Section 2.1).

Explicit numerical predictions for the inclusive jet cross sections at collider energies have been made by several authors [28,29]. The curve shown in Figure 5 was calculated by Stirling [29] using Duke and Owens structure functions [21] assuming $\Lambda_{QCD} = 0.2$ GeV and taking the momentum scale $Q = p_T$.

Process	$\sum \mid M \mid^2$	$\theta = \pi/2$
$qq' \rightarrow qq'$	$\dfrac{4}{9} \dfrac{s^2 + u^2}{t^2}$	2.22
$qq \rightarrow qq$	$\dfrac{4}{9}(\dfrac{s^2 + u^2}{t^2} + \dfrac{s^2 + t^2}{u^2}) - \dfrac{8}{27} \dfrac{s^2}{ut}$	3.26
$q\bar{q} \rightarrow q'\bar{q'}$	$\dfrac{4}{9} \dfrac{t^2 + u^2}{s^2}$	0.22
$q\bar{q} \rightarrow q\bar{q}$	$\dfrac{4}{9}(\dfrac{s^2 + u^2}{t^2} + \dfrac{t^2 + u^2}{s^2}) - \dfrac{8}{27} \dfrac{u^2}{st}$	2.59
$q\bar{q} \rightarrow gg$	$\dfrac{32}{27} \dfrac{t^2 + u^2}{tu} - \dfrac{8}{3} \dfrac{t^2 + u^2}{s^2}$	1.04
$gg \rightarrow q\bar{q}$	$\dfrac{1}{6} \dfrac{t^2 + u^2}{tu} - \dfrac{3}{8} \dfrac{t^2 + u^2}{s^2}$	0.15
$gq \rightarrow gq$	$-\dfrac{4}{9} \dfrac{s^2 + u^2}{su} + \dfrac{u^2 + s^2}{t^2}$	6.11
$gg \rightarrow gg$	$\dfrac{9}{2}(3 - \dfrac{tu}{s^2} - \dfrac{su}{t^2} - \dfrac{st}{u^2})$	30.4

Table 1: Two to two parton subprocesses. $\mid M \mid^2$ is the invariant matrix element squared. An overall factor of g^4 has been removed. The colour and spin indices are averaged (summed) over initial (final) states. All partons are assumed massless. The scattering angle in the center of mass frame is denoted by θ.

3.2   Experimental Results on Two Jets

The simplest experimental checks which can be made on the two-jet events, test the predictions that the two jets should be back-to-back in azimuth and balanced in transverse momentum. These are predictions of the parton model which remain valid in leading order QCD. The distribution in the difference ϕ_{12} in azimuthal angle between the two highest E_T jets, measured in the UA2 experiment [23], for events with $\Sigma E_T > 60$ GeV is shown in Figure 8. The distribution is strongly peaked around $\phi_{12} \simeq 180°$ showing that the jets are indeed produced preferentially back-to-back in azimuth, in excellent agreement with the parton picture and with leading order QCD.

A related study [30] focusses on the transverse momentum ($\vec{P}_T$) of the two-jet system. Figure 9 shows the distributions in the components of $\vec{P}_T$ parallel ($P_{T\parallel}$), and perpendicular ($P_{T\perp}$) to the two-jet axis. If the transverse momenta of the two jets are balanced then the transverse momentum ($\vec{P}_T$) of the two-jet system should be zero. As may be seen from Figure 9, the distributions in $P_{T\parallel}$ and $P_{T\perp}$ although peaked at

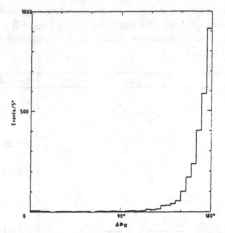

Figure 8: Separation in azimuthal angle ϕ between the two highest transverse energy jets, measured by UA2.

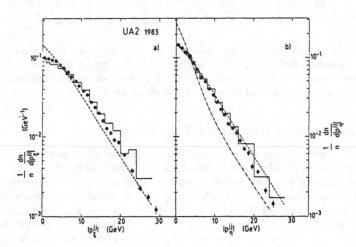

Figure 9: The distributions in $P_{T\parallel}$ and $P_{T\perp}$ for the two-jet system , measured by UA2

zero have significant widths such that $\sigma(P_{T\|}) = 9.4$ GeV and $\sigma(P_{T\perp}) = 7.1$ GeV. The extent of the imbalance which is observed, may be accounted for by a combination of the effects of calorimeter resolution and higher order QCD corrections, particularly initial state bremsstrahlung processes, a point which has been emphasised by Greco [31]. In fact, since the distribution in $P_{T\perp}$ is considerably less sensitive to the calorimeter resolution than the distribution in $P_{T\|}$, this study provides, as an important byproduct, an independent experimental check on the calorimeter resolution.

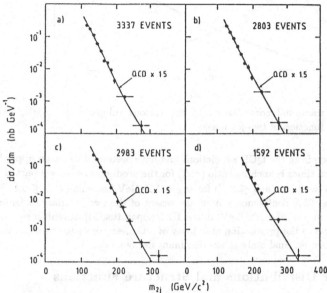

Figure 10: Two-jet mass distributions, for various $cos\theta$ intervals, measured by UA1

For a fixed cms scattering angle θ the form of the two-jet mass distribution is determined by the x-dependence of the parton distribution functions. Figure 10 shows two-jet mass distributions measured by UA1 [32] for various intervals of $cos\theta$. The available mass range at each angle is determined by the jet trigger E_T-threshold. The QCD predictions, which have been scaled up by a factor of 1.5 to fit the data better, have been computed using the structure functions of Eichten et al. [33], taking $\Lambda_{QCD} = 0.2$ GeV and assuming $Q = p_T$. Figure 11 shows the fractional contributions to the two-jet cross-section coming from the subprocesses: $gg \rightarrow gg$, $qg \rightarrow qg$ and $q\bar{q} \rightarrow q\bar{q}$ as a function of two-jet mass, based on the QCD prediction. As explained in Section 3.3, these subprocesses dominate the two jet cross-section and their relative contribution at fixed two-jet mass is very nearly independent of cms scattering angle. Notice that the dominant subprocess for $m_{2J} < 50$ GeV is gluon-gluon elastic scattering, as a result of the very large flux of gluons in the proton at small x and the relatively large subprocess cross section (see Table 1). The two-jet mass distributions shown in Figure 10 show no

146

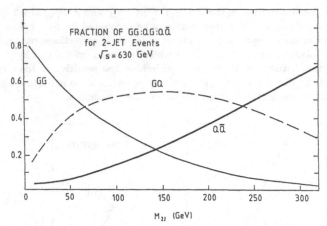

Figure 11: The fractional contribution of the various subprocesses to the two-jet cross-section as a function of two-jet mass.

significant deviation from the QCD predictions and have been used to set an upper limit on the cross-section times branching ratio ($\sigma.B$) for the production of a new particle (X) which decays into two jets. ($\sigma.B \leq 1$ nb for $m_X = 150$ GeV becoming $\sigma.B \leq 0.1$ nb for $m_X = 350$ GeV, at 95% confidence.) A measurement of the two-jet mass distribution in the mass range $m_{2J} = 70 - 100$ GeV, in the UA2 apparatus [34], reveals a significant excess of events due to the production and decay of intermediate vector bosons (W and Z) for which the two-jet final state is the dominant decay mode.

3.3 Angular Distributions and Structure Functions

A useful simplification in the theory of the two-jet cross section was introduced by Combridge and Maxwell [35]. Their idea rests on the observation that, in QCD, the subprocess cross sections for the dominant subprocesses are predicted to have very similar dependence on cms scattering angle θ. This is a consequence of the presence $1/\hat{t}$ singularities in the leading order matrix elements (Table 1) for those processes which can proceed by single gluon exchange in leading order. In QCD the dominant subprocesses are predicted to show a $(1 - cos\theta)^{-2}$ dependence closely analogous the case of, for example, Bhabha scattering in QED. In the approximation that the elastic subprocesses, ie. $gg \rightarrow gg$, $qg \rightarrow qg$ and $q\bar{q} \rightarrow q\bar{q}$ dominate, and have a common dependence on cms scattering angle θ, the sum over all incoming and outgoing parton combinations in eqn (1) may be replaced by a single term. The two-jet cross section may then be written:

$$d^3\sigma/dx_1 dx_2 dcos\theta = [F(x_1)/x_1][F(x_2)/x_2]d\hat{\sigma}/dcos\theta \qquad (6)$$

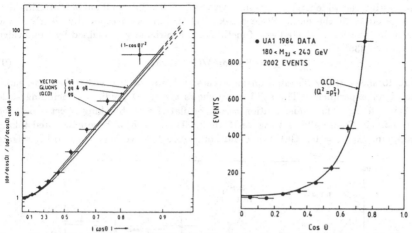

Figure 12: a) The two jet angular distribution plotted to display the power law dependence as $cos\theta \to 0$. b) The two jet angular distribution for $m_{2J} = 180 - 240$ GeV.

where $d\hat{\sigma}/dcos\theta$ is taken to be the subprocess cross section for $gg \to gg$:

$$\frac{d\hat{\sigma}}{dcos\theta} = \frac{9}{8}\frac{\pi\alpha_s^2}{2x_1x_2s}\frac{(3+cos^2\theta)^3}{(1-cos^2\theta)^2} \qquad (7)$$

and the effective structure function $F(x)$ is given by:

$$F(x) = G(x) + (4/9)[Q(x) + \bar{Q}(x)]. \qquad (8)$$

In eqn (8) $G(x)$, $Q(x)$ and $\bar{Q}(x)$ are respectively the gluon, quark and antiquark structure functions of the proton. The factor 4/9 appearing in eqn (8) reflects the relative strengths of quark-gluon and gluon-gluon couplings in QCD.

In the UA1 experiment considerable emphasis has been placed on the measurement of the two-jet angular distribution [18,36,37]. The earliest results [18] are shown in Figure 12a and are based on a sample of two-jet events for which each jet satisfied $E_T > 20$ GeV. The data are plotted so as to display directly the power law dependence as $cos\theta \to 1$. The theoretical curves shown in Figure 12a have been computed neglecting all scale-breaking effects, so that the predicted angular dependence, which is displayed separately for each of the dominant subprocesses, reflects only the explicit θ dependence of the leading order matrix elements squared (Table 1). This data demonstrates clearly the dominance of gluon exchange processes in two-jet production. In fact the data show a somewhat more rapid angular dependence than is predicted by the scaling curves and a fit to the data of the form $(1 - cos\theta)^{-n}$ for $cos\theta = 0.4 - 0.8$ gives $n = 2.38 \pm 0.10$.

More recent results on the two-jet angular distribution [37] are shown in Figure 12b for two-jet masses $m_{2J} = 180 - 240$ GeV. In Figure 12b the data are plotted on a linear scale for $cos\theta = 0 - 0.8$. The solid curve shows the QCD prediction normalised to the data. The prediction now takes account of scale-breaking effects in the strong coupling constant and in the structure functions. These effects modify the scaling curve by supressing the wide-angle scattering with respect to the small-angle scattering and are required to fit the data. Scale breaking effects are seen especially clearly if the angular distribution is plotted as a function of the variable χ as defined by Combridge and Maxwell [35]:

$$\chi = (1 + cos\theta)/(1 - cos\theta). \tag{9}$$

Figure 13 shows the χ distribution for events with $m_{2J} = 240 - 300$ GeV measured in the UA1 experiment. The solid curve shows the QCD prediction including scale breaking effects. The χ-distribution would be flat if scale-breaking effects could be neglected. (In Figure 12b and Figure 13 the QCD curves have been calculated taking $Q = p_T$ and $\Lambda_{QCD} = 0.2$ GeV.) The data of Figure 13 have been used to investigate

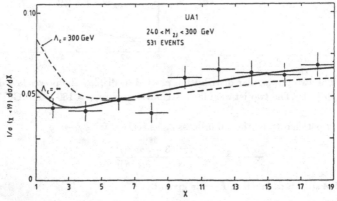

Figure 13: The distribution in the variable χ for events with $m_{2J} = 180 - 300$ GeV, measured by UA1

the possibility that the partons are composite by testing for a contact interaction as discussed by Eichten et al. [38]. A contact interaction would lead to an excess of events at wide angle ($\chi \simeq 1$), relative to the QCD prediction, as shown by the broken curve in Figure 13. The data give $\Lambda_{qq} < 415$ GeV at 95% confidence where Λ_{qq} is the energy scale of compositeness for quarks as defined in ref.[38]. Comparable limits have been given by UA2 [15] based on their measurement of the inclusive jet cross-section.

A sensitive measure of the shape of the two-jet angular distribution is obtained by computing the ratio of the wide-angle ($cos\theta = 0 - 0.6$) to the small angle ($cos\theta = 0.6 - 0.8$) scattering. The dependence of this ratio on two-jet mass is shown in Figure 14. In Figure 14 the data point plotted at $m_{2J} = 25$ GeV is based on a sample of data taken under minimum-bias conditions. Over most of the range explored this ratio is essentially constant independent of two-jet mass. The average value of R for $m_{2J} > 180$ GeV is given by $R = 0.521 \pm 0.025$[37]. It will be interesting to see if this ratio remains constant at the higher values of two-jet mass which are accessible at the FNAL Tevatron collider.

In passing we note that a jet trigger based on a fixed E_T -threshold requirement is not ideal for the study of the two-jet angular distribution. The problem is that, once such an E_T requirement has been imposed, a very restrictive mass cut has to be applied

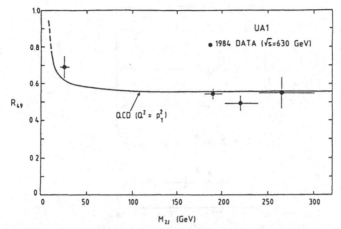

Figure 14: The ratio of wide-angle to small-angle scattering R (see text) measured by UA1, plotted as a function of two-jet invariant mass.

in order to ensure a uniform trigger efficiency over the full angular range. This means that a large fraction of the events which are accepted by the trigger are not used in the final analysis. For the purposes of this measurement, it is clearly best to implement mass and angle cuts from the start, if possible, using the saving in data taking capacity to extend the analysis to lower values of the two-jet mass.

Both UA1 and UA2 have proceeded to extract the effective structure function $F(x)$ from the two-jet data [18,30]. As a first step they extract the quantity $S(x_1, x_2)$ defined by:

$$d\sigma/dx_1 dx_2 dcos\theta = S(x_1, x_2)/(x_1 x_2) d\hat{\sigma}/dcos\theta \qquad (10)$$

If factorisation holds in x_1 and x_2 then:

$$S(x_1, x_2) = F(x_1) F(x_2) \qquad (11)$$

Figure 15 shows the ratio $S(x_1, x_2)/S(x_1, x_2 + \Delta x_2)$ measured by the UA2 collaboration [30], as a function of x_1 for various x_2 intervals setting $\Delta x_2 = 0.05$. This ratio should be independent of x_1 if factorisation holds. Figure 15 shows that factorisation holds approximately over the range studied. As it happens, this ratio is also almost independent of x_2. That this is true reflects the fact that the effective structure function at these energies is well approximated over most of the range in x, by a single exponential function.

The effective structure function $F(x)$ may be obtained from $S(x_1, x_2)$ assuming factorisation in x_1 and x_2. If the x_1, x_2 -plane is divided into n by n square bins with coordinates $x_1(i), x_2(j)$, $i, j = 1, n$ then the structure function may be projected-out using the expression:

$$F(x(k)) = \sqrt{\sum_{i=1}^{n} \sum_{j=1}^{n} S(x_1(i), x_2(j))} - \sqrt{\sum_{i \neq k}^{n} \sum_{j \neq k}^{n} S(x_1(i), x_2(j))} \qquad (12)$$

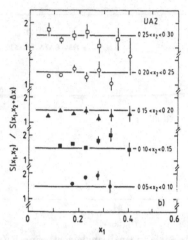

Figure 15: Test of factorisation in x_1 and x_2 (see text) : two-jet data from UA2

A recent measurement [39] of the effective structure function is shown in Figure 16, based on a sample of two-jet events from UA1, with both jets having $p_T > 70$ GeV. At such high values of Q^2 the structure function shows a roughly exponential x -dependence, and is in fair agreement with expections based on lower Q^2 deep inelastic lepton scattering data.

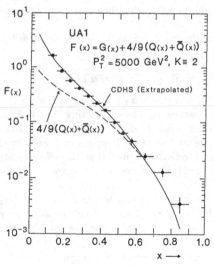

Figure 16: The effective structure function $F(x) = G(x) + 4/9(Q(x) + \bar{Q}(x))$: two jet data from UA1

3.4 The Effect of Higher Order Corrections

In QCD the definition of the scale parameter Q appearing in the theoretical formulae (for example eqn. (1)) is not fixed. It is required only to be of the order of the large scale present in the problem. For jet production this large scale is the transverse energy of the observed jet. When higher order corrections are taken into account the sensitivity of the physical prediction to the choice made for Q is expected to be reduced. In the case of two-jet production in hadron-hadron collsions the next-to-leading order corrections are extremely complicated and, although full α_s^3 corrections to parton-parton scattering processes now exist [40], they have not yet been incorporated into a phenomenological analysis. As an illustration of how the knowledge of the higher order corrections improves the prediction for the cross-section, we give a discussion of the simpler case of direct photon production, for which the effect of the higher order corrections has been analysed by Aurenche et al. [26].

The theory of direct photon production in hadron-hadron collisions closely parallels that of jet production. In particular, eqn. (1) holds also for direct photon production if the outgoing parton k is replaced by a photon and the sum over k is suppressed. In leading order of QCD the direct photon cross section is proportional to α_s. The leading order matrix elements squared are given in Table 2. The leading order QCD prediction for the direct photon cross section at a given p_T and $\sqrt{s}$, depends monotonically on the scale choice Q, as illustrated by the solid curve in Figure 17a. The analysis of

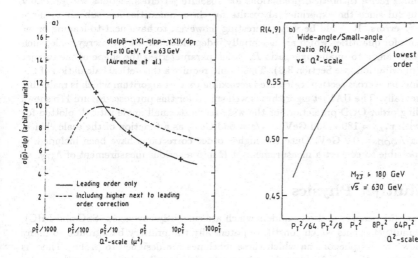

Figure 17: a) Leading order and next-to-leading order QCD predictions for direct photon production from Aurenche et al. b) Leading order QCD predictions for the wide-angle to small angle ratio R for collider jet pairs

Aurenche et al. shows that if next-to-leading order QCD effects are included, then the dependence on the scale choice Q is considerably reduced as illustrated by the broken curve in Figure 17a. Note also that the apparent rate of convergence of the perturbation series is maximised, if the scale is chosen such that $Q^2 \simeq p_T^2/20$. The theoretical curves which have been compared with the collider data (see Section 2.3) have been calculated by Aurenche et al. The curves include the complete higher order corrections, and use a scale choice $Q^2 \simeq p_T^2/20$.

Process	$\overline{\sum} \mid M \mid^2$
$q\bar{q} \rightarrow \gamma g$	$\dfrac{8}{9}\dfrac{t^2+u^2}{tu}$
$gq \rightarrow \gamma q$	$-\dfrac{1}{3}\dfrac{s^2+u^2}{su}$

Table 2: Lowest order processes for direct photon production. $\overline{\sum} \mid M \mid^2$ is the invariant matrix element squared with a factor of $g^2 e_q^2$ removed. The colour and spin indices are averaged (summed) over initial (final) states.

A similar analysis has yet to be carried out for the case of jet production. Beyond the leading order the theoretical predictions will depend on how the jets are defined experimentally: on the choice of jet cone-size, $\Delta R < 1$, for example, used in the UA1 jet algorithm. Precise theoretical predictions for absolute jet cross sections will probably not be useful since the experimental results for these quantities are subject to large systematic errors. It would be very interesting, however, to have next-to-leading order predictions for quantities which are essentially independent of the jet energy scale, such as, the wide-angle to small-angle ratio R, which parameterises the shape of the two-jet angular distribution (see Section 3.3). This would require a theoretical calculation of the two jet inclusive cross-section, calculated according to a jet algorithm which is realisable experimentally. The UA1 jet algorithm is well suited for this purpose. Figure 17b shows the leading order QCD prediction for the wide-angle to small angle ratio R plotted for events with $m_{2J} = 180 - 300$ GeV at $\sqrt{s} = 630$ GeV, as a function of the scale choice, assuming $\Lambda_{QCD} = 0.2$ GeV. Once the higher order corrections have been included, it will be possible to convert a measurement of R into a serious measurement of $\Lambda_{\overline{MS}}$.

4   Multijet Physics

In the very high energy hadron colliders which are currently envisaged (SSC and LHC), it is the multijet events which constitute potentially the primary background to many of the new physics phenomena which these machines are designed to probe. There is thus a strong motivation for extending the study of jet production at present energies beyond the study of two-jet events, to obtain as detailed an understanding as possible of multijet production mechanisms. For the case of three-jet production complete theoretical calculations exist at leading order in QCD.

4.1 Theory of the Three-Jet Cross Section

Events with three jets at large transverse energy are described in QCD by amplitudes with two incoming partons and three outgoing partons. At tree graph level the matrix elements for these processes were first calculated in refs.[42,43]. Very elegant results for the two-to-three parton scattering processes have been given by Berends et al.[44]. For a complete description it is sufficient to consider the following four processes.

$$
\begin{array}{lll}
(A) & q(p_1) + q'(p_2) & \rightarrow \quad q(p_3) + q'(p_4) + g(k) \\
(B) & q(p_1) + q(p_2) & \rightarrow \quad q(p_3) + q(p_4) + g(k) \\
(C) & q(p_a) + \overline{q}(p_b) & \rightarrow \quad g(p_1) + g(p_2) + g(p_3) \\
(D) & g(p_1) + g(p_2) & \rightarrow \quad g(p_3) + g(p_4) + g(p_5)
\end{array}
\tag{13}
$$

The momentum assignments for the partons are given in brackets. All other matrix elements for two-to-three parton amplitudes may be obtained by crossing from the above four processes.

The matrix elements squared for the processes $(A - D)$, averaged (summed) over the initial (final) colours and spins are given below. We have set the masses of the quarks equal to zero. With the momentum assignments of eqn. (13) the matrix element for process (A) is [45],

$$
\overline{\sum} \mid M^{(A)} \mid^2 = \frac{g^4 C_F}{N} \left(\frac{s^2 + s'^2 + u^2 + u'^2}{2tt'} \right) \left(2C_F \big([14] + [23] \big) + \frac{1}{N} [12; 34] \right)
\tag{14}
$$

The kinematic variables are defined as follows,

$$
s = (p_1 + p_2)^2, \quad t = (p_1 - p_3)^2, \quad u = (p_1 - p_4)^2,
$$
$$
s' = (p_3 + p_4)^2, \quad t' = (p_2 - p_4)^2, \quad u' = (p_2 - p_3)^2
\tag{15}
$$

For compactness of notation we have introduced the eikonal factor $[ij]$ which is defined as,

$$
[ij] \equiv \frac{p_i.p_j}{p_i.k \, k.p_j}
\tag{16}
$$

We have also defined the following sum of eikonal terms,

$$
[12; 34] = 2[12] + 2[34] - [13] - [14] - [23] - [24]
\tag{17}
$$

Note that this combination is free from collinear singularities. In eqn. (14) the dependence on the $SU(N)$ colour group is shown explicitly, $(C_A = N = 3, C_F = 4/3)$.

In the same notation the result for process (B) with four identical quarks may be written [45],

$$
\overline{\sum} \mid M^{(B)} \mid^2 = \frac{g^4 C_F}{N} \left(\frac{s^2 + s'^2 + u^2 + u'^2}{2tt'} \right) \left(2C_F \left([14] + [23] \right) + \frac{1}{N} [12; 34] \right)
$$

$$+\frac{g^4 C_F}{N}\left(\frac{s^2+s'^2+t^2+t'^2}{2uu'}\right)\left(2C_F\left([13]+[24]\right)+\frac{1}{N}[12;34]\right)$$

$$-\frac{2g^4 C_F}{N^2}\left(\frac{(s^2+s'^2)(ss'-tt'-uu')}{4tt'uu'}\right)\left(2C_F\left([12]+[34]\right)+\frac{1}{N}[12;34]\right)$$

$$(18)$$

To write the results for the remaining two processes we introduce a compact notation for the dot product of two momenta,

$$\{ij\}\equiv p_i\cdot p_j \tag{19}$$

Using the momentum assignments of eqn. (13) the result for process (C) may be written as [44],

$$\overline{\sum}\mid M^{(C)}\mid^2 = \frac{g^6(N^2-1)}{4N^4}\sum_{i=1}^{3}\frac{\{ai\}\{bi\}(\{ai\}^2+\{bi\}^2)}{\{a1\}\{a2\}\{a3\}\{b1\}\{b2\}\{b3\}}$$

$$\times\left[\{ab\}+N^2\left(\{ab\}-\sum_P\frac{\{a1\}\{b2\}+\{a2\}\{b1\}}{\{12\}}\right)\right.$$

$$\left.+\frac{N^4}{\{ab\}}\left(\sum_P\frac{\{a3\}\{b3\}(\{a1\}\{b2\}+\{a2\}\{b1\})}{\{23\}\{31\}}\right)\right] \tag{20}$$

The sums run over the three cyclic permutations P of the labels of the final state gluons.

Using the momentum labels of eqn. (13) the result for process (D) is [44],

$$\overline{\sum}\mid M^{(D)}\mid^2 = \frac{g^6 N^3}{240(N^2-1)}\left[\sum_P\{12\}^4\right]\left[\sum_P\{12\}\{23\}\{34\}\{45\}\{51\}\right]\left(\prod_{i<j}\{ij\}\right)^{-1} \tag{21}$$

The sums run over the 120 permutations of the momentum labels.

These matrix elements display the typical bremsstrahlung structure with the emission of soft and collinear gluons predominating. This is particularly clear from the form of the result given in eqns. (14,18) where the dominant contributions come from the region in which the eikonal factors are large. From the tree graph results one can also show that same effective structure function which is relevant for two-jet production (cf. eqn. (8)) is also valid to a very good approximation for three-jet production [46].

For three final-state (massless) partons the final-state parton configuration, at fixed cms energy, is specified by five independent variables. Two variables are required to specify how the available energy is shared between the three final-state partons, and two variables serve to fix the orientation of the three-jet system with respect to the axis defined by the colliding partons. The last variable is an overall azimuthal angle. If z_1, z_2, and z_3 are the energies of the outgoing partons scaled such that $z_1+z_2+z_3=2$ and ordered such that $z_1 > z_2 > z_3$ and θ_i is the angle between parton i and the beam direction then the subprocess differential cross section may be written:

$$\frac{d^4\hat{\sigma}}{dz_1 dz_2 d\cos\theta_1 d\psi}=\frac{1}{(1024\pi^4)}\overline{\sum}\mid M\mid^2 \tag{22}$$

In eqn. (22) the variable ψ is the angle between the plane containing jet-2 and jet-3 and the plane containing jet-1 and the axis defined by the incoming partons. The Dalitz plot variables used in the UA2 analysis are defined by $x_{ij} = m_{ij}^2/\hat{s}$, where m_{ij} is the invarant mass of any two (i, j) of the three jets, and are related to the z_i as follows:

$$x_{12} = (1 - z_3)$$
$$x_{13} = (1 - z_2) \qquad (23)$$
$$x_{23} = (1 - z_1)$$

The x_{ij} satisy the constraint: $x_{12} + x_{13} + x_{23} = 1$.

4.2 Experimental Results on Three Jets

Both UA1 and UA2 have made detailed experimental studies of three jet production. In the UA1 analysis [36], the UA1 jet algorithm has been used without modification. Cuts are applied to the three-jet sample to ensure that the events will have all three jets well separated from each other and from the beams. The cut $z_1 < 0.9$ guarantees that jet-2 and jet-3 will be separated in pseudorapidity-azimuth such that the minimum possible separation in psedorapidity-azimuth space is given by ΔR between jet-2 and jet-3 is $\Delta R = 1.2$. The average separation between the two closest jets in psedorapidity-azimuth space is given by $< \Delta R > \simeq 1.7$. The requirement $cos\theta_1 < 0.6$ and the requirement $30° <| \psi |< 150°$ guarantees that all three jets are well separated in angle from the beams.

For the UA2 three-jet analysis [16], the UA2 algorithm is modified by the inclusion of a second pass, in which primary clusters found in the first pass may be split into two (or more) secondary clusters. The secondary clusters are defined by reapplying the cluster algorithm, combining hits in adjacent calorimeter cells, if they satisy a new E_T- threshold set at 5% of the E_T of the primary cluster. This modified algorithm is found to be fully efficient in resolving jets whose axes are separated by more than 50° in angle, essentially independent of jet E_T. In the UA2 three-jet analysis, events with three (and only three) clusters satisfying a transverse energy threshold $E_T > 10$ GeV are selected if all three clusters fall in an angular acceptance region $| \eta |< 0.8$ and if the scalar sum of their transverse energies exceeds 70 GeV.

The three-jet Dalitz plot x_{12} vs. x_{23} measured by UA2 is shown in Figure 18. The absence of events at small x_{23} is due to the inability of the jet algorithm to resolve jets at small angles to each other, and the absence of events at large x_{12} is due to the requirement on the transverse energy of the third jet. Otherwise the acceptance is rather uniform over the whole plot. The increase in event density with decreasing x_{23} for fixed x_{12} reflects the tendency of final state bremsstrahlung to be produced at small angles to the parent parton. The projections of the Dalitz plot onto the x_{12} and x_{23} axes are also shown. The data are in acceptable agreement with the leading order QCD predictions and are inconsistent with the phase space distributions.

The three-jet angular distributions ($cos\theta_1$ vs ψ) measured by UA1 [36,47] are shown in Figure 19. The distribution in $cos\theta_1$ shows a pronounced forward-backward peaking, which is qualitatively similar to that which is observed for two-jet events. The distribution in $| \psi |$ shows that configurations in which jet-2 and jet-3 lie close to the plane defined by jet-1 and the beams ($| \psi |\simeq 30°, 150°$), are preferred relative to configurations for which $| \psi |\simeq 90°$. This effect reflects the tendency of initial-state bremsstrahlung

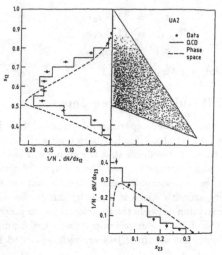

Figure 18: Three jet Dalitz plot x_{12} vs. x_{23} measured by UA2.

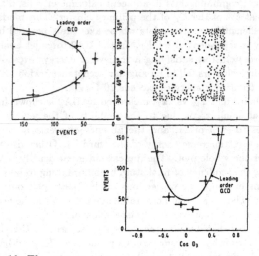

Figure 19: Three jet angular distributions measured by UA1.

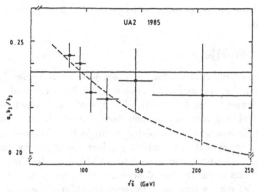

Figure 20: The value of $\alpha_s K_3/K_2$ measured by UA2 plotted as a function of three-jet (or two-jet) mass.

to be produced at small angles to the incoming partons. The projections of the scatter plot onto the $cos\theta_1$ and $|\psi|$ axes are also shown. The theoretical curves have been calculated using the leading order QCD formulae neglecting scale breaking effects. As in the case of the two-jet angular distribution, the data show a somewhat more pronounced angular dependence than is predicted by the scaling curves, and it has been shown [47] that, as in the two-jet case, the inclusion of scale breaking effects in the theoretical curves can improve the agreement with experiment.

Any attempt to determine the strong coupling constant α_s based on the relative rates of three-jet and two-jet production is frustrated by the lack of any theoretical knowledge of the next-to-leading-order QCD corrections to the three-jet cross-section. Experimenters have nonetheless proceeded to analyse the three-jet to two-jet ratio using leading order formulae introducing a factor (K_3/K_2) to represent the effect of unknown higher order QCD corrections. Figure 20 shows the value of $\alpha_s K_3/K_2$ measured by the UA2 collaboration plotted as function of three-jet (or two-jet) mass. The curve shows the expected variation of α_s over the range of masses shown. The data are not sufficiently accurate to demonstrate the variation of α_s expected in QCD. Very similar data have been published by the UA1 collaboration.

Experiment	Mass	$\alpha_s K_3/K_2$
UA1	> 150 GeV	$0.22 \pm 0.03 \pm 0.03$
UA2	> 70 GeV	$0.23 \pm 0.01 \pm 0.04$

Table 3: Experimental results on $\alpha_s K_3/K_3$ published by UA1 and UA2

The average values of $\alpha_s K_3/K_2$ measured by UA1 and UA2 for comparable angular acceptance selections are summarised in Table 3. Since any extrapolation of existing low energy measurements of α_s from e^+e^- anihilation and from deep inelastic scattering, up to collider energies would predict $\alpha_s \simeq 0.12 - 0.14$, it seems likely that higher order

QCD corrections do have an important influence on the three-jet to two-jet ratio in hadron-hadron collisions.

4.3 Theory of Multijets

The theoretical description of events containing many jets can conveniently be illustrated by considering in detail the case in which there are four jets produced. There are two mechanisms which give rise to four jet events. The first of these, which we shall refer to as the double bremsstrahlung mechanism, produces four jets because of the radiation of additional partons from a process involving a single hard scattering. The second mechanism is double parton scattering which produces four jets by the independent scatterings of two pairs of partons at two distinct values of the impact parameter. These two mechanisms can be distinguished because they have quite different kinematic structures. The four jet events produced by bremsstrahlung are expected to have a typical bremsstrahlung shape, ie. predominantly collinear and soft emission, whereas the double parton scattering events are expected to be pair-wise balanced in transverse momentum with no azimuthal correlation between the planes of the two pairs of jets.

The description of the four jet bremsstrahlung process can be treated within the context of the QCD parton model by calculating the matrix elements for the processes

$$p_1 + p_2 \rightarrow p_3 + p_4 + p_5 + p_6 \qquad (24)$$

where p_i denotes a parton of type i. The transition probabilities for all possible parton processes of this type have been calculated at tree graph level. The results for the six gluon process is given in refs. [48,49]. Processes involving four external gluons and two external quarks are studied in refs. [49,50]. Processes involving two external gluons and four external quarks are given in refs.[51]. Finally refs.[52] give the results for six quark processes. Numerical estimates of the rates of production of four or more jets are given in ref.[53,63].

The information on bremsstrahlung processes with larger numbers of radiated partons is less complete. Exact results for higher bremssstrahlung amplitudes are in general not available even at tree graph level. However for one particular helicity amplitude the exact result is known [54]. Let us denote the helicity amplitude for the scattering of n gluons with momenta $p_1, \ldots, p_n$ and helicities $\lambda_1, \ldots, \lambda_n$ by $M_n(\lambda_1, \ldots, \lambda_n)$. The momenta and helicities of the gluons are labelled as though all particles are outgoing. There are $(n+2)/2$ independent helicity amplitudes of this type. At tree graph level the three amplitudes which most violate helicity conservation are,

$$
\begin{aligned}
| M(+,+,+,+,\ldots) |^2 &= 0 \\
| M(-,+,+,+,\ldots) |^2 &= 0 \\
| M(-,-,+,+,\ldots) |^2 &= c_n(g,N) \big[(p_1.p_2)^4 \\
&\quad \times \sum_P \big((p_1.p_2)(p_2.p_3)(p_3.p_4)\ldots(p_n.p_1) \big) \big]^{-1} + O(N^{-2}) \quad (25)
\end{aligned}
$$

where $c_n(g,N) = g^{(2n-4)} N^{n-2}(N^2 - 1)/2^{n-4} n$. The sum is over all permutations P of $1 \ldots n$ and N is the number of colours. An approximate extension of the results of eqn. (25) to other helicity amplitudes is given in ref.[55]. Recursive formulae relating processes with $n + 1$ emitted gluons to processes with n emitted gluons are given in ref.[56].

Another approach to modelling multijet processes is to use a Monte Carlo program [57]. These programs contain the matrix elements appropriate for collinear and soft emission of partons, expressed as a Markov process. They can therefore be used to produce an arbitrarily high number of jets, but they are not expected to give a reliable description when the emitted partons are hard or at wide angle.

Because of the rapid growth of the number of partons at low x inside the hadron, at some energy it will become favourable to have more than one hard scattering per event. As explained above such double parton scatterings will have a distinctive kinematic structure. An exact description of events in which there is more than one active parton per hadron requires a knowledge of the correlations between the first active parton with longitudinal momentum fraction x_1 and the second active parton with longitudinal momentum fraction x_2. These parton correlation functions are not known. In order to give a crude interpretation of the information learnt from the double parton scattering we may write,

$$\sigma_{4-\mathrm{jet}} = \frac{(\sigma_{2-\mathrm{jet}})^2}{\pi R_{\mathrm{eff}}^2} \tag{26}$$

$\sigma_{2-\mathrm{jet}}$ is the measured value of the two jet cross-section evaluated at the p_T cut-off which is used to define the jets. The parameter R_{eff} depends on the correlations between partons within the proton, and in general could be x-dependent.

4.4   Results on Multijets

The UA2 Collaboration have made a preliminary study of four jet production [61,59]. The four jet events are selected using a straightforward extension of the criteria used to select three jet events. In view of the large number of independent variables required to describe a four parton configuration, it is impractical to carry out a fully comprehensive analysis of four jet final states. Instead the UA2 Collaboration have concentrated on a limited number of particular variables which seem to be especially effective in discriminating between theoretical models. Figure 21a shows the distribution in ($cos\omega_{23}^*$), the cosine of the angle between the 2nd and 3rd highest momentum jets measured in the four jet rest frame. Evidently the distribution is significantly different from the phase space curves and is consistent with the QCD 4-jet double bremsstrahlung prediction of Kunzst and Stirling [62].

In an attempt to assess the importance of the double parton mechanism relative to the double bremsstrahlung mechanism, the UA2 Collaboration have studied the pairwise correlations amongst the jets in the transverse plane. One approach is to pair-off the jets so that the scalar sum of the transverse momentum imbalance (P_T) of the two resulting jet pairs is minimised. Figure 21b shows the variable ϕ_2, the difference in azimuthal angle ϕ between the two jets of the lower E_T jet pair. The data are clearly consistent with the QCD double bremsstrahlung curve, and there is no evidence for any significant contribution due to the double parton mechanism. In terms of the parameter R_{eff} (see Section 4.3) this data yields $R_{\mathrm{eff}} > 0.6$ fm at 90 % confidence level [59]. Similar

160

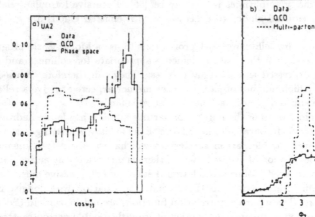

Figure 21: The distribution in a) $cos\omega_{23}^*$ and b) ϕ_2 for four jet events from the UA2 experiment.

data from the AFS collaboration at the ISR [60] gives $R_{\text{eff}} = 0.3$ fm. The latter result corresponds to a larger value of x than the collider result, suggesting that there is a greater correlation for valence quarks than there is for gluons.

Preliminary data from UA1 on the rate of production of four, five and six jet events, have recently become available [63]. Figure 22 shows the absolute rate of four, five

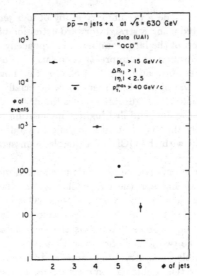

Figure 22: The rate of production of two, three, four, five and six jet events measured by UA1.

and six jet events measured in the UA1 experiment. The selection criteria for the UA1 multijet events are indicated. All the jets must satisfy $p_T > 15$ GeV, $\mid \eta \mid < 2.5$ and must be separated from each other by a pseudorapidity azimuth separation $\Delta R > 1$. In addition the highest transverse momentum jet must satisfy $p_T > 40$ GeV. The QCD predictions shown are based on the approximate matrix elements of Parke and Taylor, as discussed in Section 4.3. Although the rate of six jet events falls substantially above the theoretical prediction, the extent of the agreement between theory and experiment, up to five jets, is encouraging. More detailed experimental results on multijet production would be very valuable.

5 Fragmentation

Both the UA1 and UA2 detectors are equipped with central tracking chambers, in addition to the calorimetry, making it possible to study the fragmentation of jets into charged hadrons. The study of fragmentation at collider energies has yielded two main results. Firstly, the data show clear evidence for scale breaking effects relative to ISR energies. Secondly, significant evidence has been found in the collider data for differences in the fragmentation of quark and gluon jets. The studies which have been carried out up until now by no means exhaust the posibilities of the collider data, and it is likely that further interesting results will emerge in the future.

5.1 Fragmentation Functions

Both UA1 [64,67] and UA2 [17,23] have attempted to measure the charged hadron multiplicity of the jets produced in $p\bar{p}$ collisions. Although charged multiplicity is obtained simply by counting tracks, the measurement of the charged multiplicity for jets produced in $p\bar{p}$ collisions, is complicated by the presence of tracks coming from sources other than the jet itself. Figure 23 shows the average density of charged tracks per event, per unit azimuthal angle ϕ measured relative to the axis of the highest E_T-jet, in the UA2 experiment. The multiplicity profile of the other jet, back-to-back in azimuth with the first, is also apparent. The dot-dash line represents the corresponding track density measured in minimum bias events. Assuming that the background level within the jet is measured by the density of tracks at $90°$ to the jet axis the UA2 collaboration obtain a mean charged multiplicity per jet $< n_{ch} > = 6 - 9$, depending on the two-jet mass. This is regarded as a lower limit on the jet multiplicity and is referred to as the jet core multiplicity.

More realistic (but model dependent) estimates of the mean charged multiplicity per jet from UA1 and UA2 and are shown in Figure 24 as a function of the two-jet mass [67]. In Figure 24 the errors shown are statistical only: systematic errors are beleived to be $\simeq 15\%$ (or less). The UA1 and UA2 data agree with each other but do not fall on the extrapolation of the lower energy e^+e^- data. In fact, over the range of two-jet masses explored at the collider, the mean charged multiplicity per jet is rather independent of two-jet mass. This result is in qualitative agreement with theoretical expections: in QCD the multiplicity of gluon jets is predicted to be larger than the multiplicity of quark jets. Over the mass range plotted the proportion of gluon jets falls from $\simeq 80\%$

162

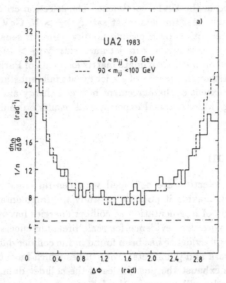

Figure 23: The distribution of charged tracks plotted as a function of the azimuthal angle ϕ measured relative to the axis of the highest E_T-jet, in the UA2 experiment.

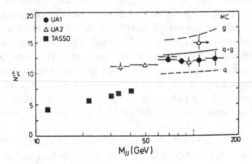

Figure 24: The mean charged multiplicity per jet measured by UA1 and UA2 plotted as a function of two-jet mass.

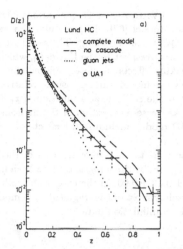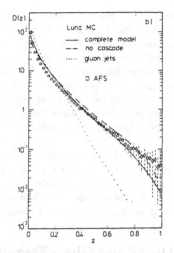

Figure 25: a) The fragmentation function measured by UA1 at the collider. b) The fragmentation function measured by the AFS collaboration at the ISR.

at 50 GeV while the proportion of quark jets grows to $\simeq$ 60% at 200 GeV.

The fragmentation function measured by UA1 [64] at the collider, plotted as a function of the fragmentation variable z, is shown in Figure 25a. (Earlier data on the z-dependence of the fragmentation function from ref.[65] were not properly corrected for smearing and energy loss effects.) Figure 25b shows the fragmentation function measured by the AFS collaboration [66] at the ISR. The collider sample comprises a mixture of quark and gluon jets in the proportion 60% : 40% with a mean jet transverse momentum $< p_T > \simeq$ 45 GeV. The ISR sample comprises quark and gluon jets

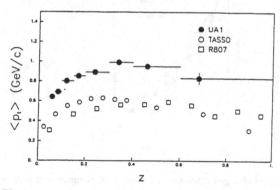

Figure 26: The mean transverse momentum measured relative to the jet axis for charged hadrons at the collider and at the ISR.

in the proportion 70% : 30% with $< p_T > \simeq 13.5$ GeV. In Figure 25 the various curves represent predictions of the LUND fragmentation model [67]. In particular the solid curves include the expected effects coming from QCD scale breaking. The collider data show a significant softening of the z distribution with respect to the ISR data, which is largely accounted for in terms of QCD scale breaking effects. A closely related effect is shown in Figure 26, which shows the mean transverse momentum, measured relative to the jet axis, for charged hadrons at the ISR and at the collider, as a function of z. The collider data show larger values for the mean transverse momentum of the hadrons in fair agreement with the predictions of the LUND model, including the effects of QCD showering.

Finally we draw attention to the observation that the apparent fraction of the jet momentum which is carried by charged particles, given by the integral of the z-weighted fragmentation function, is estimated to be $0.47 \pm 0.02 \pm 0.06$, in the case of the collider data [64]. This is somewhat lower than is predicted by most of the fragmentation models and lower than would be expected based on e^+e^- annihilation data.

5.2 Quark and Gluon Fragmentation

In the UA1 fragmentation analysis [64], a statistical separation of quark and gluon jets at fixed transverse momentum has been acheived, exploiting the difference in shape of the quark and gluon structure functions in the proton. Significant differences between quark and gluon jets are seen in the fragmentation function $D(z)$, and in the angular width of the jet, measured by plotting the charged hadron density as a function of the pseudorapidity-azimuth separation R measured relative to the jet axis. Figure 27 shows

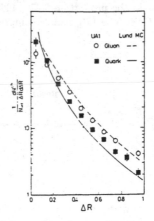

Figure 27: The density of charged tracks as a function of R measured relative to the jet axis for quark and gluon jets separately.

the density of charged tracks $1/R(dN/dR)$ plotted as a function of R for quark and gluon jets separately. The gluon jets have a lower density of tracks in the core of the jet ($\Delta R \leq 0.2$) and a higher density of tracks in the wings ($\Delta R \geq 0.2$) , relative to the quark jets. These features are in qualitative agreement with the predictions of the LUND fragmentation model [67] (see Figure 27), and demonstrate that at these energies gluon jets have a larger angular width than quark jets.

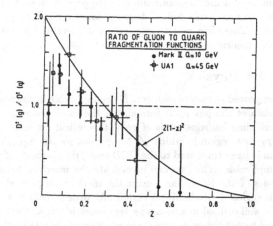

Figure 28: The ratio of gluon to quark fragmentation functions measured by UA1 and by the MARK II collaboration.

Data on the ratio of gluon and quark fragmentation functions as a function of z are shown in Figure 28. In the case of the UA1 data [64] the errors plotted include both statistical and systematic errors. It is interesting to compare the UA1 data with recent data relevant to the ratio of quark and gluon fragmentation functions published by the MARK II collaboration [68]. In the MARK II analysis, three-jet events at $\sqrt{s} = 29$ GeV are compared with two-jet events at $\sqrt{s} = 19.3$ GeV. The ratio of gluon to quark fragmentation functions based on this data are also shown in Figure 28. Within large errors the UA1 and MARK II data are consistent, and suggest that the gluon fragmentation function is softer than the quark fragmentation function by at least a factor $(1 - z)^2$. Clearly this comparison neglects scale breaking effects but these are expected to have a relatively small influence on the gluon to quark ratio in this Q^2 range.

6 The Physics of Minijets

In the spring of 1985 the SPS collider was operated in a ramping mode in which the centre of mass energy was varied continuously over a range $\sqrt{s} = 200 - 900$ GeV. The UA1 experiment took data throughout the ramping run and accumulated an integrated luminosity of about $5\mu b^{-1}$ using a minimum bias trigger. This data, together with samples of minimum bias data taken at $\sqrt{s} = 540$ GeV and $\sqrt{s} = 630$ GeV, formed the basis for an analysis of jet production as a function of beam energy [20]. Since the luminosity accumulated using the minimum bias trigger was in each case relatively modest, this analysis was necessarily focussed on the lower end of the p_T spectrum of the jets ($p_T \leq 15$ GeV). These low p_T jets generated considerable interest on the part of experimentalists and theorists alike, and have come to be known as 'minijets'.

6.1 Semihard Jet Physics

The standard QCD improved parton model, eqn. (1) is applicable in the region in which the transverse momenta of the jets p_T is formally of the same order as the centre of mass energy of the colliding hadrons. It is of great theoretical interest [69] to extend this region down in p_T to a region in which $\sqrt{s} >> p_T$ but $p_T >> \Lambda_{QCD}$. Because the transverse momentum is large compared to the QCD scale, the methods of perturbation theory will still be applicable. Thus we can investigate the interface between high p_T physics and the physics of the Regge limit using the methods of perturbation theory. We shall refer to this region as the mini-jet region. At present colliders the energy is rather too low to have well defined jets with a p_T very much less than the centre of mass energy. Because we are forced to low p_T, the contamination of the jets by the underlying minimum bias event is a serious problem. The application of the perturbative methods described below to present data is therefore not guaranteed.

In the mini-jet region the small x behaviour of the parton distribution is probed. The gluon distribution grows rapidly at small x and provides the dominant contribution. The Altarelli-Parisi (AP) equation [70] can be used to extract an asymptotic form for the gluon distribution function at small x. When $\ln(1/x)$ is large, the one loop evolution kernels in the AP equation are dominated by the poles at $x = 0$. Denoting the momentum distribution of the gluons by $G(x,t) = xg(x,t)$, at small x we obtain from the AP equation that,

$$\frac{dG(x,t)}{dt} = \frac{\alpha(t)N_c}{\pi}\int_x^1 \frac{dz}{z}G(z,t), \quad t = \ln\left(\frac{Q^2}{\Lambda_{QCD}^2}\right), \quad \alpha(t) = \frac{1}{bt} \tag{27}$$

In this case the scale Q should be identified with the transverse energy of the jet. The asymptotic solution to this equation is,

$$G(x,t) = \exp\sqrt{\frac{4N_c}{\pi b}\ln\frac{\ln Q^2/\Lambda_{QCD}^2}{\ln Q_0^2/\Lambda_{QCD}^2}\ln\frac{1}{x}}, \quad N_c = 3, \quad b = \frac{(33 - 2n_f)}{12\pi}. \tag{28}$$

Strictly speaking the AP equation is not valid in the small x region, because it sums all the leading logarithms of E_T/Λ_{QCD} but not all the leading logarithms of x. A more complete equation which treats both of these logarithms correctly has been given by Balitsky and Lipatov in ref. [74]. For values of $x > 10^{-4}$ the difference between solution to the BL equation and the full AP equation has been found numerically [71] to be less than 10%. So arguments based on the AP equation are qualitatively correct in the present energy regime.

Note that eqn. (28) shows that $G(x)$ grows at small x, whereas simple Pomeron dominance predicts that $G(x)$ tends to a constant at small x. Even if a constant behaviour is assumed at one value of Q^2 it is rapidly modified by AP evolution to give a steeper growth at small x. In the limited range of x available at present colliders it therefore more appropriate to assume the following small x behaviour [72],

$$G(x) = xg(x) \sim \frac{1}{x^\delta}, \quad \delta \sim 0.5 \tag{29}$$

Such a form of the gluon distribution predicts that mini-jet cross sections should grow approximately like

$$\frac{d\sigma}{dp_T^2} \sim \frac{s^\delta}{(q_T^2)^{2+\delta}}. \tag{30}$$

This can be interpreted as the effect of the QCD perturbative Pomeron which predicts the growth of the minijet cross-section with energy. However with a limited range in energy it is hard to turn this prediction into a reliable quantitative QCD test. This power law growth cannot continue indefinitely since it would violate the Froissart bound.

A method to disentangle the QCD Pomeron from the effects of the gluon distribution function is described in ref. [73]. Consider a two jet inclusive cross-section in which the longitudinal momentum fractions x_1, x_2 at which the parton distributions are probed is held fixed.

$$\sigma(M^2, s, x_1, x_2) = \int d^2 p_{1\,T} d^2 p_{2\,T} \frac{x_1 x_2 d\sigma}{dx_1 dx_2 d^2 p_{1\,T} d^2 p_{2\,T}} \Theta(p_{1\,T}^2 - M^2)\Theta(p_{2\,T}^2 - M^2) \tag{31}$$

By increasing the energy at fixed x_1 and x_2 the rapidity gap between the two jets grows. When the rapidity gap $Y = \ln(x_1 x_2 s/M^2)$ between the two jets becomes large, the perturbation series becomes a series in $\alpha_s Y$. At fixed x_i the effects of the gluon distributions cancel and the minijet cross-section is expected to grow asymptotically like [69],

$$\sigma(M^2, s, x_1, x_2) \sim \frac{\alpha_s^2(M^2)}{M^2} \frac{1}{\sqrt{\frac{21}{2}\zeta(3)\alpha_s(M^2)Y}} \exp(\frac{12\alpha_s(M^2)}{\pi} Y \ln 2) \tag{32}$$

For $M \sim 10$ GeV we find that $12\alpha_s \ln(2)/\pi \sim 0.5$ so that the minijet cross-section is expected to grow approximately like the square root of s.

A topic which is presently under investigation is the mechanism which limits the growth of the gluon distribution and the mini-jet cross-section. In the infinite momentum frame the Lorentz contracted proton is a disk of transverse radius r. The gluon distribution $G(x, \ln(Q^2))$ gives the number of gluons per unit of rapidity with a transverse size less than $1/Q$. At small x the number of gluons grows large. When they start to overlap, new non-perturbative effects come into play to curb the growth of the gluon distribution. A crude estimate of when this saturation begins to happen is provided by,

$$G(x, \ln(Q^2)) = \frac{\text{Area of hadron}}{\text{Area of parton}} \sim Q^2 r^2 \sim 25 \; Q^2 \; \text{GeV}^{-2}. \tag{33}$$

where $r \sim 1/m_\pi$ is the radius of the hadron. At presently attainable values of x the value of $G(x, \ln(Q^2))$ does not exceed 2 or 3, so the saturation limit is beyond the range of the present colliders. The modification of the Altarelli-Parisi equations as the saturation limit is approached is investigated in ref.[75]

6.2   Results on Minijets

In the UA1 minijet analysis [20], the UA1 jet algorithm is applied, without modification, to the low p_T jets. There is some phenomenological evidence that the UA1 jet algorithm is tolerably unbiassed for minijets provided that the transverse energy of the jets is not too low: specifically if the transverse energy of the jet satisfies $E_T > 5$ GeV. (In this context the transverse energy E_T of the jet is the vector sum over the calorimeter hits, which have been included in the jet by the jet algorithm, before any of the jet energy corrections, which are described in Section 2.1, have been applied.) Figure 29 shows the transverse energy profile plotted versus pseudorapidity for minijets with $E_T > 5$ GeV.

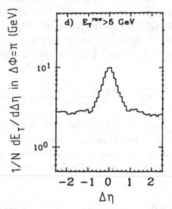

Figure 29:   The transverse energy profile versus pseudorapidity for minijets with $E_T > 5$ GeV.

Figure 29 should be compared with the jet transverse energy profile measured for high p_T jets (Figure 4). Although the core of the jet is clearly wider for low p_T jets, the base width of the jet appears to be independent of p_T. The background level in the wings of the jet is still substantial: $dE_T/d\eta \simeq 2$ GeV/(180° in ϕ). On the basis of these observations it is concluded that the minijets, found by the UA1 jet algorithm, are not fundamentally different from the high p_T jets.

Figure 30 shows the measured cross section for events with one (or more) minijets exceeding a nominal threshold $E_T > 5$ GeV, as a function of $\sqrt{s}$. The jet transverse

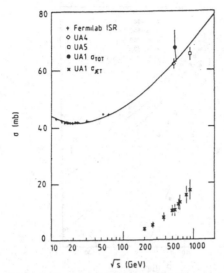

Figure 30: The cross section, measured by UA1, for events with one (or more) jets exceeding a nominal transverse energy threshold $E_T > 5$ GeV as a function of $\sqrt{s}$. The behaviour of the total cross section is also shown.

energy threshold $E_T > 5$ GeV is nominal to the extent that, as discussed above, no jet energy corrections have been applied at this stage. As is evident from the minijet pseudorapidity profile (Figure 29), for such low p_T jets the correction for energy splashing into the jet from the rest of the event is probably a substantial fraction of the jet energy itself. It is likely, therefore, that for the minijet cross section (Figure 30) the effective threshold in the jet transverse momentum is in the range $p_T \simeq 2 - 4$ GeV. From Figure 30 we conclude that the minijet cross section increases rapidly with increasing $\sqrt{s}$. The behaviour of the total cross section, as a function of $\sqrt{s}$, is shown for comparison. It is apparent that the cross section for events with minijets represents an appreciable fraction of the total cross section for centre of mass energies $\sqrt{s} \simeq 1$ TeV. The rapid increase in the jet production cross section as a function of $\sqrt{s}$ for a fixed transverse momentum threshold is predicted by perturbative QCD and is dominated by the growth of the gluon distribution function. In fact detailed QCD calculations [29,76] agree rather well with the data if the effective threshold is suitably chosen ($p_T \simeq 2$ GeV).

The inclusive jet cross section, measured in the ramping run, plotted as a function of jet transverse momentum p_T, is shown in Figure 31 for various values of the centre of mass energy $\sqrt{s} = 200, 500, 900$ GeV. The jet transverse momentum p_T, is now fully corrected for the effects of 'splash-in' etc., as described in Section 2.1. Since these corrections are much larger for the low p_T jets than for high p_T jets, considerable caution should be applied in interpreting the low p_T data points ($p_T = 5 - 10$ GeV). In ref.[20] the systematic error in the jet cross section at $p_T = 5$ GeV is estimated to be a factor of two: the errors plotted in Figure 31 are statistical only. Also shown in Figure 31 is the inclusive jet cross section at $\sqrt{s} = 63$ GeV measured by the AFS Collaboration at

170

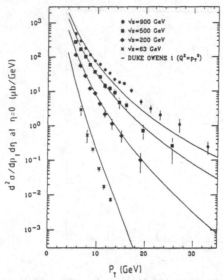

Figure 31: The inclusive jet cross section as a function of p_T for $\sqrt{s} = 200, 500, 900$ GeV measured by UA1. Data for $\sqrt{s} = 63$ GeV from the AFS Collaboration at the ISR are also shown.

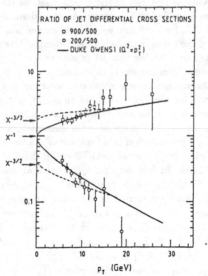

Figure 32: The ratios of jet inclusive cross sections at $\sqrt{s} = 200, 500, 900$ GeV measured by UA1. The curves indicate the possible behaviour as $p_T \rightarrow 0$ depending on the form of the gluon structure function at small x.

the ISR [77].

In Figure 31 the curves represent QCD predictions, computed consistently [29] with the predictions for the high-p_T jets shown in Figure 5, ie. using Duke and Owens structure functions, taking $\Lambda_{QCD} = 0.2$ GeV and assuming $Q^2 = p_T^2$. In this case the predictions have been scaled up by a factor of two to match the data better. Although qualitatively correct, these QCD predictions seem to fail quantitatively. In particular the $\sqrt{s} = 900$ GeV data lie above the theoretical curve for $p_T > 10$ GeV, while the ISR data tend to fall below, suggesting that the energy dependence is more rapid than predicted in this version of the theory.

A partial cancellation of systematic errors is achieved by computing the ratio of inclusive jet cross sections for different beam energies. Figure 32 shows the cross section ratios $\sqrt{s} = 900$ GeV : $\sqrt{s} = 500$ GeV and $\sqrt{s} = 200$ GeV : $\sqrt{s} = 500$ GeV plotted as a function of p_T computed from the data of Figure 31. In this case the errors shown include estimates for the systematic errors. The behaviour of these ratios in the limit $p_T \to 0$ is a measure of the behaviour of the gluon distribution at small x, as discussed in Section 6.1. Naive pomeron dominance gives a bremsstrahlung distribution ($g(x) = 1/x$) and leads to a cross section ratio of unity as $p_T \to 0$, as indicated by the solid curves in Figure 32. If we instead choose a gluon distribution function given by eqn. (29) with $\delta = 0.5$ the cross-sections are predicted by eqn. (30) to be approximately in the ratio of beam energies as indicated by the broken curves. The present data seem to favour a $1/x$ behaviour, but lower p_T 's, or in practice higher beam energies, are needed to settle the question.

172

References

[1]   M. Banner et al. , *Phys. Lett.***118B**(1982)203.

[2]   G. Arnison et al., *Phys. Lett.***123B**(1983)115.

[3]   B.Alper et al., *Phys. Lett.***B44**(1973)521;
        M.Banner et al. *Phys. Lett.***B44**(1973)537;
        F.W.Busser et al., *Phys. Lett.***B46**(1973)471.

[4]   P. Darriulat, Ann. Rev. Nucl. Part. Sci. 30 (1980) 159

[5]   L. Di Lella, Ann. Rev. Nucl. Part. Sci. 35 (1985) 107.

[6]   C. De Marzo et al., *Phys. Lett.***112B**(1982)173;
        C. De Marzo et al., *Nucl. Phys.***B211**(1983)375.

[7]   B. Brown et al., *Phys. Rev. Lett.***49**(1982)711.

[8]   G. Arnison et al., CERN EP/82-122 (unpublished).

[9]   G. Arnison et al., *Zeit. Phys.***C36**(1987)33.

[10]   P. Bagnaia et al., *Phys. Lett.***138B**(1984)430.

[11]   J. A. Appel et al., *Phys. Lett.***165B**(1985)441.

[12]   R. Ansari et al., *Zeit. Phys.***C36**(1987)175.

[13]   A. L. Angelis et al., *Phys. Lett.***126B**(1983)132;
        T. Akesson et al., *Phys. Lett.***128B**(1983)354.

[14]   G. Arnison et al., *Phys. Lett.***132B**(1983)214.

[15]   J. A. Appel et al., *Phys. Lett.***160B**(1985)349.

[16]   J. A. Appel et al., *Zeit. Phys.***C30**(1986)341.

[17]   P. Bagnaia et al., *Phys. Lett.***144B**(1984)291.

[18]   G. Arnison et al., *Phys. Lett.***136B**(1984)294.

[19]   G. Arnison et al., *Phys. Lett.***172B**(1986)461.

[20]   G. Arnison et al., CERN-EP/88-29 (submitted to Nucl. Phys. B).

[21]   D.W. Duke and J.F. Owens *Phys. Rev. Lett.***D30**(1984)49.

[22]   W. Furmanski and H. Kowalski, *Phys. Lett.***B148**(1984)247.

[23]   P. Bagnaia et al., *Zeit. Phys.***C20**(1983)117.

[24]   J. A. Appel et al., *Phys. Lett.***176B**(1986)239;
        P. Hansen, Private Communication.

[25] C. Albajar et al., CERN-EP/88-45 (submitted to Phys Lett.B).

[26] P. Aurenche et al., *Phys. Lett.***140B**(1984)87;
P. Aurenche et al., *Nucl. Phys.***B297**(1988)661.

[27] B. L. Combridge, J. Kripfganz and J. Ranft, *Phys. Lett.***70B**(1977)234.

[28] R. Horgan and M. Jacob *Nucl. Phys.***B179**(1981)441.

[29] W. J. Stirling, Proceedings of the 6th Topical Workshop on $p\bar{p}$ Collider Physics :
Aachen (1986). Edited by K. Eggert.

[30] P. Bagnaia et al., *Phys. Lett.***144B**(1984)283.

[31] M. Greco, *Zeit. Phys.***C26**(1985)567.

[32] G. Arnison et al., CERN-EP/88-54(submitted to Phys. Lett. B).

[33] E.J. Eichten et al., *Rev. Mod. Phys.***56**(1984)579.

[34] R. Ansari et al., *Phys. Lett.***186B**(1987)452.

[35] B. L. Combridge and C. J. Maxwell, *Nucl. Phys.***239B**(1984)428.

[36] G. Arnison et al., *Phys. Lett.***158B**(1985)494.

[37] G. Arnison et al., *Phys. Lett.***177B**(1986)244.

[38] E. J. Eichten et al., *Phys. Rev. Lett.***50**(1983)811.

[39] S. Li, Proceedings of the International Europhysics Conference on High Energy
Physics : Uppsala, Sweden. June 1987. Edited by O. Botner.

[40] R. K. Ellis and J. C. Sexton, *Nucl. Phys.***B282**(1987)642.

[41] J. F. Owens, *Rev. Mod. Phys.***59**(1987)465

[42] Z. Kunszt, *Nucl. Phys.***164**(1980)45.

[43] T. Gottschalk and D. Sivers, *Phys. Rev.***21**(1980)102.

[44] F. A. Berends et al, *Phys. Lett.***103B**(1981)124.

[45] R.K. Ellis, G. Marchesini and B. R. Webber, *Nucl. Phys.***286**(1987)643.

[46] B.L. Combridge and C. J. Maxwell, *Phys. Lett.***151B**(1985)299;
F. Halzen and P. Hoyer, *Phys. Lett.***130B**(1983)326.

[47] E. J. Buckley, Ph. D. Thesis, RAL T 029 (1986)

[48] S. J. Parke and T. R. Taylor, *Nucl. Phys.***269**(1986)410
J. F. Gunion and J. Kalinowski, *Phys. Rev.***34**(1986)2119;
F. A. Berends and W. Giele, *Nucl. Phys.***294**(1987)700;
M. Mangano, S. J. Parke and Z. Xu, *Nucl. Phys.***298**(1988)653.

[49]   Z. Kunszt, *Nucl. Phys.***271**(1986)333.

[50]   S. J. Parke and T. R. Taylor, *Phys. Rev.***35**(1987)313;
       M. Mangano and S. J. Parke, *Nucl. Phys.***299**(1988)673.

[51]   J. F. Gunion and Z. Kunszt, *Phys. Lett.***159B**(1985)167;
       Z. Xu, D-H. Zhang and L. Chang, *Nucl. Phys.***291**(1987)392

[52]   J. F. Gunion and Z. Kunszt, *Phys. Lett.***176B**(1986)163;
       R. Kleiss, *Nucl. Phys.***241**(1984)61

[53]   Z. Kunszt, *Phys. Lett.***145B**(1984)132;
       Z. Kunszt and W. J. Stirling, *Phys. Lett.***171B**(1986)307

[54]   S. J. Parke and T. R. Taylor, *Phys. Rev. Lett.***56**(1986)2459.

[55]   C. J. Maxwell, *Phys. Lett.***192B**(1987)190.

[56]   F. A. Berends and W. Giele, Leiden preprint Print-88-0100 (1987).

[57]   B. R. Webber, Ann. Rev. Nucl. Part. Sci. 36 (1986) 253;

[58]   B. Humpert, *Phys. Lett.***131B**(1983)461;
       N. Paver and D. Treleani, Zeit. Phys. C28 (1985) 187.

[59]   K.Meier, Proc. of the Workshop on QCD Hard Hadronic Processes, St Croix,
       Virgin Islands (1987).

[60]   T. Akesson et al, Zeit. Phys. C34 (1987) 163.

[61]   J. R. Hansen, Proceedings of the XXXIII International Conference on High Energy
       Physics : Berkeley, California (1986). Edited by S. C. Loken.

[62]   Z. Kunszt and W. J. Stirling , CERN-TH 4351/86 (1986).

[63]   Z. Kunszt, Proceedings of the Workshop on Physics at Future Accelerators : La
       Thuille ; CERN 87-07.

[64]   G. Arnison et al., *Nucl. Phys.***B286**(1986)253.

[65]   G. Arnison et al., *Phys. Lett.***132B**(1983)223.

[66]   T. Akesson et al., *Zeit. Phys.***C30**(1986)27.

[67]   P. Ghez and G. Ingleman, *Zeit. Phys.***C33**(1987)465.

[68]   A. Petersen et al., *Phys. Rev.***55**(1985)1954.

[69]   L. V. Gribov, E. M. Levin and M. G. Ryskin, Phys. Rep. 100 (1983) 1

[70]   G. Altarelli and G. Parisi, Nucl. Phys. B126 (1977) 298.

[71]   J. Kwiecinski, Z. Phys. C29 (1985) 561.

[72]   J. C. Collins, in Proceedings of the SSC workshop UCLA (1986).

[73]   A. H. Mueller and H. Navelet, Nucl. Phys. B282 (1987) 727.

[74]   Ya. Ya. Balitsky and L. N. Lipatov, Sov. J. Nucl. Phys. 28 (1978) 822.

[75]   A. H. Mueller and J. Qiu, Nucl. Phys. B268 (1986) 427.

[76]   G. Pancheri and Y. Srivastava, *Phys. Lett.*182B(1986)199.

[77]   T. Akesson et al., *Phys. Lett.*123B(1983)133.

PHYSICS OF THE INTERMEDIATE VECTOR BOSONS

G. Altarelli and L. Di Lella
CERN, Geneva, Switzerland

1. INTRODUCTION

The conversion of the CERN 450 GeV proton synchrotron (SPS) into a proton–antiproton collider was originally proposed in 1976 [1] as a fast and relatively cheap way to produce and detect the weak Intermediate Vector Bosons (IVB), $W^\pm$ and Z, by achieving hadronic collisions at an energy large enough to provide observable rates. The properties of such particles had been predicted already in the 60's in the framework of the so-called Standard Model of the unified electroweak theory developed by Glashow, Salam and Weinberg [2]; however, the interest in this theory arose only some years later, following the proof of renormalizability [3] and the first experimental observation of neutrino interactions mediated by Z-exchange [4]. In particular, this experiment obtained a measurement of the weak mixing angle, which allowed a quantitative prediction of the IVB mass values.

The CERN Collider project was approved in 1978 and the first $\bar{p}p$ collisions at a total centre-of-mass energy ($\sqrt{s}$) of 546 GeV were observed in 1981. The decay $W \to e\nu$ was first observed [5, 6] among data collected at the end of 1982, and the decay $Z \to e^+e^-$ [7, 8] and $Z \to \mu^+\mu^-$ [7] were observed a few months later. At present, following two more data-taking runs in 1984 and 1985 at a slightly increased centre-of-mass energy ($\sqrt{s} = 630$ GeV), samples of ~ 250 $W \to e\nu$ and ~ 30 $Z \to e^+e^-$ events are available from each of the two major experiments (UA1 and UA2), making possible a quantitative comparison of IVB properties with the predictions of the Standard Model.

In this article, we first describe the Standard Model of the unified electroweak theory (Section 2), and we use this theoretical framework to derive the IVB mass values and their decay properties (Sections 3 and 4). In Section 5, we discuss the IVB production properties, taking into account the effect of QCD corrections. Finally, after a description of the two major experiments and of the technique used to detect neutrinos (Sections 6 and 7), in Section 8 we discuss the large amount of data collected until the end of 1985, and compare the experimental results with Standard Model expectations.

2. THE STANDARD MODEL OF THE ELECTROWEAK INTERACTIONS

In this section, we summarize the structure of the standard electroweak Lagrangian [2] and specify the couplings of $W^\pm$ and Z, the intermediate vector bosons (IVBs).

For this discussion we split the Lagrangian into two parts by separating the Higgs boson couplings:

$$\mathcal{L} = \mathcal{L}_{symm} + \mathcal{L}_{Higgs}. \tag{2.1}$$

We start by specifying $\mathcal{L}_{\text{symm}}$, which involves only gauge bosons and fermions:

$$\mathcal{L}_{\text{symm}} = -\tfrac{1}{4} \sum_{A=1}^{3} F_{\mu\nu}^A F^{A\mu\nu} - \tfrac{1}{4} B_{\mu\nu} B^{\mu\nu} + \bar{\psi}_L i\gamma^\mu D_\mu \psi_L + \bar{\psi}_R i\gamma^\mu D_\mu \psi_R . \qquad (2.2)$$

This is the Yang–Mills Lagrangian for the gauge group $SU(2) \otimes U(1)$ with fermion matter fields. Here

$$B_{\mu\nu} = \partial_\mu B_\nu - \partial_\nu B_\mu$$

and

$$F_{\mu\nu}^A = \partial_\mu W_\nu^A - \partial_\nu W_\mu^A - g\, \epsilon^{ABC}\, W_\mu^B W_\nu^C \qquad (2.3)$$

are the gauge antisymmetric tensors constructed out of the gauge field B_μ associated with $U(1)$, and W_μ^A corresponding to the three $SU(2)$ generators; ϵ^{ABC} are the group structure constants [see Eqs. (2.6)] which, for $SU(2)$, coincide with the totally antisymmetric Levi-Civita tensor. The normalization of the $SU(2)$ gauge coupling g is therefore specified by Eq. (2.3).

The fermion fields are described through their left-hand and right-hand components:

$$\psi_{L,R} = [(1 \mp \gamma_5)/2]\psi , \qquad \bar{\psi}_{L,R} = \bar{\psi}[(1 \pm \gamma_5)/2] . \qquad (2.4)$$

In the Standard Model the left and right fermions have different transformation properties under the gauge groups. Thus mass terms for fermions (of the form $\bar{\psi}_L \psi_R + \text{h.c.}$) are forbidden in the symmetric limit. In particular, all ψ_R are singlets in the minimal Standard Model. But for the moment, by $\psi_{L,R}$ we mean a column vector, including all fermions in the theory which span a generic reducible representation of $SU(2) \otimes U(1)$.

The covariant derivatives $D_\mu \psi_{L,R}$ are explicitly given by

$$D_\mu \psi_{L,R} = [\partial_\mu + ig \sum_{A=1}^{3} t_{L,R}^A W_\mu^A + ig'\tfrac{1}{2}Y_{L,R}B_\mu]\psi_{L,R}, \qquad (2.5)$$

where $t_{L,R}^A$ and $\tfrac{1}{2}Y_{L,R}$ are the $SU(2)$ and $U(1)$ generators, respectively, in the reducible representations $\psi_{L,R}$. The commutation relations of the $SU(2)$ generators are given by

$$[t_L^A, t_L^B] = i\,\epsilon^{ABC}\, t_L^C,$$

$$[t_R^A, t_R^B] = i\,\epsilon^{ABC}\, t_R^C. \qquad (2.6)$$

We use the normalization [in the fundamental representation of $SU(2)$],

$$\text{Tr } t^A t^B = \tfrac{1}{2}\delta^{AB} \qquad (2.7)$$

which gives, for doublets, $t^3 = \pm\tfrac{1}{2}$. The electric charge generator Q (in units of e, the positron charge) is given by

$$Q = t_L^3 + \tfrac{1}{2}Y_L = t_R^3 + \tfrac{1}{2}Y_R . \qquad (2.8)$$

Note that the normalization of the $U(1)$ gauge coupling g' in Eq. (2.5) is now specified.

All fermion couplings to the gauge bosons can directly be derived from Eqs. (2.2) and (2.5). The charged-current (CC) couplings are the simplest. From

$$g(t^1 W^1_\mu + t^2 W^2_\mu) = g\{[(t^1 + it^2)/\sqrt{2}][(W^1_\mu - iW^2_\mu)/\sqrt{2}] + \text{h.c.}\} = g\{[(t^+ W^-_\mu)/\sqrt{2}] + \text{h.c.}\}, \quad (2.9)$$

where $t^\pm = t^1 \pm it^2$, $W^\pm = (W^1 \pm iW^2)/\sqrt{2}$, we obtain the vertex

$$V_{\bar\psi\psi W} = g\bar\psi\gamma_\mu [(t^+_L/\sqrt{2})(1-\gamma_5)/2 + (t^+_R/\sqrt{2})(1+\gamma_5)/2]\psi W^-_\mu + \text{h.c.} \quad (2.10)$$

In the neutral current (NC) sector, the photon A_μ and the mediator Z_μ of the weak neutral current are orthogonal and normalized linear combinations of B_μ and W^3_μ:

$$A_\mu = \cos\theta_w\, B_\mu + \sin\theta_w\, W^3_\mu,$$

$$Z_\mu = -\sin\theta_w\, B_\mu + \cos\theta_w\, W^3_\mu, \quad (2.11)$$

Equations (2.11) define the weak mixing angle θ_w. The photon is characterized by equal couplings to left and right fermions with a strength equal to the electric charge. Recalling Eq. (2.8) for the charge matrix Q, we immediately obtain

$$g \sin\theta_w = g' \cos\theta_w = e, \quad (2.12)$$

or equivalently,

$$\text{tg}\, \theta_w = g'/g, \quad (2.13)$$

$$e = gg'/\sqrt{g^2 + g'^2}.$$

Once θ_w has been fixed by the photon couplings, it is a simple matter of algebra to derive the Z couplings, with the result

$$\Gamma_{\bar\psi\psi Z} = (g/2\cos\theta_w)\bar\psi\gamma_\mu [t^3_L(1-\gamma_5) + t^3_R(1+\gamma_5) - 2Q\sin^2\theta_w]\psi Z^\mu, \quad (2.14)$$

where $\Gamma_{\bar\psi\psi Z}$ is a notation for the vertex. In the minimal Standard Model, $t^3_R = 0$ and $t^3_L = \pm\frac{1}{2}$.

In order to derive the effective four-fermion interactions that are equivalent at low energies to the CC and NC couplings given in Eqs. (2.10) and (2.14), we anticipate that large masses, as experimentally observed, are provided for $W^\pm$ and Z by $\mathcal{L}_{\text{Higgs}}$. For left–left CC couplings, when the momentum transfer squared can be neglected with respect to m^2_W in the propagator of Born diagrams with single W exchange, from Eq. (2.10) we can write

$$\mathcal{L}^{CC}_{\text{eff}} \simeq (g^2/8m^2_W)\, [\bar\psi\gamma_\mu(1-\gamma_5)t^+_L\,\psi][\bar\psi\gamma^\mu(1-\gamma_5)t^-_L\,\psi]. \quad (2.15)$$

By specializing further in the case of doublet fields such as ν_e–e^- or ν_μ–μ^-, we obtain the tree-level relation of g with the Fermi coupling constant G_F measured from μ decay ($G_F = 1.16637(2) \times 10^{-5}$ GeV^{-2}):

$$G_F/\sqrt{2} = g^2/8m^2_W. \quad (2.16)$$

By recalling that $g \sin\theta_w = e$, we can also cast this relation in the form

$$m_W = \mu_{Born}/\sin\theta_w, \tag{2.17}$$

with

$$\mu_{Born} = [\pi\alpha/\sqrt{2}\, G_F]^{1/2} \simeq 37.281\ \text{GeV}, \tag{2.18}$$

where α is the fine structure constant of QED ($\alpha \equiv e^2/4\pi = 1/137.036$).

In the same way, for neutral currents we obtain in Born approximation from Eq. (2.14) the effective four-fermion interaction given by

$$\mathcal{L}_{eff}^{NC} = \sqrt{2}\, G_F \varrho_0 \bar\psi \gamma_\mu\, [...]\, \psi \bar\psi \gamma^\mu\, [...]\, \psi, \tag{2.19}$$

where

$$[...] \equiv t_L^3 (1 - \gamma_5) + t_R^3 (1 + \gamma_5) - 2Q \sin^2\theta_w \tag{2.20}$$

and

$$\varrho_0 = m_W^2/m_Z^2 \cos^2\theta_w. \tag{2.21}$$

All couplings given in this section are obtained at tree level and are modified in higher orders of perturbative theory. In particular, the relations between m_W and $\sin\theta_w$ [Eqs. (2.17) and (2.18)] and the observed values of ϱ ($\varrho = \varrho_0$ at tree level) in different NC processes, are altered by computable electroweak radiative corrections, as discussed in Section 3.

The gauge-boson self-interactions can be derived from the $F_{\mu\nu}$ term in $\mathcal{L}_{symm}$, by using Eq. (2.11) and $W_\mu^\pm = (W^1 \pm iW^2)/\sqrt{2}$. Defining the three-gauge-boson vertex as in Fig. 1, we obtain

$$\Gamma_{W^- W^+ V} = ig_{W^- W^+ V}\, [g_{\mu\nu}(q-p)_\lambda + g_{\mu\lambda}(p-r)_\nu + g_{\nu\lambda}(r-q)_\mu], \tag{2.22}$$

with

$$g_{W^- W^+ \gamma} = g \sin\theta_w = e,$$

$$\tag{2.23}$$

$$g_{W^- W^+ Z} = g \cos\theta_w.$$

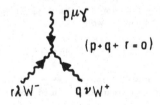

Fig. 1   The three-gauge-boson vertex

A number of four-gauge-boson vertices of order g^2 are also present, but they will not be reproduced here.

We now turn to the Higgs sector [9] of the electroweak Lagrangian: $\mathcal{L}_{\text{Higgs}}$ is specified by the gauge principle and the requirement of renormalizability to be

$$\mathcal{L}_{\text{Higgs}} = (D_\mu \phi)^\dagger (D^\mu \phi) - V(\phi^\dagger \phi) - \bar{\psi}_L \Gamma \psi_R \phi - \bar{\psi}_R \Gamma^\dagger \psi_L \phi^\dagger, \qquad (2.24)$$

where ϕ is a column vector including all Higgs fields; it transforms as a reducible representation of the gauge group. The quantities Γ (which include all coupling constants) are matrices that make the Yukawa couplings invariant under the Lorentz and gauge groups. The potential $V(\phi^\dagger \phi)$, symmetric under SU(2) $\otimes$ U(1), contains, at most, quartic terms in ϕ so that the theory is renormalizable. Spontaneous symmetry breaking is induced if the minimum of V—which is the classical analogue of the quantum mechanical vacuum state (both are the states of minimum energy)— is obtained for non-vanishing ϕ values. Precisely, we denote the vacuum expectation value (v.e.v.) of ϕ, i.e. the position of the minimum, by v:

$$\langle 0 | \phi(x) | 0 \rangle = v \neq 0. \qquad (2.25)$$

The fermion mass matrix is obtained from the Yukawa couplings by replacing $\phi(x)$ by v:

$$M = \bar{\psi}_L \mathfrak{M} \psi_R + \bar{\psi}_R \mathfrak{M}^\dagger \psi_L, \qquad (2.26)$$

with

$$\mathfrak{M} = \Gamma \cdot v. \qquad (2.27)$$

In the minimal Standard Model, where all left fermions ψ_L are doublets and all right fermions ψ_R are singlets, only Higgs doublets can contribute to fermion masses. There are enough free couplings in Γ, so that one single complex Higgs doublet is indeed sufficient to generate the most general fermion mass matrix. It is important to observe that by a suitable change of basis we can always make the matrix $\mathfrak{M}$ Hermitian, γ_5-free, and diagonal. In fact, we can make separate unitary transformations on ψ_L and ψ_R according to

$$\psi'_L = U \psi_L, \qquad \psi'_R = V \psi_R \qquad (2.28)$$

and consequently

$$\mathfrak{M} \rightarrow \mathfrak{M}' = U^\dagger \mathfrak{M} V. \qquad (2.29)$$

This transformation does not alter the general structure of the fermion couplings in $\mathcal{L}_{\text{symm}}$. For quarks, the Cabibbo–Kobayashi–Maskawa [10] unitary transformation relates the mass eigenstates d, s, and b to the CC eigenstates d', s', b', i.e. the states coupled by W emission to u, c, and t, respectively. The NC is then automatically diagonal in flavour at tree level (GIM mechanism [11]). In the case of leptons, if the neutrinos are massless clearly there is no mixing.

If only one Higgs doublet is present, the change of basis that makes M diagonal will at the same time diagonalize also the fermion–Higgs Yukawa couplings. Thus, in this case, no flavour-changing neutral Higgs exchanges are present. This is not true, in general, when there are several Higgs doublets. But one Higgs doublet for each electric charge sector—i.e. one doublet coupled only to u-type quarks, one doublet to d-type quarks, one doublet to charged leptons—would also be all right [12], because the mass matrices of fermions with different charges are diagonalized separately. For several Higgs doublets it is also possible to generate CP violation by complex phases in the Higgs couplings [13]. In the presence of six quark flavours, this CP violation mechanism is not necessary. In fact, at the moment, the simplest model with only one Higgs doublet seems adequate for describing all observed phenomena.

We recall that the Standard Model, with N fermion families with the observed quantum numbers, is automatically free of γ_5 anomalies [14] owing to cancellation of quarks with lepton loops.

We now consider the gauge-boson masses and their couplings to the Higgs. These effects are induced by the $(D_\mu\phi)^\dagger(D^\mu\phi)$ term in $\mathcal{L}_{Higgs}$ [Eq. (2.24)], where

$$D_\mu\phi = [\partial_\mu + ig \sum_{A=1}^{3} t^A W_\mu^A + ig'(Y/2) B_\mu]\phi. \tag{2.30}$$

Here t^A and $\frac{1}{2}Y$ are the SU(2) $\otimes$ U(1) generators in the reducible representation spanned by ϕ. Not only doublets but all non-singlet Higgs representations can contribute to gauge-boson masses. The condition that the photon remains massless is equivalent to the condition that the vacuum is electrically neutral:

$$Q|v\rangle = (t^3 + \frac{1}{2}Y)|v\rangle = 0. \tag{2.31}$$

The charged W mass is given by the quadratic terms in the W field arising from $\mathcal{L}_{Higgs}$, when $\phi(x)$ is replaced by v. We obtain

$$m_W^2 W_\mu^+ W^{-\mu} = g^2 |(t^+ v/\sqrt{2})|^2 W_\mu^+ W^{-\mu}, \tag{2.32}$$

whilst for the Z mass we get [recalling Eq. (2.11)]

$$\frac{1}{2}m_Z^2 Z_\mu Z^\mu = |[g \cos\theta_w t^3 - g' \sin\theta_w (Y/2)]v|^2 Z_\mu Z^\mu, \tag{2.33}$$

where the factor of $\frac{1}{2}$ on the left-hand side is the correct normalization for the definition of the mass of a neutral field. By using Eq. (2.31), relating the action of t^3 and $\frac{1}{2}Y$ on the vacuum v, and Eqs. (2.13), we obtain:

$$\frac{1}{2}m_Z^2 = (g \cos\theta_w + g' \sin\theta_w)^2 |t^3 v|^2 = (g^2/\cos^2\theta_w) |t^3 v|^2. \tag{2.34}$$

For Higgs doublets

$$\phi = \begin{pmatrix} \phi^+ \\ \phi^0 \end{pmatrix}, \quad v = \begin{pmatrix} 0 \\ v \end{pmatrix}, \tag{2.35}$$

we have

$$|t^+ v|^2 = v^2,$$
$$|t^3 v|^2 = \frac{1}{4} v^2, \tag{2.36}$$

so that

$$m_W^2 = \tfrac{1}{2}g^2v^2, \qquad m_Z^2 = \tfrac{1}{2}g^2v^2/\cos^2\theta_w. \qquad (2.37)$$

Note that by using Eq. (2.16) we obtain

$$v = 2^{-3/4}\,G_F^{-1/2} = 174.1\,\text{GeV}. \qquad (2.38)$$

It is also evident that for Higgs doublets

$$\varrho_0 = m_W^2/m_Z^2\cos^2\theta_w = 1. \qquad (2.39)$$

This relation is typical of one or more Higgs doublets and would be spoiled by the existence of Higgs triplets, etc. In general,

$$\varrho = \sum_i [(t_i)^2 - (t_i^3)^2 + t_i]v_i^2 \Big/ \sum_i 2(t_i^3)^2 v_i^2 \qquad (2.40)$$

for several Higgses with v.e.v.'s v_i, weak isospin t_i, and z-component t_i^3. These results are valid at the tree level and are modified by calculable electroweak radiative corrections, as discussed in Section 3.

If only one Higgs doublet is present, then the fermion–Higgs couplings are in proportion to the fermion masses. In fact, from the Yukawa couplings $g_{\phi\bar{f}f}$ ($\bar{f}_L\phi f_R$ + h.c.), the mass m_f is obtained by replacing ϕ by v, so that $g_{\phi\bar{f}f} = m_f/v$.

With only one complex Higgs doublet, three out of the four Hermitian fields are removed from the physical spectrum by the Higgs mechanism and become the longitudinal modes of W^+, W^-, and Z. The fourth neutral Higgs is physical and should be found. If more doublets are present, two more charged and two more neutral Higgs scalars should be around for each additional doublet.

Finally, the couplings of the physical Higgs H to the gauge bosons can be simply obtained from $\mathcal{L}_{\text{Higgs}}$, by the replacement

$$\phi(x) = \begin{pmatrix} \phi^+(x) \\ \phi^0(x) \end{pmatrix} \to \begin{pmatrix} 0 \\ v + (H/\sqrt{2}) \end{pmatrix}, \qquad (2.41)$$

[so that $(D_\mu\phi)^\dagger(D^\mu\phi) = \tfrac{1}{2}(\partial_\mu H)^2 + \ldots$], with the result

$$\mathcal{L}|H,W,Z| = g^2\,(v/\sqrt{2})\,W_\mu^+ W^{-\,\mu}\,H + (g^2/4)\,W_\mu^+ W^{-\,\mu}H^2$$
$$+ [(g^2\,v\,Z_\mu Z^\mu)/(2\sqrt{2}\cos^2\theta_w)]H + [g^2/(8\cos^2\theta_w)]\,Z_\mu Z^\mu H^2. \qquad (2.42)$$

We have thus completed our summary of the standard electroweak theory and of the $W^\pm$, Z couplings.

3. RADIATIVE CORRECTIONS

In the minimal standard electroweak theory, we have seen that, at tree level, three — in principle, different — quantities, all identified with $\sin^2\theta_w$, coincide:

i) The quantity $\sin^2\theta_w = e^2/g^2$ [see Eq. (2.12)], which specifies the ratio between the positron electric charge and the SU(2) gauge coupling. This is experimentally obtained through Eq. (2.16) by measuring m_W [see Eqs. (2.17) and (2.18)]:

$$\sin^2\theta_w = (\pi\alpha/\sqrt{2}\,G_F)\,(1/m_W^2) = \mu_{Born}^2/m_W^2. \tag{3.1}$$

ii) The quantity $\sin^2\theta_w$ derived from the gauge-boson mass matrix [see Eqs. (2.21) and (2.39)] by the relation

$$\varrho_0 = m_W^2/m_Z^2\cos^2\theta_w = 1. \tag{3.2}$$

iii) The parameter $\sin^2\theta_w$ which appears in the NC effective interaction [see Eqs. (2.19) and (2.20)]. Schematically,

$$\mathcal{L}_{eff} = \sqrt{2}\,G_F\varrho_0\,(J_3 - 2\sin^2\theta_w\,J_{em})^2. \tag{3.3}$$

The equality at tree level of these three 'definitions' of $\sin^2\theta_w$ is the signature of the minimal Standard Model. Nearly all conceivable departures from the minimal Standard Model remove the degeneracy of the above 'definitions' of $\sin^2\theta_w$. Thus precise measurements comparing NC couplings with measurements of weak gauge-boson masses are crucial tests of the Standard Model and could well lead to the discovery of new physics.

The relations (3.1), (3.2) and (3.3) are also modified, in a calculable way, by radiative corrections [15]. In particular, at tree level

$$\varrho_0 = m_W^2/m_Z^2\cos^2\theta_w = \varrho_{\nu N} = \varrho_{\nu e} = \varrho_{eq} = ..., \tag{3.4}$$

i.e. ϱ is the same for all NC processes and is related in a simple way to W and Z masses. When radiative corrections are included in Eq. (3.3), $\varrho_0 \rightarrow \varrho_0 + \delta\varrho$ and $\sin^2\theta_w \rightarrow k\sin^2\theta_w$, with different $\delta\varrho$ and k for each channel, so that in particular

$$m_W^2/m_Z^2\cos^2\theta_w \neq \varrho_{\nu N} \neq \varrho_{\nu e} \neq \varrho_{eq} .... \tag{3.5}$$

It is useful to fix the definition of $\sin^2\theta_w$ by imposing [16] that, at all orders,

$$\varrho \equiv m_W^2/m_Z^2\cos^2\theta_w = \varrho_0, \tag{3.6}$$

with $\varrho_0 = 1$ for doublet Higgs bosons. Thus by including radiative corrections, we can write

$$\varrho_{\nu N} = \varrho_0(1 + \delta\varrho_{\nu N}),$$

$$\varrho_{\nu e} = \varrho_0(1 + \delta\varrho_{\nu e}), \text{ etc.} \tag{3.7}$$

With the same definition of $\sin^2\theta_w$, also the relation between $\sin^2\theta_w$ and m_W^2 is modified. The corrected expression of Eq. (3.1) can be conveniently written in the form

$$\sin^2\theta_w = [\mu_{Born}^2/(1-\Delta r)]\,(1/m_W^2), \tag{3.8a}$$

or equivalently

$$\sin^2\theta_w \cos^2\theta_w = [\mu_{Born}^2/(1-\Delta r)] (1/m_Z^2), \qquad (3.8b)$$

where μ_{Born} is given in Eq. (3.1).

In order to measure ϱ_0 accurately from the NC data, or to extract $\sin^2\theta_w$ precisely from a measurement of m_w^2, we must compute the radiative corrections $\delta\varrho_{\nu N}$, $\delta\varrho_{\nu e}$, ..., and Δr which appear in Eqs. (3.7) and (3.8). However, even assuming that we have three fermion families and one Higgs doublet, the radiative corrections still depend sensitively on m_t and m_H, i.e. the masses of the t-quark and of the Higgs particle, respectively. Usually, in most existing computations, one makes a guess: typically, $m_t \simeq 45$ GeV/c^2 and $m_H \approx m_Z$. With these values a result for ϱ_0 is obtained from the data, which according to two very recent analyses [17] is given by

$$(\varrho_0)_{exp} = \begin{array}{l} 0.998 \pm 0.0086 \text{ (Amaldi et al.)}, \\ 1.001 \pm 0.007 \quad \text{(Costa et al.)}. \end{array} \qquad (3.9)$$

Clearly even if the minimal electroweak theory is correct, $(\varrho_0)_{exp}$ could still deviate from one, simply because the input values of m_t and m_H are not correct. In particular, $\varrho_0 - 1$ is quadratically increasing with m_t and logarithmically varying with m_H. Since $(\varrho_0)_{exp}$ — obtained from the data when the radiative corrections are computed with 'light' m_t and m_H — is remarkably close to one, it can be deduced that the Higgses are (at least dominantly) doublets and that an upper bound on m_t can be given (which is [17] around $m_t \leqslant 200$ GeV/c^2 for any $m_H \leqslant 1$ TeV/c^2). Also, for 'light' m_t and m_H, the value of Δr in Eqs. (3.8) is essentially coincident with $[\alpha(m_W)/\alpha(m_e)] - 1$, i.e. it is determined by the running of the QED coupling from the defining scale of order m_e (the electron mass) up to a scale of order m_W. A precise calculation for $m_t = 35$ GeV/c^2, $m_H = 100$ GeV/c^2, and $m_Z = 93$ GeV/c^2 leads to [18]

$$\Delta r = 0.071 \pm 0.0013. \qquad (3.10)$$

For increasing m_t the value of Δr varies dramatically as is shown [18] in Fig. 2 for different values of the Higgs mass. If we use the value of Δr given in Eq. (3.10), together with the UA1 and UA2 measurements of m_W combined (see subsections 8.3 and 8.11), we obtain from Eqs. (3.8):

$$(\sin^2\theta_w)_{m_W} = 0.229 \pm 0.003 \text{ (stat.)} \pm 0.007 \text{ (syst.)}. \qquad (3.11)$$

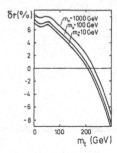

Fig. 2   The electroweak correction factor δr as a function of m_t [18]

This value is in very good agreement with the measurement of $\sin^2\theta_w$ from NC data. The most precise NC experiments are those on νN deep-inelastic scattering [19]. Combining the corresponding data by the CDHS and CHARM experiments at CERN, and by the FMMF and CCFRR experiments at Fermilab, we obtain (including electroweak radiative corrections),

$$(\sin^2\theta_w)_{\nu N} = 0.233 \pm 0.003 \pm 0.005. \tag{3.12}$$

The agreement between these two determinations of $\sin^2\theta_w$ is the most precise test of the electroweak theory available at present. Given the behaviour of Δr with m_t shown in Fig. 2, the agreement would be spoiled for too large values of m_t (with the definition of $\sin^2\theta_w$ adopted here, the radiative corrections to νN deep-inelastic scattering are less sensitive to m_t [20]). This argument leads to an independent confirmation that $m_t \leqslant 200$ GeV/c^2.

In future, when m_Z will be precisely measured at LEP and the SLC, it will be possible to test the standard electroweak theory by comparing m_Z with the ratio m_W/m_Z measured by collider experiments at the CERN p$\bar{\text{p}}$ Collider and the FNAL Tevatron. For this, we modify Eqs. (3.8) by using the definition of $\sin^2\theta_w = 1 - (m_W^2/m_Z^2)$ (obtained with $\varrho_0 = 1$ as appropriate for Higgs doublets) according to

$$m_Z^2 = [\mu_0^2/(1 - \Delta r)]/[(m_W^2/m_Z^2)(1 - m_W^2/m_Z^2)] \tag{3.13}$$

The relation between m_W and m_Z [18, 21, 22] implied by Eq. (3.13) is shown in Fig. 3 for different values of m_t and m_H.

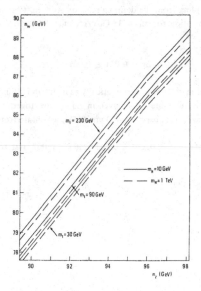

Fig. 3 The W mass versus the Z mass, including the effect of radiative corrections for different values of the Higgs and the t-quark masses [22].

4. W AND Z DECAYS

In this section we briefly summarize the W and Z decay properties in the Standard Model.

The partial widths of W and Z into a massless fermion–antifermion pair, including first-order strong and electroweak radiative corrections, are given by

$$\Gamma(W \rightarrow f\bar{f}') = N(G_F m_W^3/6\pi\sqrt{2})(1 + \delta_f^W) \tag{4.1}$$

$$\Gamma(Z \rightarrow f\bar{f}) = N(G_F \varrho_0 m_Z^3/24\pi\sqrt{2})[1 + (1 - 4|Q_f| \sin^2\theta_w)^2](1 + \delta_f^Z), \tag{4.2}$$

where $\varrho_0 = 1$ for doublet Higgses; δ_f^W and δ_f^Z are the electroweak radiative corrections, given explicitly, for example, in Refs. [18] and [21]; and

$$N = 1 \qquad \text{for leptons,}$$
$$N = 3(1 + \alpha_s/\pi + ...) \qquad \text{for quarks.} \tag{4.3}$$

Here α_s is the QCD coupling. We can define α_s according to the $\overline{MS}$ prescription [23]. In this case the second-order correction is also known [23]. From experimental determinations of Λ_{QCD} [24], we obtain a value of $\alpha_s(m_W)/\pi$ in the range

$$\alpha_s(m_W)/\pi \simeq 0.036 \pm 0.005. \tag{4.4}$$

In the case of W decays formula (4.1) refers to the sum of all down-quarks associated with a given up-quark. For a particular down-quark, q', a factor $|V_{qq'}|^2$ would appear, where $V_{qq'}$ is the relevant term of the Cabibbo-Kobayashi-Maskawa matrix [10] ($\Sigma_{q'}|V_{qq'}|^2 = 1$ by unitarity). Note that by writing $\Gamma_{Born} \approx G_F M^3$, instead of $\Gamma_{Born} \approx g^2 M \approx \alpha M/\sin^2\theta_w$, we make the Born approximation more precise (i.e. $\delta_f^{W,Z}$ smaller). In fact the radiative corrections (which result in a scale dependence) are large on both α and $\sin^2\theta_w$. On the other hand, there are no leading logs in the scale dependence of the Fermi coupling G_F [25]; $\delta_f^{W,Z}$ are very small [15, 21] ($\delta_f \leq 0.1\%$) if $m_t \leq m_Z$, and there are no additional heavy-fermion generations [provided the physical values of $m_{W,Z}$ and $\sin^2\theta_w$ are inserted in Eqs. (4.1) and (4.2)]. The electroweak radiative corrections become considerably larger if m_t is large or if weak isospin multiplets with large mass splittings exist [15, 21]. With the exception of this case, for example, we obtain

$$\Gamma(W \rightarrow e\nu) = 233 \, (m_W/81 \, \text{GeV}/c^2)^3 \, \text{MeV}/c^2, \tag{4.5}$$

$$\Gamma(Z \rightarrow e^+e^-) = 86 \, (m_Z/92 \, \text{GeV}/c^2)^3 \, \text{MeV}/c^2, \tag{4.6}$$

$$\Gamma(Z \rightarrow \nu_e\bar{\nu}_e) = 170 \, (m_Z/92 \, \text{GeV}/c^2)^3 \, \text{MeV}/c^2. \tag{4.7}$$

Note that the dependence of $\Gamma(Z \rightarrow ee)$ on $\sin^2\theta_w$ is very small for $\sin^2\theta_w \simeq 0.23$. If a careful determination of the partial widths is required for quark modes, one should implement the dependence of $\sin^2\theta_w$ on m_Z given by Eq. (3.8b).

For $W \rightarrow t\bar{b}$ and $Z \rightarrow t\bar{t}$ (and perhaps $Z \rightarrow b\bar{b}$) the quark-mass corrections cannot be neglected. The widths in Born approximation are then modified as follows [26]:

$$\Gamma_{\text{Born}}(W \to t\bar{b}) = (G_F m_W^3/2\pi\sqrt{2})\{[(m_W^2 - m_t^2 - m_b^2)^2 - 4m_b^2 m_t^2]/m_W^4\}^{1/2}$$

$$\times \{1 - (m_t^2 + m_b^2)/2m_W^2 - \tfrac{1}{2}[(m_t^2 + m_b^2)/m_W^2]^2\}$$

$$\simeq (G_F m_W^3/2\pi\sqrt{2})(1 - \epsilon)^2(1 + \epsilon/2), \tag{4.8}$$

$$\Gamma_{\text{Born}}(Z \to Q\bar{Q}) = (G_F \varrho_0 m_Z^3/8\pi\sqrt{2})[\beta^3 + (1 - 4|Q_Q|\sin^2\theta_w)^2 \tfrac{1}{2}\beta(3 - \beta^2)] \tag{4.9}$$

with $\epsilon = m_t^2/m_W^2$ and $\beta = [1 - (4m_Q^2/m_Z^2)]^{1/2}$. The QCD corrections of order α_s are also modified with respect to the massless case and are given in Ref. [27].

The leptonic branching ratios are independent of m_W and m_Z (at fixed $\sin^2\theta_w$). We obtain

$$BR(W \to e\nu) \simeq 0.089, \qquad m_t \simeq 40 \text{ GeV/c}^2,$$

$$\tag{4.10}$$

$$BR(W \to e\nu) \simeq 0.109, \qquad m_t > m_W - m_b,$$

$$BR(Z \to e^+e^-) \simeq 0.034, \tag{4.11}$$

$$BR(Z \to \sum_{i = e,\mu,\tau} \nu_i\bar{\nu}_i) \simeq 0.20, \tag{4.12}$$

if $m_t > m_Z/2$ and $\sin^2\theta_w \simeq 0.23$.

The total widths Γ_W and Γ_Z are given to a good approximation by summing up the rates for $V \to f\bar{f}$ discussed above, because the rare decays of W/Z in the Standard Model are really rare and can be neglected [28]. For example, for $m_t > m_W - m_b$, we obtain

$$\Gamma_W \equiv \Gamma(W \to e\nu)/BR(W \to e\nu) \simeq 2.14(m_W/81 \text{ GeV/c}^2)^3 \text{ GeV/c}^2.$$

5. PRODUCTION PROPERTIES OF $W^\pm$ AND Z

The total cross-sections and distributions for the production of the W and Z bosons can be reliably computed in perturbative QCD [23]. To lowest order the production is determined by the parton process $q + \bar{q} \to V$, where $V \equiv W^\pm$ or Z. The next-order corrections in α_s are due to the process $q + \bar{q} \to V + g$ and ${}^{(\bar{q})} + g \to {}^{(q)} + V$, and so on, order by order in α_s. The total cross-sections, integrated over the rapidity y and the transverse momentum p_T are given by [29]

$$\sigma_V = N_V \int (dx_1 dx_2/x_1 x_2)\left\{[q(x_1, m_V^2)\bar{q}(x_2, m_V^2) + (1 \leftrightarrow 2)]\right.$$

$$\times [\delta(1 - \tau/x_1 x_2) + 2/3\pi\,\alpha_s(m_V^2)\,\theta(1 - \tau/x_1 x_2)2f_q(\tau/x_1 x_2)]$$

$$+ \left([q(x_1, m_V^2) + \bar{q}(x_1, m_V^2)]\,g(x_2, m_V^2) + (1 \leftrightarrow 2)\right)$$

$$\left. \times [\alpha_s(m_V^2)/4\pi]\,\theta(1 - \tau/x_1 x_2 f_g)\,(\tau/x_1 x_2)\right\} + O(\alpha_s^2), \tag{5.1}$$

where $\tau = m_V^2/s$ ($\sqrt{s}$ is the total centre-of-mass energy of the incoming hadrons), $0 \leqslant x_1, x_2 \leqslant 1$, and N_V is

$$N_{W^\pm} = \pi^2\alpha/3s\sin^2\theta_w \tag{5.2}$$

$$N_Z = \pi^2\alpha/12s \sin^2\theta_w \cos^2\theta_w. \tag{5.3}$$

[For comparison, $N(\gamma^*) = 4\pi^2\alpha/3s$ is the normalization for Drell–Yan production of charged lepton pairs.]

The appropriate combinations of quark flavours are needed in the initial state. Precisely, for W^+ production,

$$q_1\bar{q}_2 \to (u_1\bar{d}_2 + c_1\bar{s}_2)\cos^2\theta_C + (u_1\bar{s}_2 + c_1\bar{d}_2)\sin^2\theta_C, \tag{5.4}$$

$$(q_1 + \bar{q}_1)g_2 \to (u_1 + c_1 + \bar{d}_1 + \bar{s}_1)g_2, \tag{5.5}$$

where θ_C is the Cabibbo angle [10]; and similarly for W^-. For Z production:

$$q_1\bar{q}_2 \to \sum_f n_f^Z q_{f_1}\bar{q}_{f_2}, \tag{5.6}$$

$$(q_1 + \bar{q}_1)g \to \sum_f n_f^Z [q_{f_1} + \bar{q}_{f_1}]g_2, \tag{5.7}$$

with

$$n_f^Z = [1 + (1 - 4|Q_f|\sin^2\theta_w)^2]. \tag{5.8}$$

The kernels $f_{q,g}(z)$ are given by

$$f_q(z) = (\tfrac{1}{2} + \tfrac{2}{3}\pi^2)\,\delta(1 - z) - 3 - 2z + (1 + z^2)\,[\ln(1-z)/(1-z)]_+ + 3/2(1-z)_+ \tag{5.9}$$

$$f_g(z) = (\tfrac{9}{2}z^2 - 5z + \tfrac{3}{2} + [z^2 + (1-z)^2]\ln(1-z). \tag{5.10}$$

The 'plus' definition applies to the integration range from zero to one and is specified as follows:

$$\int_0^1 dz\, g(z)[f(z)]_+ \equiv \int_0^1 dz\, [g(z) - g(1)]\, f(z), \tag{5.11}$$

where $g(z)$ is any regular function in $z = 1$. To be completely explicit, this means in practice:

$$\int (dx_1 dx_2/x_1 x_2)\, \theta[1 - (\tau/x_1 x_2)]\, h(x_1, x_2)\, f(\tau/x_1 x_2)_+ = \int_\tau^1 (dx_1/x_1) \int_{\tau/x_1}^1 (dz/z)\, h(x_1, \tau/x_1 z)\, f(z)_+$$

$$= \int_\tau^1 (dx_1/x_1) \int_0^1 dz\, [(1/z)\, h(x_1, \tau/x_1 z) - h(x_1, \tau/x_1)]\, f(z) - \int_\tau^1 (dx_1/x_1) \int_0^{\tau/x_1} (dz/z)\, h(x_1, \tau/x_1 z)\, f(z)$$

$$= \int_\tau^1 (dx_1/x_1) \int_{\tau/x_1}^1 dz\, [(1/z)\, h(x_1, \tau/x_1 z) - h(x_1, \tau/x_1)]\, f(z) - \int_\tau^1 (dx_1/x_1)\, h(x_1, \tau/x_1) \int_0^{\tau/x_1} dz\, f(z). \tag{5.12}$$

In the last expression the integrands are regular at both ends of the x_1, z integrations.

A similar formula can also be written for $d\sigma/dy$ (or $d\sigma/dx_F$), where y (or x_F) is the V rapidity (or the Feynman variable). The corrections of order α_s are somewhat complicated. The complete formula is given in Refs. [29] and [30], and will not be repeated here. It turns out that $f_q(z)$ corresponds to a rather large correction, whilst $f_g(z)$ is quite small. The corrections of order α_s can be described as a rescaling of the lowest-order result by a factor called, by tradition, the 'K factor' (K is a function of Q^2, with $Q \equiv m_{\bar{q}q}$, and τ). Since the dependence on Q^2 and τ is mild over the range of most of the actual experiments on Drell–Yan production, K is often taken approximately as a constant. At fixed target energies, the K factor for Drell–Yan processes increases the lowest-order cross-section by a dangerously large amount, and resummation techniques must be invoked in an attempt to control the perturbation series. At collider energies their size is reduced because the coupling constant is smaller, and for the production of weak intermediate bosons they lead to a correction of about 30%. The total cross-section is therefore more reliably predicted by perturbation theory at these energies, with a smaller theoretical error on the overall normalization. Note that the bulk of the cross-section arises from $x_1 \approx x_2 \approx \sqrt{\tau}$. For W or Z production at the CERN $p\bar{p}$ Collider, $\sqrt{\tau} \approx 0.15$, a value where the quark and gluon densities are under good control (at the FNAL Tevatron, $\sqrt{\tau} = 0.04$–0.05). Thus W and Z production in the range $\sqrt{s} \simeq 0.54$ to 2 TeV is very suitable for theoretical calculations in the framework of QCD. The main uncertainty arises from the Q^2 scale appearing in the order α_s terms. In fact, once the order α_s corrections are computed, the lowest-order term is completely defined. That is, one needs a precise definition of the lowest-order term in order to determine the appropriate corresponding corrections. Thus in Eq. (5.1) the quark densities are, for example, those given by the electroproduction structure function F_2 at the scale m_V^2 (that is, $q^2 = -m_V^2$, where q_μ is the space-like momentum transfer of leptoproduction). However, the Q^2 scale—where the running coupling and the densities are to be evaluated for estimating the order α_s^2 term—is free, because the difference between two choices is of order α_s^2. Of course, a natural choice is $Q^2 \approx m_V^2$ also in this case. And this is what was done in Eq. (5.1). However, one could argue in favour of choosing for Q^2 the average p_T^2 value of the W/Z, $\langle p_T^2 \rangle$. Such a choice results in a sizeable difference because $\langle p_T^2 \rangle \ll M^2$. This ambiguity must be taken into account when the theoretical errors are estimated.

An updated evaluation [31] of $\sigma_W \equiv \sigma_{W^+} + \sigma_{W^-}$ and σ_Z is presented in Table 1 for $\sin^2\theta_w = 0.229$, $m_W = 80.8$ GeV/c^2, and $m_Z = 92.0$ GeV/c^2. Experimentally one measures $\sigma\cdot$BR, where BR is the branching ratio into the electron modes. This quantity depends on the t-quark mass, especially in

Table 1

Values of W ($= W^+ + W^-$) and Z production cross-sections in $p\bar{p}$ collisions
for $\sin^2\theta_w = 0.229$, $m_W = 80.8$ GeV/c^2, $m_Z = 92.0$ GeV/c^2

$\sqrt{s}$ (TeV)	σ^W (nb)	σ^Z (nb)
0.54	$4.3^{+1.3}_{-0.6}$	$1.4^{+0.4}_{-0.2}$
0.63	$5.4^{+1.6}_{-0.9}$	$1.7^{+0.5}_{-0.3}$
1.6	$17^{+4.0}_{-2.5}$	$5.1^{+1.2}_{-0.8}$
1.8	$19^{+5.0}_{-3.3}$	$5.8^{+1.6}_{-1.0}$
2.0	$21^{+6.0}_{-4.0}$	$6.4^{+1.9}_{-1.2}$

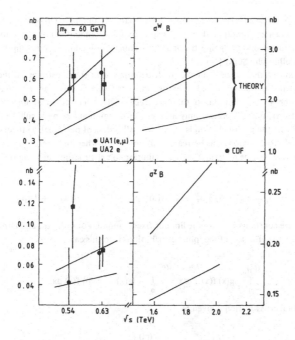

Fig. 4   A comparison of the theoretical predictions on $\sigma \cdot BR$ versus the experimental data [31].

the W case [see Eq. (4.10)]. A comparison of the theoretical prediction with the data is presented in Fig. 4 [31] (these data will be discussed in Section 8).

A particularly important quantity is the ratio $R' = \sigma_W/\sigma_Z$. A substantial part of the theoretical error is cancelled in the ratio. For the same values of $\sin^2\theta_w$, m_W, and m_Z as quoted above, one finds [31]

$$\sigma_W/\sigma_Z = 3.28 \pm 0.15, \qquad \sqrt{s} = 0.63 \, \text{TeV}. \tag{5.13}$$

The ratio R' plays a crucial role in limiting the number of neutrino flavours N_ν (or, in general, of non-minimal contributions to Γ_W/Γ_Z) from W/Z physics in collider experiments. This method will be discussed in detail in subsection 8.14.

The prediction of the boson transverse-momentum distribution is more subtle, since effects of all orders need to be taken into account. Renormalization-group-improved perturbation theory is valid when the transverse momentum p_T is of the same order as the vector mass m_V. The large p_T tail of the distribution was one of the early predictions of the QCD-improved parton model. As p_T becomes less than m_V, so that $\Lambda \ll p_T \ll m_V$, a new scale is present in the problem and large terms of order $\alpha_s^n(p_T^2) \times \ln^m (m_V^2/p_T^2)$, $(m \leqslant 2n - 1)$, occur: this forces the consideration of all orders in n. These terms are characteristic of a theory with massless vector gluons. Fortunately, in the leading double-logarithmic approximation (DLA: $m = 2n - 1$), these terms can be reliably resummed. This resummation was first attempted by Dokshitzer et al. [32], and was subsequently modified and consolidated by Parisi et al. [33]. A consistent framework for going beyond the leading double-logarithmic approximation has been indicated by Collins and Soper [34].

The combination of these results on the p_T distribution with the constraint on the area of the distribution provided by the integrated cross-section at $O(\alpha_s)$, allows an essentially complete reconstruction of the p_T distribution to that accuracy [30]. There is some uncertainty due to non-perturbative effects, but it is present only in a restricted region at low p_T at collider energies. It is interesting to recall why the p_T distribution is sensitive to all orders of perturbation theory in α_s, whilst its integral, the total cross-section, can be reasonably described when only terms up to order α_s are included. Schematically, for the $q\bar{q}$ term, the p_T distribution in perturbation theory at order α_s is of the form

$$d\sigma/dp_T^2 \simeq \sigma_0(1+A)\,\delta(p_T^2) + B[\ln(m_V^2/p_T^2)/p_T^2]_+ \; + \; C/(p_T^2)_+ \; + \; D(p_T^2), \tag{5.14}$$

where A, B, and C are constants in p_T (in the limit of fixed coupling) of order α_s, and $D(p_T)$ is a regular function in $p_T = 0$. The definition of the 'plus' distributions is in this case,

$$\int_0^{(p_T^2)_{max}} g(x)\, f(x)_+ \, dx = \int_0^{(p_T^2)_{max}} [g(x) - g(0)]\, f(x)\, dx. \tag{5.15}$$

In particular, the total cross-section is given by

$$\sigma = \sigma_0(1+A) + \int_0^{(p_T^2)_{max}} D(x)\, dx \tag{5.16}$$

because the integral over the whole p_T^2 range of the B and C terms vanishes owing to Eq. (5.15) (with $g = 1$). The corresponding resummed result is given by

$$d\sigma/dp_T^2 \simeq D(p_T^2) + \int (d^2b/4\pi)\,[\exp(-i\,\vec{p}_T\,\vec{b})]\,\sigma_0(1+A)\,\exp S(b), \tag{5.17}$$

where

$$S(b) = \int_0^{(p_T^2)_{max}} dk^2\,\{B[\ln(m_V^2/k^2)/k^2] + (C/k^2)\}\,[J_0(bk) - 1]. \tag{5.18}$$

The Bessel function $J_0(bk)$ originates from the angular integration in d^2k:

$$\int_0^{2\pi} d\phi\, e^{\pm iz\cos\phi} = 2\pi\, J_0(z), \tag{5.19}$$

and the -1 in the $(J_0 - 1)$ factor is from the 'plus' definition, Eq. (5.15). The b-transform trick is introduced in order to ensure $\vec{p}_T$ conservation in multiple gluon emission. (This is explained, for example, in Ref. [23].)

When the integration over all values of p_T^2 is performed in order to derive the total cross-section, a $\delta^{(2)}(\vec{b})$ is produced in Eq. (5.17):

$$\sigma = \int_0^{(p_T^2)_{max}} D(x)\, dx + \sigma_0 \int d^2b\, \delta^2(\vec{b})\, (1+A)\, \exp S(b) = \sigma_0(1+A) + \int_0^{(p_T^2)_{max}} D(x)\, dx, \qquad (5.20)$$

where the last step follows because $S(0) = 0$. Thus the whole tower of exponentiated logarithms decouples from σ.

Actually, $S(b)$ is known to order α_s^2. The main $O(\alpha_s^2)$ term was computed in Ref. [35], and was later confirmed [36] by using the results of Ref. [37]. In the $\overline{MS}$ prescription for α_s, one obtains

$$B = +\tfrac{4}{3}\, [\alpha_s(k^2)/\pi][1 + T\, \alpha_s(k^2) + \ldots], \qquad (5.21)$$

$$C = -2\, [\alpha_s(k^2)/\pi] + \ldots, \qquad (5.22)$$

with

$$T = 1/2\pi\, [67/6 - \pi^2/2 - 5n_f/9]_{n_f = 5} \simeq 0.55. \qquad (5.23)$$

The $O(\alpha_s^2)$ term in C is also known [38], but being less important it will not be reported here. The functions A and D are given explicitly in Ref. [30].

A complete analysis of the p_T distributions for W and Z production at Collider and Tevatron energies ($\sqrt{s} \approx 0.54$–2 TeV) was given in Refs. [30] and [39], including also an evaluation of the theoretical error. Examples of theoretical p_T distributions for W production at $\sqrt{s} = 546$ GeV are displayed in Fig. 5.

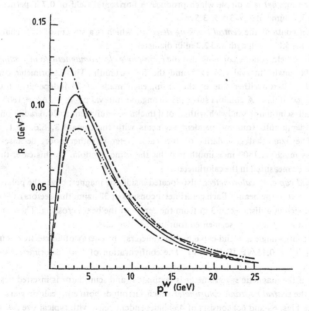

Fig. 5 The ratio $R = [(d\sigma/dp_T^W\, dy)/(d\sigma/dy)]$ at W rapidity $y = 0$, as given by Altarelli et al. [30] for four different choices of structure functions and definitions of the Q^2 scale.

6. THE DETECTORS

A typical experimental apparatus at a hadron or e^+e^- collider consists, in general, of many successive layers of different kinds of detectors.

Moving outwards from the region where the beams collide, one finds first a tracking detector which serves to reconstruct charged-particle tracks. Such a detector may be located in a magnetic field, which provides a measurement of the charged-particle momenta.

Following the tracking detector, one finds an electromagnetic calorimeter with large angular coverage, which is used to measure the energy of secondary electrons and photons. Such a device contains a high-Z material, and is thick enough to absorb almost completely the electromagnetic showers produced by high-energy electrons or photons.

Just behind the electromagnetic calorimeter, a hadronic calorimeter is installed, in general, to absorb hadronic showers associated with strongly interacting particles, such as π, K, or nucleons, and to measure their energies.

A calorimeter is generally subdivided into many independent cells, in order to cope with the high multiplicity of secondary particles. Such a property is often called granularity.

If the calorimeters are thick enough, no known long-lived particles, except for muons and neutrinos, can escape. Muon detectors, consisting in general of large-area tracking chambers, can be used therefore as the outermost shell of the apparatus. To measure muon momenta, a magnetic field is required. Such a field could either be the same field as is used in the central region, or it could be provided by magnetized iron, or both.

The UA1 experiment (Fig. 6) contains all the features described above. It is based on a general-purpose magnetic detector [40], which was designed to achieve an almost complete solid-angle coverage down to polar angles of 0.2° with respect to the beams, using both track detection and calorimetry. The *magnet* is a dipole which produces a horizontal field of 0.7 T perpendicular to the beam axis over a volume of $7 \times 3.5 \times 3.5 \text{ m}^3$.

The magnet contains the *central tracking detector,* which is a system of drift chambers filling a cylindrical volume 5.8 m in length and 2.3 m in diameter.

The magnetic field volume contains also the *large-angle electromagnetic (e.m.) calorimeter,* which covers the polar angle interval 25°–155° and the full azimuth. This calorimeter consists of two half-cylindrical shells on either side of the beam, each made of 24 independent total-absorption counters called 'gondolas'. A gondola subtends an angular interval $\Delta\theta = 5°$, $\Delta\phi = 180°$. It consists of a multilayer lead/scintillator sandwich with a total thickness equivalent to 26.6 radiation lengths (r.l.) subdivided in depth into four independent segments with thickness 3.3, 6.6, 10.1, and 6.6 r.l., respectively. The azimuthal granularity of this calorimeter is rather poor, because two particles separated by as much as 180° in azimuth may hit the same gondola. In this case, their individual energies cannot be measured in the calorimeter.

The two *end-cap e.m. calorimeters,* also located inside the magnet, cover the polar angle interval 5°–25° with respect to the beams. Each calorimeter consists of 32 azimuthal sectors of lead/scintillator multilayer sandwich at a distance of 3 m from the centre of the beam crossing. These counters have a total thickness of 27 r.l. and are segmented four times in depth.

In both the large-angle and end-cap e.m. calorimeters, the energy of an electron is measured with a resolution $\sigma_E/E = 0.15/\sqrt{E}$ (E in GeV). The configuration of both calorimeters can be seen in Figs. 6b and 6c.

The yoke of the magnetic spectrometer is laminated, and scintillator is inserted between the iron plates to form the *central hadronic calorimeter* which surrounds both e.m. calorimeters. The hadronic calorimeter (see Figs. 6a and 6c) consists of 450 independent cells, with typical size $\Delta\theta \times \Delta\phi$ equal to

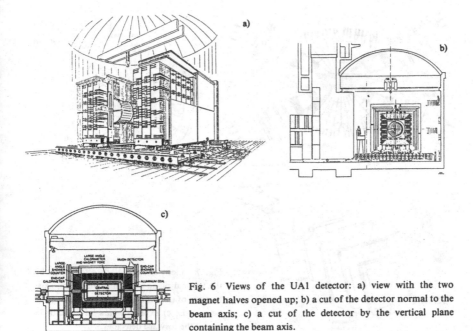

Fig. 6 Views of the UA1 detector: a) view with the two magnet halves opened up; b) a cut of the detector normal to the beam axis; c) a cut of the detector by the vertical plane containing the beam axis.

$15° \times 18°$ in the central region and $5° \times 10°$ in both forward regions. Their thickness is ≈ 5 and ≈ 7 absorption lengths, respectively. The energy resolution is typically $\sigma_E/E \approx 0.8/\sqrt{E}$ (E in GeV).

Muon detectors surround the magnet yoke (see Figs. 6a and 6c). They consist of two planes of drift chambers covering the polar angle interval $5° < \theta < 175°$ and the full azimuth, separated by a distance of 60 cm. Each of these planes is made of 50 drift chambers of $\approx 4 \times 6$ m^2. Each chamber consists of four layers of drift tubes (maximum drift space 7 cm), which define two orthogonal coordinates. By knowing the position of the interaction vertex and the muon-track parameters in these chambers, it is possible to obtain an independent measurement of the muon momentum with a relative precision of $\approx 20\%$ through the determination of its deflection in the magnet yoke.

In addition to the detectors just described, more calorimeters and track detectors located along the beam pipe on both sides of the spectrometer cover the region of polar angles from 5° down to $\approx 0.2°$ with respect to the beam line (see Fig. 6b).

The UA2 detector (Fig. 7a) consists of three parts [41]:

i) The *vertex detector*, which is a system of cylindrical chambers to reconstruct charged-particle tracks in a region without magnetic field. This detector covers the polar angle interval $20° < \theta < 160°$. A 'preshower' counter, consisting of a 1.5 r.l. thick tungsten cylinder followed by a multiwire proportional chamber (MWPC) with cathode strip readout and pulse-height measurement on the wires, is located just behind the last chamber, covering the central region $40° < \theta < 140°$. As we shall see later, this device is essential for electron identification.

ii) The *central calorimeter*, which is a system of 240 independent counters covering the full azimuth (see Fig. 7b) and the polar angle interval $40° < \theta < 140°$. Each counter has an angular acceptance

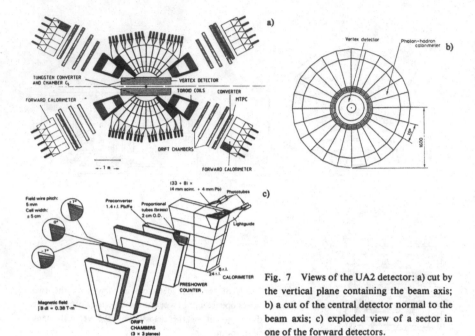

Fig. 7 Views of the UA2 detector: a) cut by the vertical plane containing the beam axis; b) a cut of the central detector normal to the beam axis; c) exploded view of a sector in one of the forward detectors.

$\Delta\theta \times \Delta\phi = 10° \times 15°$ and consists of a first e.m. section (lead/scintillator) 17 r.l. thick, followed by two independent hadronic sections (iron/scintillator), each 2 absorption lengths thick. The energy resolution is $\sigma_E/E \approx 0.15/\sqrt{E}$ for e.m. showers, and $\approx 0.32E^{-1/4}$ for hadronic showers (E in GeV).

iii) The two *forward detectors*, covering the polar angle intervals $20° < \theta < 37.5°$ and $142.5° < \theta < 160°$ and the full azimuth. Each detector consists of 12 sectors in which a toroidal magnetic field is generated by 12 coils equally spaced in azimuth (the field integral is ≈ 0.38 T·m). Following the magnetic field volume, each sector contains nine drift-chamber planes, which are used to measure the charged-particle momenta together with the information from the vertex detector. A preshower counter, consisting of a 1.5 r.l. thick lead/iron plate followed by four layers of proportional tubes, is located after these chambers. Energy measurement, limited to e.m. showers only, is performed with a calorimeter containing ten independent counters per sector ($\Delta\theta \times \Delta\phi = 3.5° \times 15°$ per counter). Each counter is a lead/scintillator multilayer sandwich, subdivided in depth into two independent sections, 24 and 6 r.l. thick. Energy resolution is similar to that of the central calorimeter. A view of a sector of a forward detector is shown in Fig. 7c.

No muon detector is implemented in the UA2 experiment.

7. MISSING TRANSVERSE MOMENTUM

The presence of one or more non-interacting particles, such as neutrinos, among the outgoing secondaries, can be detected, at least in principle, by measuring the total momentum vector associated with all observed final-state particles. Such a vector should be zero within the measurement errors, if no high-energy particle escapes detection.

In practice, even in the case of a detector with very large angular coverage, the presence of the machine vacuum chamber prevents the observation of particles emitted at angles smaller than $\lesssim 0.5°$. The total momentum carried by these particles may be large and undergoes important fluctuations from one event to the other, thus making the total longitudinal momentum a useless quantity. Such undetected particles, however, carry a negligibly small transverse momentum; hence a measurement of the total transverse momentum in the event can still be effectively used to detect the presence of invisible high-p_T particles.

The missing transverse momentum vector, $\vec{p}_T^{miss}$, is defined as

$$\vec{p}_T^{miss} = -\vec{p}_T^\mu - \sum_i \vec{p}_T^i , \qquad (7.1)$$

where $\vec{p}_T^\mu$ is the total transverse momentum carried by muons (if any), and $\vec{p}_T^i$ is a vector with magnitude given by the energy deposited in the i-th cell of the calorimeter, and directed from the event vertex to the cell centre. In Eq. (7.1), the sum is extended to all calorimeter cells.

Obviously, the measurement of $\vec{p}_T^{miss}$ requires as large an angular coverage as possible (this property is often called hermeticity), and this is indeed the case for the UA1 apparatus, with the exception of two holes at central rapidities, having the shape of an azimuthal sector of $\pm 4°$ with respect to the vertical axis, which are needed to let through the signals from the various detectors contained in the magnet. The main consequence of these holes is that, in most cases, if $\vec{p}_T^{miss}$ makes an angle of less than $\pm 15°$ with the vertical axis, its measurement is not reliable. After excluding such events from a sample of purely hadronic final states, it is found [42] that each component of $\vec{p}_T^{miss}$ has a Gaussian distribution centred at zero, with a r.m.s. value $\sigma = 0.5(\Sigma_i E_T^i)^{1/2}$, where E_T^i is the transverse energy (in GeV) observed in the i-th calorimeter cell. Hence the distribution of $|\vec{p}_T^{miss}|^2$ has the form $\exp(-|\vec{p}_T^{miss}|^2/2\sigma^2)$.

As described in Section 6, the UA2 apparatus does not detect any particles below an angle of 20° to the beams, and measures only partially hadronic jets emitted at polar angles between 20° and 40°. As a consequence, in a certain fraction of events a large value of p_T^{miss} may result from instrumental effects, such as partially contained or undetected high-p_T jets. In the case of two-jet events, the probability of losing one of the jets is found to decrease from $\approx 10\%$ at $p_T(jet) = 15$ GeV, to $\approx 2\%$ at $p_T(jet) = 40$ GeV. We also recall that there is no muon detector in UA2.

8. W AND Z DETECTION

8.1 General Considerations

For both IVBs, $W^\pm$ and Z, a large fraction of the decay modes (about 70%) consists of $q\bar{q}$ pairs which appear as two high-p_T hadronic jets (see Section 4). In searching for such configurations we expect a background from the continuum of high-p_T two-jet events which results from hard-parton scattering [43]. Such a background is approximately two orders of magnitude larger than the IVB signal, and it cannot be easily reduced experimentally because there is no way to distinguish jets from W and Z decay from those which result from hard-parton scattering. A measurement of the two-jet invariant mass with excellent resolution (of the order of the natural width of the IVBs) would certainly help to identify a peak superimposed on the smooth mass distribution of the background. However, in both UA1 and UA2, jet energies are measured using conventional calorimetry, and the resolution so obtained for the two-jet invariant mass in the region of the IVB masses is ≈ 8 GeV/c^2, a value barely adequate for the event sample available at present (see subsection 8.15).

For these reasons, both experiments have chosen to detect the IVBs by identifying their leptonic decays:

$$\left. \begin{array}{l} W^{\pm} \to e^{\pm} \nu_e(\bar{\nu}_e) \\ \\ Z \to e^+ e^- \end{array} \right\} \text{ in both UA1 and UA2,}$$

$$\left. \begin{array}{l} W^{\pm} \to \mu^{\pm} \nu_{\mu}(\bar{\nu}_{\mu}) \\ \\ Z \to \mu^+ \mu^- \end{array} \right\} \text{ in UA1 only,}$$

for which the backgrounds are much lower, as we shall see later, in spite of the smaller branching ratios.

The decision to detect electrons instead of muons in the UA2 experiment was based on the possibility of measuring electron energies with good precision using compact e.m. calorimeters (typically only ≈ 25 cm thick), whereas muon detection implies large magnetic field volumes for momentum measurement and the use of thick iron absorbers.

When searching for the decay $Z \to e^+ e^-$, or $Z \to \mu^+ \mu^-$, the Z mass can be directly determined by measuring the energies (or momenta) of the two leptons and the angle α between their directions:

$$m^2 = 4E_1 E_2 \sin^2 \alpha/2 .$$

The mass resolution σ_m can be estimated from a knowledge of the measuring errors σ_{E1}, σ_{E2}, and σ_{α}:

$$\sigma_m/m = \frac{1}{2}[(\sigma_{E1}/E_1)^2 + (\sigma_{E2}/E_2)^2 + (\sigma_{\alpha}/\tan \alpha/2)^2]^{1/2} .$$

The opening angle α is measured using tracking devices, and its error $\Delta\alpha$ can easily be made as low as 0.01 rad, or less. Its effect on σ_m/m is then negligible, because at Collider energies α is generally larger than 90°.

Electron energies are measured in calorimeters with a resolution $\sigma_E/E \approx 0.15/\sqrt{E}$ (E in GeV, see Section 6), giving

$$\sigma_m/m \approx 0.15/\sqrt{m} ,$$

where m is in GeV/c^2 for a symmetric decay configuration. For m ≈ 90 GeV/c^2 the mass resolution is $\sigma_m \approx 1.4$ GeV/c^2, which is comparable with the expected natural width (see Section 4).

For the decay $Z \to \mu^+ \mu^-$, muon momenta p_1 and p_2 replace the electron energies. They are determined by measuring the muon deflection in the magnetic field with a resolution which is typically $\sigma_p/p \approx 0.5\%$ p (p in GeV/c) in the UA1 detector (there is only a 0.6 mm sagitta in a 1 m long track from a 45 GeV/c particle). In this case we obtain

$$\sigma_m/m \approx (1.8 \times 10^{-3})m ,$$

where m is in GeV/c^2, which corresponds to a mass resolution $\sigma_m \approx 14$ GeV/c^2 at m ≈ 90 GeV/c^2 and is a much worse value than in the $e^+ e^-$ case. Furthermore, σ_m/m increases linearly with m, whereas for e^+e^- pairs it decreases as $m^{-1/2}$, showing the superiority of the e^+e^- channel in searching for possible new particles with masses larger than those of the IVBs.

The detection of the decay mode $W \to \ell\nu$, where ℓ is either an electron or a muon, is based on a completely different method, because the neutrino cannot be detected. In the W rest frame the lepton energy is just $m_W/2$. If we neglect the W transverse momentum p_T^W, the lepton transverse momentum $p_T^\ell = (m_W/2) \sin \theta^*$ (where $\sin \theta^*$ is the decay angle with respect to the beam direction in the W rest frame) is Lorentz invariant, and the p_T spectrum can be obtained from the decay angular distribution $f(\theta^*)$ by a simple change of variables:

$$dn/dp_T \propto [1 - (2p_T/m_W)^2]^{-1/2} f[\sin^{-1} (2p_T/m_W)].$$

The singularity at $p_T = m_W/2$ arises only from the change of variables and is often called the Jacobian peak. The W transverse motion and the W natural width have the effect of smearing this peak. However, if the mean value of p_T^W is much smaller than $m_W/2$, the Jacobian peak is still the most distinctive feature of the charged-lepton p_T distribution in the decay $W \to \ell\nu$.

In addition to the peculiar shape of the charged-lepton p_T distribution, $W \to \ell\nu$ decays are characterized by a large missing transverse momentum, which results from the presence of a high-p_T neutrino. Obviously, the p_T^{miss} distribution will also show a Jacobian peak.

8.2 The Decay $W \to e\nu$

In a search for high-p_T electrons at the Collider, the main background results from hadronic jets consisting of one or more high-p_T π^0's and one charged particle, which could be either a charged jet fragment or an electron from photon conversion. Such a configuration is hard to distinguish from a genuine electron in the UA1 and UA2 detectors.

As a consequence, both experiments use somewhat similar electron identification criteria, which aim at a high rejection against jets while maintaining a reasonably high efficiency for electrons: These criteria require the energy deposition in the calorimeter to match that expected from an isolated electron, which implies that a very large fraction of events containing electrons inside or near a jet are rejected. In the following, we describe the main cuts used to select an inclusive electron sample.

i) The transverse energy E_T associated with the energy cluster in the calorimeter is required to exceed a given threshold.

ii) The shower energy leakage into the hadronic compartment of the calorimeter is required to be small.

iii) In the UA1 experiment the longitudinal development of the electromagnetic shower, which is measured over four samples, is required to be compatible with that expected for an electron. In the UA2 experiment, the lateral shower profile is required to be small, using the small cell size of the calorimeters.

iv) The presence of a charged-particle track pointing to the energy cluster in the calorimeter is required.

v) In the UA1 experiment the momentum measurement is used to ensure that the track transverse momentum is larger than 7 GeV/c (or compatible with 15 GeV/c within 3σ). In UA2, a magnetic field exists only in the two forward regions $20° < \theta < 37.5°$ with respect to the beams. In these regions the track momentum is required to match the particle energy, as measured in the calorimeter, within 4σ.

vi) A distinctive feature of the UA2 detector is the presence of a 'preshower' counter (see Section 6) in front of the calorimeters. This counter is used to verify that the electromagnetic shower is initiated in the converter and is also associated with the measured charged-particle track, as expected for electrons. This is done in practice by requiring that the observed signal exceeds a threshold corresponding to several minimum-ionizing particles, and, furthermore, that its position in space matches the track impact point within the space resolution of the counter itself (a few millimetres).

vii) Finally, an explicit isolation criterion is applied. In UA1 this is done by requiring only a limited amount of transverse energy (typically less than 3 GeV) associated with charged particles and calorimeter cells contained in a cone of $\approx 40°$ half-angle around the electron track. In UA2 this cone has a half-angle of typically $\approx 15°$.

The combination of all these cuts is estimated to be $\approx 75\%$ efficient for isolated electrons in both experiments.

Figure 8 shows, for UA1 [44] and UA2 [45], examples of distributions of events containing at least one electron candidate in the (p_T^e, p_T^ν) plane, where p_T^e is the measured electron transverse momentum, and $p_T^\nu \equiv p_T^{miss}$. In the high-$p_T^e$ region ($p_T^e \gtrsim 25$ GeV/c), the signal from $W \rightarrow e\nu$ ($p_T^\nu \approx p_T^e$) is clearly visible above the background of misidentified hadrons, which is dominant at low p_T^e. The events with high-p_T^e values and small p_T^ν result mostly from $Z \rightarrow e^+ e^-$ decays, and will be discussed in subsection 8.10.

Examples of p_T^ν distributions are shown in Fig. 9. They demonstrate clearly that the rejection power against background of a cut on p_T^ν is much larger in UA1 than in UA2. The UA1 data (Fig. 9a) show two well-separated classes of events: those with $p_T^\nu > 15$ GeV/c, which show the characteristic Jacobian structure expected from $W \rightarrow e\nu$ decay; and those with $p_T^\nu < 15$ GeV/c, which are mostly misidentified hadronic final states. In the latter class the non-zero value of p_T^ν is the effect of the p_T^{miss} resolution. The simple requirement $p_T^\nu > 15$ GeV/c is sufficient, therefore, to select an almost pure sample of $W \rightarrow e\nu$ decays. A typical p_T^e distribution for events satisfying such a requirement is shown in Fig. 10a for a partial event sample, together with the estimated background and signal contributions.

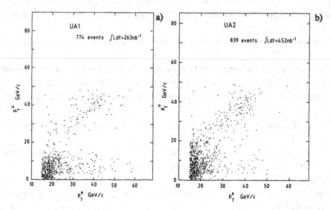

Fig. 8   Examples of distributions of the events containing at least one electron candidate in the p_T^e, p_T^ν plane: a) UA1 data; b) UA2 data.

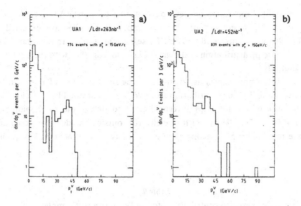

Fig. 9 Examples of p_T^ν distribution for events containing at least one electron candidate with $p_T^e >$ 15 GeV/c: a) UA1 data; b) UA2 data.

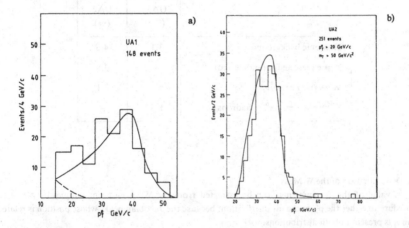

Fig. 10 a)Example of a p_T^e distribution for electron candidates in events having $p_T^\nu >$ 15 GeV/c (UA1); b) The p_T^e distribution for $W \rightarrow e\nu$ candidates satisfying the UA2 selection criteria (see text). Broken curve: estimated hadronic background. Full curves: expected distributions for values of m_W as given by Eqs. (8.2a) and (8.2b).

The separation between the two classes of events is much less clear in the UA2 data (Fig. 9b), as a result of the non-Gaussian tails of the p_T^{miss} resolution in the UA2 experiment. In the UA2 data, a cut on p_T^ν is not sufficient to reduce the hadronic background to a tolerable level. For this reason, the W sample is restricted to the events containing an electron candidate of $p_T^e > 20$ GeV/c. For these events, a transverse mass m_T, with the property $m_T \leqslant m_W$, is evaluated using the formula

$$m_T^2 = 2p_T^e p_T^\nu (1 - \cos \Delta\phi) \quad , \tag{8.1}$$

where $\Delta\phi$ is the azimuthal separation between $\vec{p}_T^e$ and $\vec{p}_T^\nu$. Of all the W $\rightarrow$ eν decays for which the electron falls within the acceptance of the UA2 apparatus, 80% are expected to satisfy the requirements $p_T^e > 20$ GeV/c, $m_T > 50$ GeV/c^2. These criteria, applied to the total UA2 event sample, select 251 events, whose p_T^e distribution is shown in Fig. 10b, together with the estimated background and signal contributions.

Background contributions, given as fractions of the total number of events in the sample, are listed in Table 2. We note that the contribution from W $\rightarrow$ $\tau\nu_\tau$, followed by $\tau \rightarrow$ e$\nu_e\nu_\tau$, is larger in UA1 than in UA2, because of the lower p_T^e threshold used to define the final sample. However, the contribution from Z $\rightarrow$ e$^+$e$^-$ decays with one electron outside the detector acceptance is larger in UA2, because the probability of detecting both electrons is only $\approx$ 60% in UA2, whilst it is nearly 100% in UA1.

Table 2

Background contributions to the W $\rightarrow$ eν event samples
(in percent of the number of events in the sample)

	UA1 (%)	UA2 (%)
Hadronic background	3.1	4.6
Z $\rightarrow$ e (detected) e (undetected)	–	3.8
W $\rightarrow$ $\tau\nu_\tau$, $\tau \rightarrow$ e$\nu_e\nu_\tau$	5.3	2.1
W $\rightarrow$ $\tau\nu_\tau$, $\tau \rightarrow$ $\nu_\tau\pi^0$ + hadrons	1.6	–

8.3 Determination of the W Mass

A value of the W mass, m_W, can be extracted from the W $\rightarrow$ eν event samples by a best-fit procedure to either the p_T^e or the m_T distribution, because the Jacobian peak, whose position is related to m_W, is present in both distributions.

In practice, although p_T^e is measured, in general, more precisely than m_T, the p_T^e distribution is sensitive to the W transverse momentum, p_T^W, whose distribution has large theoretical uncertainties [30], which cannot be reduced using the present experimental data. For this reason, both experiments use the m_T distribution, which is rather insensitive to the W transverse motion.

In the UA1 experiment [44], a sample of 149 events, for which both p_T^e and p_T^ν exceed 30 GeV/c, is selected from the total event sample, which corresponds to a total integrated luminosity of L = 706 nb^{-1}. The m_T distribution for this sample, which is background-free, is shown in Fig. 11a.

In UA2, the m_T distribution of the events containing an electron candidate of $p_T^e > 20$ GeV/c is shown in Fig. 11b.

A Monte Carlo simulation is used to generate m_T distributions for different values of m_W, and the most probable value of m_W is found by a maximum-likelihood fit to the experimental distributions. The large backgrounds in the UA2 sample (see Fig. 11b) have a negligible effect on the mass

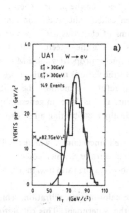

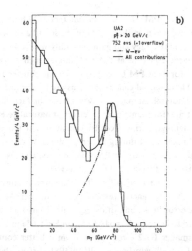

Fig. 11 Electron–neutrino transverse mass distribution for the UA1 (a) and UA2 (b) event samples. The full curves are expectations for values of m_W as given by Eqs. (8.2a) and (8.2b). The broken curve in Fig. 11b represents the $W \rightarrow e\nu$ contribution to the UA2 event sample.

determination, because they are predominantly at small m_T, and the best-fit value of m_W depends mainly on the upper edge of the distribution.

The results of the fit are

$$\text{UA1 [44]:} \quad m_W = 82.7 \pm 1.0 \, (\text{stat.}) \pm 2.7 \, (\text{syst.}) \, \text{GeV}/c^2; \tag{8.2a}$$

$$\text{UA2 [45]:} \quad m_W = 80.2 \pm 0.8 \, (\text{stat.}) \pm 1.3 \, (\text{syst.}) \, \text{GeV}/c^2. \tag{8.2b}$$

The best-fit curves, which correspond to these values, are superimposed on the experimental m_T distributions in Fig. 11.

In both experiments, the fit is performed using the Standard Model value for the W width, $\Gamma_W \approx 2.6 \, \text{GeV}/c^2$ (see Section 4). By using Γ_W as a second free parameter, it is possible to obtain an upper limit on Γ_W. The results from the two experiments are $\Gamma_W < 5.4 \, \text{GeV}/c^2$ (UA1 [44]) and $< 7 \, \text{GeV}/c^2$ (UA2 [45]), at the 90% confidence level (CL).

The systematic error in Eqs. (8.2a) and (8.2b) reflects the uncertainty on the absolute energy scale of the calorimeter, and it will be discussed in detail in the next subsection. It is quoted separately because it cancels in the ratio m_W/m_Z.

8.4 Mass Scale Uncertainty

The m_W measurement discussed in the previous subsection and the measurement of the Z mass, m_Z, from the decay $Z \rightarrow e^+ e^-$ (see subsection 8.11), involve energy measurements in the calorimeter.

It is possible, at least in principle, to calculate the energy lost by a particle of a given energy in the calorimeter scintillator. However, what is finally measured in both UA1 and UA2 is the total electric charge collected at the anodes of several photomultipliers (PMs), which results from the conversion of energy loss into scintillation light, followed by photoelectric effect and amplification by secondary emission in the PM. The conversion factors which describe these two steps cannot be predicted, because they depend on the quality of the scintillator and of the optical coupling to the PM, and on the PM quantum efficiency and gain. The only way to know the ratio between the particle energy and the electric charge collected at the PM anode (such a ratio is often called calibration constant), is to measure it for each PM by exposing the calorimeter to test beams of electrons and pions of known energies. These special calibration runs are performed before assembling the detector at the Collider.

An additional complication arises from the fact that the calorimeter properties and PM gains may vary with time (for example, as an effect of natural deterioration of the various materials or because of radiation damage during Collider operation). It is necessary, therefore, to monitor the calibration constant for each PM during the entire duration of the experiment. This is done, in general, by using artificial light flashes and radioactive sources, and also by periodically recalibrating a few calorimeter modules using test beams of known energies.

There are two kinds of calibration uncertainties: a cell-to-cell uncertainty, which results in a deterioration of the energy resolution; and an overall energy scale uncertainty, which corresponds to a systematic error in the unit of energy (this is why such an error cancels in the ratio m_W/m_Z). The latter is estimated to be $\pm 3.2\%$ in UA1 [44] and $\pm 1.6\%$ in UA2 [45].

The smaller UA2 error results from a better control of the calorimeter calibration, which was performed for all calorimeter cells in 1981, before the start of the experiment. This calibration has then been repeated periodically for a fraction of the calorimeter modules. A recalibration of 39 calorimeter cells performed in July 1986 is consistent with the quoted uncertainty. On the contrary, in the UA1 experiment only a few calorimeter modules were initially calibrated, and no recalibration has been made because the large size of the calorimeter modules makes them cumbersome and practically impossible to transport to a test-beam area.

It should be noted that a contribution of $\approx 0.7\%$ to the quoted UA2 energy scale uncertainty results from the uncertainty on the energy of the test beams used to calibrate the calorimeters.

The calorimeter calibration represents a less severe problem in experiments with e^+e^- machines, where elastic scattering and two-jet production via the reaction $e^+e^- \rightarrow q\bar{q}$ provide calibration standards *in situ*. No known source of mono-energetic particles or jets, on the contrary, is available at hadron colliders.

8.5 Charge Asymmetry in the Decay W → eν

At the energies of the CERN $p\bar{p}$ Collider, W production is dominated by $q\bar{q}$ annihilation involving at least one valence quark or antiquark. As a consequence of the V − A coupling, the helicity of the quarks (antiquarks) is −1 (+1) and the W is almost fully polarized along the $\bar{p}$ beam. At higher energies (e.g. at the Fermilab Tevatron Collider) the contribution from $q\bar{q}$ annihilation with both partons belonging to the sea is important, and the W polarization is greatly reduced.

Similar helicity arguments applied to W → eν decay predict that the leptons (e^- or ν_e) should be preferentially emitted opposite to the direction of the W polarization, and antileptons (e^+ or $\bar{\nu}_e$) along it. More precisely, the angular distribution of the charged lepton in the W rest frame has the form $dn/d(\cos \theta^*) \propto (1 + \cos \theta^*)^2$, where θ^* is the e^+ (e^-) angle in the W rest frame, measured with respect to a direction parallel (antiparallel) to the W polarization. This axis coincides with the direction of the incident $\bar{p}$ (p) beam only if the W transverse momentum p_T^W is zero. For $p_T^W \neq 0$ the initial parton directions are not known, and the Collins–Soper convention [46] is used to define θ^*.

A further complication arises from the fact that p_L^ν is not measured, and the condition that the invariant mass of the eν pair be equal to m_W gives two solutions for p_L^ν or p_L^W. The UA1 analysis [44]

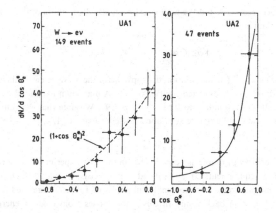

Fig. 12 The UA1 and UA2 distributions of the product $q \cos \theta_e^*$ for electrons ($q = -1$) and positrons ($q = +1$) from $W \rightarrow e\nu$ decay. The angle θ_e^* is measured with respect to the incident $\bar{p}$ beam.

retains only those events for which one solution is unphysical and the charge sign is unambiguously determined (149 events). In UA2 [45], the solution corresponding to the smaller value of $|p_L^W|$ is chosen.

Figure 12a shows the $\cos \theta^*$ distribution from UA1, corrected for the detector acceptance.

In the UA2 experiment [45], 47 events containing an electron candidate with $p_T^e > 20$ GeV/c and $m_T > 40$ GeV/c^2 are detected in the forward regions, where the magnetic spectrometers allow a measurement of the electric charge. The background to this sample from misidentified hadrons is estimated to be 4.2 ± 0.3 events. The background-subtracted $\cos \theta^*$ distribution is shown in Fig. 12b, where it is compared with the expected $(1 + \cos \theta^*)^2$ form, distorted by the effect of the detector acceptance.

Both distributions of Fig. 12 agree well with the expected $(1 + \cos \theta^*)^2$ form. It must be noted, however, that this result cannot distinguish between $V - A$ and $V + A$. In the latter case, all helicities change sign and the angular distribution remains the same. In order to separate the two alternatives, a direct measurement of the lepton helicity would be needed.

It has been shown [47] that, for a particle of arbitrary spin J, one expects

$$\langle \cos \theta^* \rangle = \langle \lambda \rangle \langle \mu \rangle / J(J + 1),$$

where $\langle \mu \rangle$ and $\langle \lambda \rangle$ are, respectively, the global helicity of the production system (in our case, $u\bar{d}$ or $d\bar{u}$) and of the decay system ($e^-\bar{\nu}_e$ or $\nu_e e^+$). For $V - A$ one has $\langle \mu \rangle = \langle \lambda \rangle = -1$, which, together with the assignment $J = 1$, leads to a maximal value $\langle \cos \theta^* \rangle = 0.5$. The UA1 result [44] is

$$\langle \cos \theta^* \rangle = 0.43 \pm 0.07,$$

in agreement with the maximal helicity states at production and decay, and with the assignment $J = 1$. For $J = 0$ one obviously expects $\langle \cos \theta^* \rangle = 0$; for any other spin value $J \geqslant 2$, $|\langle \cos \theta^* \rangle| \leqslant 1/6$. Hence the UA1 result demonstrates that the W boson has indeed spin 1.

If x is the ratio between the A and V couplings, the angular distribution of the e^- (e^+) with respect to incident p ($\bar{p}$) direction has the form

$$dn/d\cos\theta^* \propto (1 + \cos\theta^*)^2 - 2\alpha\cos\theta^*,$$

where $\alpha = [(1 - x^2)/(1 + x^2)]^2$. The distribution $d^2n/dp_T^e d\theta_e$ for the 47 W $\rightarrow$ eν events observed in UA2 [45] is consistent with $\alpha = 0$, as expected for $|V/A| = 1$. A maximum likelihood fit gives $\alpha < 0.35$ (90% confidence level), corresponding to $0.51 < |x| < 1.97$. We note that the value of α does not provide any information about the sign of x, or about the choice of x or 1/x.

8.6 The Decay W $\rightarrow$ $\mu\nu$

The observation of high-p_T muons is possible only in the UA1 experiment, since there is no muon detector in UA2. The requirements of an isolated track consistent with a high-p_T muon ($p_T^\mu >$ 15 GeV/c) and of a missing transverse momentum in excess of 15 GeV/c are used to select W $\rightarrow$ $\mu\nu$ candidates [44, 48]. The main background in this search arises from medium-energy K$^\pm$ mesons, which decay into $\mu\nu$ at angles such that the decay kink on the track occurs in a direction opposite to the magnetic deflection. In these cases the track curvature appears smaller than it is in reality, thus simulating a high-momentum muon. To reject this background, which is absent in the W $\rightarrow$ eν channel, a careful visual inspection of the track is performed in order to identify the K $\rightarrow$ $\mu\nu$ decay kink, using a high-resolution graphics display.

A sample of 57 W $\rightarrow$ $\mu\nu$ events is selected from the full data sample, with negligible background contamination [44]. Since the momentum measurement error is Gaussian in 1/p, and not in p, the distribution of the variable $1/m_T$, rather than m_T, is more convenient to display the Jacobian peak structure expected from W $\rightarrow$ $\mu\nu$ decay (see Fig. 13). We recall that in UA1 the momentum resolution for a 40 GeV/c muon is $\approx \pm 20\%$, whereas it is $\approx \pm 2.5\%$ for a 40 GeV/c electron as measured in the calorimeter.

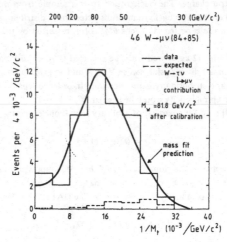

Fig. 13 Inverse transverse mass distribution for muon–neutrino pairs in UA1. The variable $1/m_T$ is chosen instead of m_T, because the error resulting from momentum resolution is Gaussian in $1/m_T$ and not in m_T.

It must also be noted that the number of events in the W $\rightarrow \mu\nu$ sample is $\approx \frac{1}{3}$ of that of the W $\rightarrow$ eν sample. This is due to the incomplete angular coverage of the muon trigger, which was not operational over the whole solid angle during data taking.

A fit to the distribution of Fig. 13, using the expected shape of the $1/m_T$ distribution, distorted by the effects of momentum resolution, provides an independent determination of m_W. The result is [44]

$$m_W = 81.8 \, ^{+6.0}_{-5.3} \, (\text{stat.}) \pm 2.6 \, (\text{syst.}) \, \text{GeV/c}^2,$$

where the systematic error results from uncertainties in the corrections for track distortions in the central detector. This result agrees with the determination from the W $\rightarrow$ eν channel [Eq. (8.2)], but is affected by a much larger statistical error.

8.7 The Decay W $\rightarrow \tau\nu_\tau$

The decay chain W $\rightarrow \tau\nu_\tau$, $\tau \rightarrow \nu_\tau$ + hadrons is expected to result in purely hadronic events with large p_T^{miss}. Furthermore, since the τ energy is generally much larger than the τ mass, hadrons from τ decay will appear as a highly collimated jet with low charged-particle multiplicity (we remind the reader [49] that the decay $\tau \rightarrow \nu$ + hadrons represents $\approx$ 64% of all τ decays, and of these decays, the fraction with five or more charged particles amounts to less than 1%).

A search for purely hadronic events with large values of p_T^{miss} has been performed in the UA1 experiment [50] from a data sample corresponding to a total integrated luminosity of 715 nb^{-1}. The main selection criteria are:

i) $p_T^{\text{miss}} > 15 \, \text{GeV/c}$;

ii) $p_T^{\text{miss}}/0.7(\Sigma_i E_T^i)^{1/2} > 4$ (as discussed in Section 7, this requirement ensures that the large p_T^{miss} value is not an effect of the p_T^{miss} resolution);

iii) the presence of at least one jet with transverse energy $E_T > 12 \, \text{GeV}$;

iv) absence of muons or electrons in the event.

A total of 57 events satisfy these requirements. In this sample, W $\rightarrow \tau$ events are identified using the following three variables:

i) the fraction F of the jet energy contained in a cone with half-aperture $R = (\Delta\phi^2 + \Delta\eta^2)^{1/2} = 0.4$, where $\Delta\phi$ ($\Delta\eta$) is the azimuthal (pseudorapidity) separation with respect to the jet axis (in UA1, jets are defined by a cone with $R = 1$ and thus F is a measure of the jet collimation);

ii) the separation $r = (\Delta\phi^2 + \Delta\eta^2)^{1/2}$ between the charged-particle track with the highest momentum and the jet axis;

iii) the charged-particle multiplicity, ν_{ch}, in a cone with $R = 0.4$ around the jet axis.

For each event, a τ likelihood, L_τ, is defined as

$$L_\tau = \ln[(dn/dF)(dn/dr)(dn/d\nu_{\text{ch}})],$$

where dn/dF, dn/dr, and $dn/d\nu_{\text{ch}}$ are the expected distributions for W $\rightarrow \tau$ decay, as obtained by Monte Carlo simulations, calculated at the measured values of the three variables F, r, and ν_{ch}.

The L_τ distribution for the 56 events is shown in Fig. 14, where it is compared with the expected distribution from W $\rightarrow \tau$ decay (normalized to the expected number of W $\rightarrow \tau$ events), and also with the distribution measured for ordinary jets from a sample of two-jet events. The cut $L_\tau > 0$, which is satisfied by 78% of all hadronic τ decays, and which rejects 89% of the ordinary jets, selects a sample of 32 events. Figure 15 shows the m_T distribution for these events, together with the expected distribution for a W mass of 82.7 GeV/c^2 (in this case m_T is evaluated from Eq. (8.1) using $\vec{p}_T^{\text{miss}}$ and

208

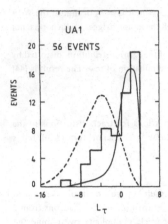

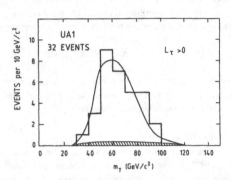

Fig. 14 Distribution of the τ-likelihood for the UA1 data (56 events, solid histogram). The solid curve is a Monte Carlo expectation for W $\rightarrow$ $\tau\nu$ $\rightarrow$ $\nu\bar{\nu}$ + hadrons, absolutely normalized. The broken curve is the τ-likelihood, as measured for hadronic jets in UA1, normalized to 56 events.

Fig. 15 Transverse mass distribution for events passing the τ selection criteria in UA1 (32 events, solid histogram). The solid curve is a Monte Carlo prediction for W $\rightarrow$ $\tau\nu$ $\rightarrow$ $\nu\bar{\nu}$ + hadrons, normalized to the expected number of 28.7 events. The shaded area represents the expected hadronic background (2.7 events).

$\bar{p}_T^{jet}$ in the place of $\bar{p}_T^e$ and $\bar{p}_T^\mu$). A fit to this distribution gives m_W = 89 ± 3 (stat.) ± 6 (syst.) GeV/c^2, in agreement with the measurements from W $\rightarrow$ eν and W $\rightarrow$ $\mu\nu$ decay.

8.8 W Production Cross-Section and e-μ-τ Universality

The cross-sections for inclusive W production, followed by the decay W $\rightarrow$ $\ell\nu$, σ_W^ℓ (where ℓ ≡ e, μ or τ), are computed in a straightforward way from the number of events in the samples, after subtracting the various background contributions and correcting for the geometrical acceptance of the detector, the efficiency of the selection criteria, and the machine luminosity.

Experimental values of σ_W^ℓ, as measured by UA1 [44] and UA2 [51], are listed separately in Table 3 for $\sqrt{s}$ = 546 and 630 GeV.

The quoted systematic uncertainties arise mainly from the uncertainties on the total luminosity (± 15% in UA1, and ± 8% in UA2 which benefits from the measurement of the total cross-section by UA4 [52] in the same interaction).

From the ratios of σ_W^μ/σ_W^e and σ_W^τ/σ_W^e, which are not affected by uncertainties on the machine luminosity, the UA1 Collaboration [44] obtains the following values for the ratios of weak coupling constants:

$$g_\mu/g_e = 1.01 \pm 0.07 \text{ (stat.)} \pm 0.04 \text{ (syst.)},$$
$$g_\tau/g_e = 1.01 \pm 0.11 \text{ (stat.)} \pm 0.06 \text{ (syst.)},$$

where the constants g describe the strength of the $(\ell\nu)$ current coupling to the W. These results provide direct evidence for e–μ–τ universality at $Q^2 = m_W^2$.

Table 3
W cross-sections (nb)

		$\sqrt{s} = 546$ GeV	$\sqrt{s} = 630$ GeV
	σ_W^e	0.55 ± 0.08 (stat.) ± 0.09 (syst.)	0.63 ± 0.04 (stat.) ± 0.10 (syst.)
UA1 [44]	σ_W^μ	0.56 ± 0.18 (stat.) ± 0.12 (syst.)	0.63 ± 0.08 (stat.) ± 0.11 (syst.)
	σ_W^τ		0.63 ± 0.12 (stat.) ± 0.11 (syst.)
UA2 [51]	σ_W^e	0.61 ± 0.10 (stat.) ± 0.07 (syst.)	0.57 ± 0.04 (stat.) ± 0.07 (syst.)
Theory [39]	σ_W^ℓ	$0.46^{+0.14}_{-0.07} \pm 0.06$	$0.57^{+0.18}_{-0.10} \pm 0.07$

8.9 Longitudinal Momentum of the W

The W longitudinal momentum distribution is expected to reflect the structure functions of the incident annihilating partons, which are predominantly u- and d- quarks (see Section 5). If we neglect the W transverse momentum, the W longitudinal momentum, p_L^W, measured with respect to the incident proton direction, is equal to $\sqrt{s}(x_q - x_{\bar{q}})/2$, where x_q and $x_{\bar{q}}$ are the fractional momenta of the two annihilating partons. Unfortunately, the measurement of p_L^W requires the simultaneous measurement of both the charged lepton and neutrino longitudinal momentum, and the latter, as explained in Section 7, is not measurable. It can be calculated, however, from the constraint

$$m_W^2 = (E_\ell + E_\nu)^2 - (\vec{p}_\ell + \vec{p}_\nu)^2,$$

where E_ℓ and $\vec{p}_\ell$ are the energy and momentum of the charged lepton, respectively. This equation gives two solutions for the neutrino longitudinal momentum, and hence for p_L^W. However, in a large fraction of the events (of the order of ~ 75% in the UA1 experiment [44]), only one solution is possible, because the other one is either unphysical or inconsistent with the total energy measured in the event. In the ambiguous cases, the solution corresponding to the smaller value of $|p_L^W|$ is chosen.

Figure 16 shows the distribution of $|x_W| = 2|p_L^W|/\sqrt{s}$, as measured in UA1 [44] (a) and UA2 [51] (b) for $W \to e\nu$ events. The distributions are not very sensitive to $\sqrt{s}$ (see Fig. 16a), and agree well with the theoretical expectation, based on the structure function parametrization of Ref. [53].

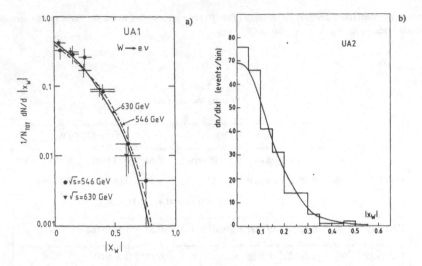

Fig. 16 Distribution of the variable $|x_W| = 2|p_L^W|/\sqrt{s}$, as measured in UA1 (a) and UA2 (b). The curves represent theoretical predictions, based on the structure function parametrization of Ref. [53].

8.10 Transverse Momentum of the W

The W transverse momentum $\vec{p}_T^W$ is equal and opposite to the total transverse momentum carried by all hadrons produced in association with the W itself. Figure 17 shows the p_T^W distribution, as measured by UA1 [54] and UA2 [51] for a sample of 323 W $\rightarrow$ eν or W $\rightarrow$ $\mu\nu$ events. A QCD prediction [30], also shown in Fig. 17, agrees with the data over the full p_T^W range. The W bosons produced with high p_T^W are expected to recoil against one or more jets, as a result of higher-order QCD effects, and such jets are indeed observed experimentally.

As shown in Fig. 17, the two UA1 events with the highest p_T values correspond to a rate in reasonable agreement with QCD expectations. However, both events contain a high-mass two-jet system recoiling against the W. Such a configuration is rather unlikely, as illustrated in Fig. 18, where m_{jj}, the two-jet invariant mass, is plotted against m_{Wjj}, the invariant mass of the W + two jets system, for all W + two jets events of the UA1 sample (only jets with $E_T > 7$ GeV are considered for this plot). A comparison with a QCD Monte Carlo calculation [55], which includes terms of order α_s^2, represented by equal-probability contours in Fig. 18, shows that the two events (designated as A and B in Fig. 18) lie in a region where the probability is only a few per cent.

A list of measured parameters for the two events is given in Table 4. Within measurement errors, the m_{jj} values for the two events is consistent with either m_W or m_Z, and also m_{Wjj} is consistent with a unique value around 290 GeV/c^2. It is tempting, therefore, to interpret these events in terms of WW or WZ production. However, the production of pairs of gauge bosons by standard mechanisms is much too low to be observable in the present event sample.

An alternative mechanism, first suggested by Denegri [56], considers QCD production of a heavy quark-antiquark pair $Q\bar{Q}$, with $m_Q > m_W$. In such a case the decay $Q \rightarrow q + W$, where q is a lighter quark, would result in a final state containing two W bosons. However, as recently discussed [57], also

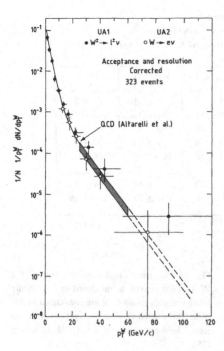

Fig. 17 Distribution of p_T^W for both $W \to e\nu$ and $W \to \mu\nu$ events in UA1, and for $W \to e\nu$ events in UA2. The curve represents a QCD prediction [30], and the shaded band shows the theoretical uncertainty in the region of high p_T^W.

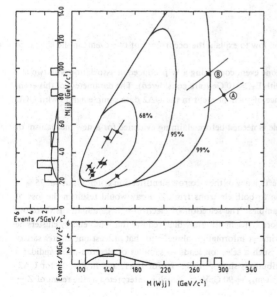

Fig. 18 Correlation between the W-jet–jet invariant mass and the two-jet invariant mass for 10 W-jet–jet events in UA1 (3 $W \to \mu\nu$ and 7 $W \to e\nu$ decays). Only jets having $E_T > 7$ GeV are considered in this plot. The contours are second-order QCD predictions [55], distorted by the jet energy resolution, containing the event fraction as indicated. Events A and B are emphasized in the plot.

Table 4

Parameters of UA1 events A (W $\to \mu\nu$)

and B (W $\to e\nu$), and of the UA2 event (W $\to e\nu$)

		A (UA1)	B (UA1)	UA2
p_T	Charged lepton	22 ± 3	19 ± 1	35 ± 3
(GeV/c)	Neutrino	85 ± 11	99 ± 14	55 ± 6
	Jet 1	72 ± 11	90 ± 13	38 ± 5
	Jet 2	20 ± 4	17 ± 4	24 ± 3
	W boson	82 ± 12	105 ± 14	36 ± 5
m	Jet 1–jet 2	82 ± 10	97 ± 12	54 ± 10
(GeV/c^2)	W–jet 1–jet 2	299 ± 28	279 ± 28	151 ± 16

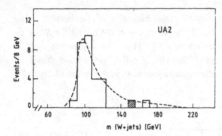

Fig. 19 Invariant mass distribution of the W + jet(s) system, as measured by UA2. Only jets having $E_T > 10$ GeV are considered in this plot. The shaded event is the only one with two jets. The broken curve is a QCD prediction [55].

in this case the expected event rate is too low to explain the occurrence of two events in the UA1 data sample.

In the UA2 data sample [51], only one event containing a W produced in association with two jets is observed (in this analysis, only jets with $E_T > 10$ GeV are considered). The parameters of this event are also listed in Table 4. The occurrence of such an event in the UA2 data sample agrees with QCD expectations (see Fig. 19).

Clearly, a much larger event sample is needed before claiming evidence for a new mechanism of boson-pair production.

8.11 The Z $\to e^+e^-$ Event Samples

As observed in subsection 8.1, the efficiency of the electron identification criteria is about 75% in both experiments, and their application to both electrons from Z decay would result in the loss of about 50% of the Z $\to e^+e^-$ event samples. The selection of electron-pair candidates is therefore performed by using less selective but more efficient criteria which require that both energy clusters be compatible with an electron from calorimeter information alone, and that at least one cluster satisfy the full electron identification criteria. Such a selection leads to samples of electron-pair candidates whose invariant mass m_{ee} has the distributions shown in Fig. 20 for UA1, and in Fig. 21 for UA2. Both distributions show a clear peak near $m_{ee} \approx 90$ GeV/c^2, which is interpreted as the result of Z $\to$

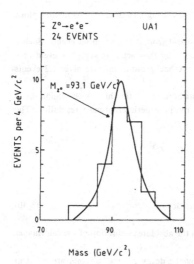

Fig. 20 Invariant mass distribution for e^+e^- pairs in the UA1 data sample, shown above a mass value of 70 GeV. The curve represents the best fit to the data [see Eq. (8.3a)].

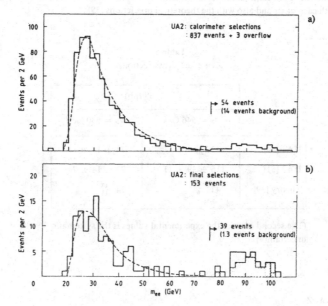

Fig. 21 Invariant mass distribution for e^+e^- pairs in the UA2 data sample: a) only calorimetric selections are applied; b) at least one electron candidate is required to satisfy all electron identification criteria. The broken lines show the estimated hadronic background. The hatched region shows the m_{ee} distribution of the 25 events used in the evaluation of the Z mass [see Eq. (8.3b)].

e^+e^- decay. Such a peak is visible in the UA2 sample even using calorimeter information alone (Fig. 21a).

The low-mass continuum in Fig. 21b is mostly due to background from two-jet events. Such a background is negligible under the Z peak, as demonstrated by the fact that very few events are observed between the low-mass continuum and the peak. A background estimate gives 1.3 events under the Z peak for the distribution of Fig. 21b.

A value of m_Z can be obtained from the m_{ee} distributions in the high-mass region by a maximum-likelihood fit of a Breit-Wigner shape distorted by the experimental mass resolution. The UA1 result is [44]

$$m_Z = 93.1 \pm 1.0 \,(\text{stat.}) \pm 3.1 \,(\text{syst.}) \,\text{GeV/c}^2. \tag{8.3a}$$

The corresponding UA2 result is [45]

$$m_Z = 91.5 \pm 1.2 \,(\text{stat.}) \pm 1.7 \,(\text{syst.}) \,\text{GeV/c}^2. \tag{8.3b}$$

In both cases the systematic error reflects the uncertainty in the absolute calibration of the calorimeter energy scale (see subsection 8.4).

Cross-sections for inclusive Z production followed by the decay $Z \to e^+e^-$, σ_Z^e, are listed in Table 5. Within the rather large statistical and systematic errors, the results of the two experiments are consistent with each other and also with the theoretical predictions [39].

Table 5

$Z \to e^+e^-$ cross-sections

	σ_Z^e (pb)	
	$\sqrt{s} = 546$ GeV	$\sqrt{s} = 630$ GeV
UA1 [44]	$42^{+33}_{-20} \pm 6$	$74 \pm 14 \pm 11$
UA2 [51]	$116 \pm 39 \pm 11$	$73 \pm 14 \pm 7$
Theory [39]	43^{+13}_{-6}	54^{+17}_{-11}

(The second error in the experimental values is the systematic uncertainty)

8.12 The Decay $Z \to \mu^+\mu^-$

By requiring the simultaneous presence of two muons, each having $p_T^\mu > 15$ GeV/c and such that their invariant mass $m_{\mu\mu}$ exceeds 40 GeV/c^2, a total of 19 events are selected in the UA1 experiment [44]. The $m_{\mu\mu}$ distribution of these events (Fig. 22) clearly shows the peak expected from $Z \to \mu^+\mu^-$ decay. Such a peak is much broader than that measured in the e^+e^- decay channel, because of the

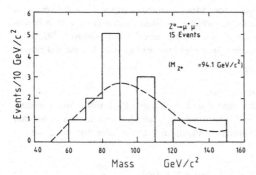

Fig. 22 Invariant mass distribution for $\mu^+\mu^-$ pairs in UA1, before momentum balance. Four events having $m_{\mu\mu} > 160$ GeV are not shown. The curve represents the best fit to the data.

poor momentum resolution in UA1 (see subsection 8.1). Since the resolution is Gaussian in $1/p$, and not in p, the measured $m_{\mu\mu}$ distribution from $Z \to \mu^+\mu^-$ decay has an asymmetric shape, with a tail extending to large mass values (four of the 19 events are not shown in Fig. 22 because the measured $m_{\mu\mu}$ value exceeds 160 GeV/c^2).

A fit to the peak of Fig. 22 gives

$$m_Z = 94.1 \, {}^{+8.4}_{-6.6} \,(\text{stat.}) \pm 2.8 \,(\text{syst.}) \, \text{GeV/}c^2,$$

where the systematic error reflects uncertainties mostly associated with distortions in the tracking detector.

A more precise result can be obtained by requiring that the dimuon transverse momentum be equal and opposite to the total transverse momentum carried by all other particles in the event (this method assumes that there is no real neutrino or other undetected particle in the event). In this case the fit gives [44]

$$m_Z = 90.7 \, {}^{+5.2}_{-4.8} \,(\text{stat.}) \pm 3.2 \,(\text{syst.}) \, \text{GeV/}c^2.$$

The cross-section for inclusive Z production, followed by the decay $Z \to \mu^+\mu^-$, is measured to be [44]

$$\sigma_Z^\mu = 67 \pm 17 \,(\text{stat.}) \pm 11 \,(\text{syst.}) \, \text{pb}$$

at $\sqrt{s} = 630$ GeV. A comparison of this value with the measured value of σ_Z^e (see Table 5) provides a test of μ-e universality in Z decay:

$$g_\mu/g_e = 1.03 \pm 0.03 \,(\text{syst.}),$$

where g_μ (g_e) is the Z coupling constant to the muon (electron) weak neutral current.

8.13 Charge Asymmetry in the Decay $Z \to e^+e^-$

Since the Z coupling to charged leptons is almost purely axial (see Section 4), the charge asymmetry in the decay angular distribution is expected to be small. Such an angular distribution should be completely symmetric in the case $\sin^2\theta_w = 0.25$ (pure axial coupling), and is thus sensitive to the value of $\sin^2\theta_w$.

Using 21 $Z \rightarrow e^+e^-$ and 12 $Z \rightarrow \mu^+\mu^-$ decays for which the lepton charge sign is well determined, the UA1 Collaboration finds an angular distribution which is symmetric within errors [44]. A best fit gives $\sin^2\theta_w = 0.24^{+0.03}_{-0.04}$, with a rather large statistical error. This method of measuring $\sin^2\theta_w$, which is totally independent of the values of m_W and m_Z, is at present limited by statistics, and will become interesting in the future.

8.14 The Z Width and the Number of Neutrino Species

Within the context of the Standard Model, the value of the Z width, Γ_Z, is related to the number of fermion doublets for which the decay $Z \rightarrow f\bar{f}$ is kinematically allowed. Under the assumption that for any additional fermion family only the neutrino is significantly less massive than $m_Z/2$, we can write

$$\Gamma_Z = \sum_\ell \Gamma_Z(\ell\bar{\ell}) + \sum_q \Gamma_Z(q\bar{q}) + N_\nu \Gamma_Z(\nu\bar{\nu}),$$

where $\ell \equiv e, \mu, \tau$, $q \equiv u, d, c, s, t, b$, and N_ν is the number of neutrino species.

A value of Γ_Z can be obtained, in principle, by the same maximum-likelihood fit as used to determine m_Z. However, the expected width for three fermion families ($\Gamma_Z \approx 2.5$ GeV/c^2) is comparable with the experimental mass resolution achieved in UA1 and UA2. Under these circumstances the determination of Γ_Z depends critically on a precise knowledge of both the measurement errors and the shape of the experimental resolution. In the present experiments, given also the small event samples available, this method does not provide meaningful results:

$$\Gamma_Z < 5.2 \text{ GeV/c}^2 \text{ (90\% CL)} \qquad \text{from UA1 [44]}$$

and

$$\Gamma_Z = 2.7 \pm 2.0 \text{ (stat.)} \pm 1.0 \text{ (syst.) GeV/c}^2;$$

$$\Gamma_Z < 5.6 \text{ GeV/c}^2 \text{ (90\% CL)} \qquad \text{from UA2 [45]}.$$

A model-dependent method, first proposed by Cabibbo [58], which does not depend on the mass resolution, consists in measuring the ratio $R = \sigma_W^\ell/\sigma_Z^\ell$ ($\ell \equiv$ e or μ), which is related to the ratio of total widths, Γ_Z/Γ_W, by the identity

$$R \equiv [\sigma_W \cdot BR(W \rightarrow e\nu)]/[\sigma_Z \cdot BR(Z \rightarrow ee)] = R'[\Gamma(W \rightarrow e\nu)/\Gamma(Z \rightarrow e^+e^-)](\Gamma_Z/\Gamma_W), \quad (8.4)$$

where $R' = \sigma_W/\sigma_Z$ is the ratio between the cross-sections for W and Z inclusive production already discussed in Section 4. The measured values of R are:

$$\text{UA1 [59]:} \quad R = 9.1^{+1.7}_{-1.2};$$

$$R < 11.5 \text{ (90\% CL)}; \qquad\qquad (8.5a)$$

$$R < 12.2 \text{ (95\% CL)};$$

$$\text{UA2 [45]:} \quad R = 7.2^{+1.7}_{-1.2};$$

$$R < 9.5 \text{ (90\% CL)}; \qquad\qquad (8.5b)$$

$$R < 10.4 \text{ (95\% CL)}.$$

In these results, the errors are dominated by statistics, because the value of the integrated luminosity cancels out. The UA1 result is obtained by combining both electron and muon data.

In Eq. (8.4) both R' and the ratio of the leptonic partial widths depend on $\sin^2\theta_w$, explicitly through the neutral-current couplings and implicitly via the W and Z masses. However, their product is constant to within 1% over the range $\sin^2\theta_w = 0.229 \pm 0.005$. The theoretical calculation of R' gives the most serious uncertainty. Calculations based on all published sets of structure functions give values of R' ranging between the two extreme values 3.08 and 3.50. Recent estimates give R' = 3.28 ± 0.15 [31]; 3.25 ± 0.20 [39]; 3.36 ± 0.09 [60]; and 3.41 ± 0.08 [61]. Choosing the value R ' = 3.28 ± 0.22, with a rather conservative error, we rewrite Eq. (8.4) in the numerical form

$$\Gamma_Z/\Gamma_W = (0.114 \pm 0.008)R, \qquad (8.6)$$

where the error reflects the theoretical uncertainty on R' as quoted above. Using R as in Eqs. (8.5a) and (8.5b) we evaluate

$$\text{UA1:} \quad \Gamma_Z/\Gamma_W = 1.04^{+0.19}_{-0.14}\,(\text{stat.}) \pm 0.07\,(\text{theor.}) \;;$$

$$\text{UA2:} \quad \Gamma_Z/\Gamma_W = 0.82^{+0.19}_{-0.14}\,(\text{stat.}) \pm 0.06\,(\text{theor.}).$$

These results can be statistically combined [62] to give

$$\Gamma_Z/\Gamma_W = 0.96^{+0.14}_{-0.10}\,(\text{stat.}) \pm 0.07\,(\text{theor.}), \qquad (8.7a)$$

$$\Gamma_Z/\Gamma_W < 1.20 \pm 0.08\,(\text{theor.})\,(95\%\ \text{CL}), \qquad (8.7b)$$

where the theoretical error reflects the uncertainty on R'.

The ratio Γ_Z/Γ_W is sensitive both to the number of neutrino types, N_ν, and to the mass of the top quark, m_t. The results are summarized in Fig. 23, which shows the expected variation of Γ_Z/Γ_W with

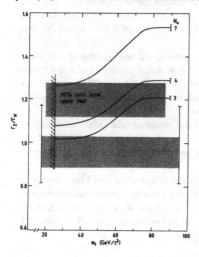

Fig. 23 The value of Γ_Z/Γ_W displayed with both statistical and theoretical uncertainties [see Eq. (8.7a)]. Also shown is the 95% CL upper limit with its theoretical uncertainty [Eq. (8.7b)]. The solid lines are the expected variations of Γ_Z/Γ_W with the top-quark mass, m_t, for 3, 4, and 7 neutrino types. The error bars at the ends of these curves represent the $\sin^2\theta_w$ uncertainty. The shaded region marks the lower limit on m_t ($m_t > 25$ GeV) as determined by e^+e^- experiments.

m_t if the number of light neutrino species is 3, 4, or 7. The error bars at the end of each line show the effect of the uncertainty on $\sin^2 \theta_w$. Also shown in Fig. 23 is the combined result on Γ_Z/Γ_W [Eq. (8.7a)], together with the 95% CL upper limit [Eq. (8.7b)]. The following conclusions can be drawn from Fig. 23:

  i) for $N_\nu = 3$, the data provide no bound for m_t;

  ii) $N_\nu \leqslant 6$ independent of m_t;

  iii) if $m_t \gtrsim 75$ GeV/c^2, $N_\nu \geqslant 4$ is excluded. Hence it is not possible to have at the same time a heavy t quark and more than three neutrino types.

Finally, we note that the method just discussed will become obsolete as soon as the forthcoming Z factories (SLC, LEP) begin operation, because Γ_Z will be directly measured with a precision of ± 30 MeV/c^2 at these machines [63]. At that moment, Eq. (8.6) will be used to determine Γ_W with a precision far superior to any other possible method.

8.15 The Decays W, Z → Two Jets

As we have seen in Section 4, the W and Z bosons are expected to decay into $q\bar{q}$ pairs with calculable rates. These decays can be observed, in general, as two-jet final states. Examples of these decays are W$^+$ → u$\bar{\text{d}}$ and c$\bar{\text{s}}$, and Z → u$\bar{\text{u}}$, d$\bar{\text{d}}$, c$\bar{\text{c}}$, s$\bar{\text{s}}$, and b$\bar{\text{b}}$. Decays involving the t quark, if kinematically allowed, are expected to result in more than two jets because of the heavy t-quark mass.

Under the assumption that the decays which are kinematically allowed involve only the fermions belonging to the three known families, the branching ratios into two jets are expected to be between 55% and 66% for the W, and between 62% and 69% for the Z, depending on m_t. However, in spite of such sizeable values, the observation of these decays is made difficult by the QCD background of two-jet final states from hard-parton collisions [43]. In the region of the W and Z masses, this background consists mostly of gluon jets. This does not help, however, because at present there is no known method for distinguishing a gluon jet from a quark jet on an event-by-event basis.

Obviously, the signal-to-noise ratio depends on the resolution of the two-jet mass, m_{jj}, because the QCD background has a smooth and rapidly falling distribution, whereas the two-jet invariant-mass distribution from the decays of the W and Z is peaked at the mass values of these particles. For a mass resolution $\sigma_m \approx 8$ GeV/c^2, which is relevant for the present UA1 and UA2 calorimeters, the signal-to-noise ratio is expected to be of the order of $1/100$. In the ideal case of a perfect mass resolution it is dominated by the natural width of the W and Z bosons, and becomes $\approx 1/10$.

The signal-to-noise ratio can be optimized by suitably choosing the rapidity interval over which the two jets are observed, because the angular distribution for W and Z decay into $q\bar{q}$ has the form $1 + \cos^2 \theta^*$, whilst for hard collisions it is strongly peaked at forward angles [43]. The signal-to-noise ratios quoted above correspond to the choice $|\eta| < 0.85$, which is well matched to the acceptance of the UA2 central calorimeter and is very close to the optimal value.

We note that the good agreement between the measured and predicted values of the cross-sections for W and Z production in p$\bar{\text{p}}$ collisions provides evidence for the validity of the Standard Model couplings of the intermediate bosons to $q\bar{q}$ pairs, which have been used to calculate the branching ratio values quoted above.

Results from a search for the decay of the W and Z particles into two jets have been reported by UA2 [64]. The data were collected at $\sqrt{s} = 630$ GeV and correspond to a total integrated luminosity of 730 nb^{-1}. The trigger for this search requires the presence of two jets at opposite azimuth in the central calorimeter, with a transverse energy threshold for each jet low enough (typically between 12.5

and 20 GeV) to ensure full efficiency in the W and Z mass regions. For lower masses, a measure of the trigger efficiency was obtained by comparing data taken at different thresholds.

The main concern of this search is to ensure that the jet energies are measured as well as possible. A source of measurement errors is the limited acceptance of the calorimeter, which covers the pseudorapidity interval $|\eta| < 1$ over the full azimuth. Jets close to the calorimeter edges, which might have been contained only partially, are rejected by requiring that the jet axis satisfy the condition $|\eta| < 0.85$.

To define the jet energy, a cone is constructed around the jet axis, and the energies deposited in the calorimeter cells within the cone are added together. We note that a large cone aperture ensures full jet containment, but adds to the jet the energies of the spectators which happen to fall within the cone. Hence for large cone apertures we expect, in general, E(cone) > E(jet). This inequality is particularly dangerous in the presence of a background which decreases rapidly as the energy of the jet increases, because low-energy jets, which are produced with large cross-sections, may be shifted to higher energies and therefore overwhelm the real high-energy jets which are produced with smaller cross-sections.

For small cone apertures we expect, in general, that the reverse inequality holds: E(cone) < E(jet). It is possible, therefore, to find a cone aperture which optimizes the jet energy resolution. If ω is the half-aperture of the cone, the best resolution is found over the range $0.2 < \cos \omega < 0.6$, and the value $\cos \omega = 0.6$, which corresponds to the narrowest cone, is used.

Other important criteria for ensuring good mass resolution are the following:

i) Jets with more than 4 GeV deposited in the region between two cones with $\cos \omega = 0.5$ and $\cos \omega = 0.7$ are rejected. This ensures that the jet is well contained within the cone.

ii) The hadronic shower associated with each jet must be well contained in depth. In the UA2 central calorimeter, each cell consists of three independent compartments, and a jet is rejected if the energy fraction deposited in the third compartment exceeds a given threshold (typically set at 40%).

iii) The total transverse momentum of the two-jet system must satisfy the condition $p_T < 24$ GeV/c. This condition rejects events with a large p_T imbalance, which might be due to measurement errors.

All these requirements remove 34% of the events from the original sample. A further reduction of approximately 40% is obtained by requiring that the energy of all other particles not contained in the cones does not exceed 10 GeV in the central calorimeter and 6 GeV in the two forward ones. This condition ensures that the energy associated with the spectators is not too large, because it is conceivable that the jet energy measurement is less precise in 'noisy' than in 'quiet' events.

The selection criteria just described are meant to provide optimal energy resolution. However, it is plausible that they have had the effect of enriching the final sample in quark jets, because gluon radiation by gluons is more probable than gluon radiation by quarks, and, as a consequence, gluon jets are likely to be more affected than quark jets by the selection criteria.

The mass resolution for the final sample is estimated to be of the order of 8 GeV/c^2 for two-jet masses of 80 GeV/c^2, and 9 GeV/c^2 for masses of 90 GeV/c^2.

Figure 24 shows the two-jet mass distribution for the final sample. In order to remove most of the QCD fall-off, the data are multiplied by the factor m_{jj}^5. An almost constant statistical error is obtained for all data points by increasing the bin size as m_{jj} increases.

Given the estimated mass resolution quoted above, the W and Z peaks are not resolved. However, a broad bump is visible in the expected region of the spectrum of Fig. 24.

220

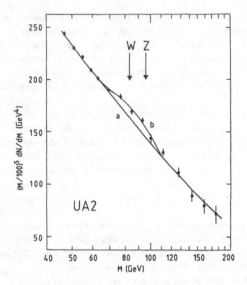

Fig. 24 Two-jet invariant mass distribution for events passing the UA2 selection criteria (see text). The distribution has been multiplied by the factor $(M/100)^5$ to make it less steep. The curves are results of the best fits to the QCD background alone (curve a), and by adding to it two Gaussians describing W and Z decays (curve b).

A maximum-likelihood fit to these data is made by using the form

$$d\sigma/dm_{jj} = A[m_{jj}^{-\alpha}\exp(-\beta m_{jj}) + xS(m_{jj})], \qquad (8.8)$$

where the first term in the square brackets is meant to parametrize the QCD background. The second term (the signal) is the sum of two Gaussians and has the form

$$S(m_{jj}) = \frac{1}{\sqrt{2\pi}}\left\{\frac{1}{\sigma_W}\exp\left[-\frac{(m_{jj}-m_W)^2}{\sigma_W^2}\right] + \frac{1}{3\sigma_Z}\exp\left[-\frac{(m_{jj}-m_Z)^2}{\sigma_Z^2}\right]\right\},$$

where the factor 1/3 in front of the second Gaussian reflects the ratio of the Z to the W inclusive cross-section, as expected from the Standard Model.

In the fit, the free parameters are α, β, x, and m_W. For the mass resolution, fixed values of $\sigma_W = 8$ GeV/c^2 and $\sigma_Z = 9$ GeV/c^2 are used, as quoted above. The Z mass is taken to be $m_Z = 1.14m_W$, in agreement with the mass measurements from the leptonic decay modes. For each set of parameters, the constant A in Eq. (8.8) is adjusted to provide the appropriate normalization.

The best-fit value of m_W is 82 ± 3 GeV/c^2, in agreement with the value obtained from the decay $W \to e\nu$, and the integral of the signal term in Eq. (8.8) is found to correspond to 632 ± 190 events. This number is consistent with the number of events expected from the Standard Model after correcting for the acceptance and the efficiency of the selection criteria. It has a statistical significance which corresponds to 3.3 standard deviations. The result of the fit is shown as curve (b) in Fig. 24.

If the signal term is suppressed in Eq. (8.8), one obtains a poor fit to the data ($\chi^2 = 21.1$ for 12 degrees of freedom). If, however, the data points contained in the interval $65 < m_{jj} < 105$ GeV/c^2 are excluded, the fit becomes good ($\chi^2 = 4.7$ for 7 degrees of freedom), as shown by curve (a) in Fig. 24.

Other fits, using either a different parametrization of the QCD background or different data samples obtained by small changes of the selection criteria, always give a non-zero signal with a statistical significance which varies between 2 and 3.5 standard deviations.

This result is encouraging, in the sense that the data suggest the presence of a structure in the two-jet mass distribution, with the properties expected from the decay of the W and Z particles into $q\bar{q}$ pairs. However, the statistical significance of this structure is far from providing overwhelming evidence, and more data are needed to confirm it.

9. CONCLUSION

In the previous section we have summarized the experimental study of a rather large number of physical properties of the weak intermediate vector bosons. The data accumulated and analyzed so far provide us with a description of these physical states which is already rich and accurate enough (within the limits imposed by the size of the available event samples) to derive some important conclusions. The emerging profile of the W and Z states—mass values, charge asymmetry in W $\rightarrow$ eν, decay modes, production cross-sections, p_T and p_L distributions, etc. — is in all details in complete agreement with the predictions of the Standard Model, which were reviewed in Sections 2-5 of this article.

Whilst the study of W and Z physics has strongly enhanced our confidence in the present theoretical framework, it is true that much work remains to be done. In particular, more precise tests, obtained with larger event samples, might reveal signals of new physics. Work is being continued at the improved CERN p$\bar{p}$ Collider and at the FNAL Tevatron, in parallel and complementary to LEP/SLC experiments. Together with the search for the top-quark, the study of the W and Z properties, in particular a precise determination of the ratio m_W/m_Z, remains the most promising area for collider physics at present, even in the most pessimistic case that no signal of new physics is directly observed.

222

REFERENCES

[1] Rubbia, C., McIntyre, P. and Cline, D., Proc. Int. Neutrino Conference, Aachen, 1976, eds. H. Faissner, H. Reithler and P. Zerwas (Vieweg, Braunschweig, 1977), p. 683.

[2] Glashow, S.L., Nucl. Phys. **22**, 579 (1961).
Weinberg, S., Phys. Rev. Lett. **19**, 1264 (1967).
Salam, A., Proc. 8th Nobel Symposium, Aspenäsgården, 1968, ed. N. Svartholm (Almqvist and Wiksell, Stockholm, 1968), p. 367.

[3] 't Hooft, G., Nucl. Phys. **B35**, 167 (1971).

[4] Hasert, F.J. et al., Phys. Lett. **46B**, 121 and 138 (1973).

[5] Arnison, G. et. al. (UA1 Collab.), Phys. Lett. **122B**, 103 (1983).

[6] Banner, M. et al. (UA2 Collab.), Phys. Lett. **122B**, 476 (1983).

[7] Arnison, G. et al. (UA1 Collab.), Phys. Lett. **126B**, 398 (1983).

[8] Bagnaia, P. et al. (UA2 Collab.), Phys. Lett. **129B**, 130 (1983).

[9] Higgs, P.W., Phys. Lett. **12**, 132 (1964); Phys. Rev. Lett. **13**, 508 (1964); Phys. Rev. **145**, 1156 (1966).
Anderson, P.W., Phys. Rev. **130**, 439 (1963).
Englert, F. and Brout, R., Phys. Rev. Lett. **13**, 321 (1964).
Guralnik, G.S., Hagen C.R. and Kibble, T.W.B., Phys. Rev. Lett. **13**, 585 (1964).

[10] Cabibbo, N., Phys. Rev. Lett. **10**, 531 (1963).
Kobayashi M. and Maskawa, M., Progr. Theor. Phys. **49**, 652 (1973).

[11] Glashow, S.L., Iliopoulos, J. and Maiani, L., Phys. Rev. **D2**, 1285 (1970).

[12] Glashow, S.L. and Weinberg, S., Phys. Rev. **D15**, 1968 (1977).
Paschos, E.A., Phys. Rev. **D15**, 1966 (1977).

[13] Lee, T.D., Phys. Rep. **9C**, 143 (1974).
Weinberg, S., Phys. Rev. Lett. **37**, 657 (1976).

[14] Adler, S.L., Phys. Rev. **177**, 2426 (1969).
Bell, J.S. and Jackiw, R., Nuovo Cimento **51**, 47 (1969).
See also Bouchiat, C., Iliopoulos, J. and Meyer, Ph., Phys. Lett. **38B**, 519 (1972).
Gross, D. and Jackiw, R., Phys. Rev. **D6**, 477 (1972).

[15] Marciano, W.J., Phys. Rev. **D20**, 274 (1979).
Antonelli, F. et al., Phys. Lett. **91B**, 90 (1980).
Veltman, M., Phys. Lett. **91B**, 95 (1980).

[16] Sirlin, A., Phys. Rev. **D22**, 971 (1980).
Marciano, W. and Sirlin, A., Phys. Rev. **D22**, 2695 (1980).

[17] For a recent summary of electroweak radiative corrections and a comparison with experiment, see: Amaldi, U. et al., Phys. Rev. **D36**, 1385 (1987).
Costa, G. et al., Nucl. Phys. **B297**, 244 (1988).

[18] Jegerlehner, F., Z. Phys. **C32**, 425 (1986).
Bardin, Yu.D., Riemann, S. and Riemann, T., Z. Phys. **C32**, 121 (1986).

[19] Abramowicz, H. et al. (CDHS Collab.), Phys. Rev. Lett. **57**, 298 (1986).
Allaby, J.V. et al. (CHARM Collab.), Z. Phys. **C36**, 611 (1987).
Bogert, D. et al. (FMMF Collab.), Phys. Rev. Lett. **55**, 1969 (1985).
Reutens, P. et al. (CCFRR Collab.), Phys. Lett. **152B**, 404 (1985).

[20] Stuart, R.G., Z. Phys. **C34**, 445 (1987).

[21] Consoli, M., LoPresti, S. and Maiani, L., Nucl. Phys. **B223**, 474 (1983).

[22] Lynn, B.W. and Stuart, R.G., Nucl. Phys. **B253**, 216 (1985).

[23] See, for example, Altarelli, G., Phys. Rep. **81**, 1 (1982).

[24] See, for example, Wu, S.L., e^+e^- interactions at high energy, Proc. Int. Symp. on Lepton and Photon Interactions at High Energies, Hamburg, 1987 (North-Holland, Amsterdam, 1988), p. 39.

[25] See, for example, Antonelli, F. and Maiani, L., Nucl. Phys. **B186**, 269 (1981).
Bellucci, S., Lusignoli, M. and Maiani, L., Nucl. Phys. **B189**, 2645 (1981).
For a pedagogical introduction, see also Altarelli, G., Acta Phys. Austriaca, Suppl. **24**, 229 (1982).

[26] See, for example, Ellis, J., Gaillard, M.K., Girardi, G. and Sorba, P., Annu. Rev. Nucl. Part. Sci. **32**, 48 (1982).

[27] Alvarez, T., Leites, A. and Terron, J., preprint DESY 87–105 (1987).
See also Güsken, S., Kühn, J.H. and Zerwas, P.M., Phys. Lett. **155B**, 185 (1986);
Kühn, J.H. et al., Nucl. Phys. **B272**, 560 (1986).

[28] See, for example, Franco, E., *in* Physics at LEP (CERN 86–02, Geneva, 1986), p. 187.

[29] Altarelli, G., Ellis, R.K. and Martinelli, G., Nucl. Phys. **B143**, 521 (1978), (E) **B146**, 544 (1978), and **B147**, 461 (1979).
Kubar-André, J. and Paige, F.E., Phys. Rev. **D19**, 221 (1979).
Kubar-André, J., Le Bellac, M., Meunier, J.L. and Plaut, G. Nucl. Phys. **B157**, 251 (1980).

[30] Altarelli, G., Ellis, R.K., Greco, M. and Martinelli, G., Nucl. Phys. **B246**, 12 (1984).

[31] Diemoz, M., Ferroni, F., Longo, E. and Martinelli, G., Z. Phys. **C39**, 21 (1988).
Diemoz, M. private communication.

[32] Dokshitzer, Yu.L., Dyakonov, D.I. and Troyan, S.I., Phys. Lett. **78B**, 290 (1978); Phys. Rep. **58**, 269 (1980).

[33] Parisi, G. and Petronzio, R., Nucl. Phys. **B154**, 427 (1979).
Curci, G., Greco, M. and Srivastava, Y., Phys. Rev. Lett. **43**, 434 (1979); Nucl. Phys. **B159**, 451 (1979).

[34] Collins, J. and Soper, D.E., Nucl. Phys. **B139**, 381 (1981), **B194**, 445 (1982), and **B197**, 446 (1982).
Collins, J., Soper, D.E. and Sterman, G., preprint CERN–TH.3923/84 (1984).

[35] Kodaira, J. and Trentadue, L., Phys. Lett. **112B**, 66 (1982) and **123B**, 335 (1983).

[36] Davies, C.T. and Stirling, W.J., Nucl. Phys. **B244**, 337 (1984).
Davies, C.T., Webber, B.R., and Stirling, W.J., preprint CERN–TH.3987/84 (1984).
See also Davies, C.T. and Webber, B., Z. Phys. **C24**, 133 (1984).

[37] Ellis, R.K., Martinelli, G. and Petronzio, R., Nucl. Phys. **B211**, 106 (1983).

[38] Davies, C.T. and Stirling, W.J., preprint CERN–TH.3853 (1984).

[39] Altarelli, G., Ellis, R.K. and Martinelli, G., Z. Phys. **C27**, 617 (1985) and Phys. Lett. **151B**, 457 (1985).

[40] Astbury, A. et al. (UA1 Collab.), Phys. Scr. **23**, 397 (1981).

[41] Mansoulié, B., Proc. Moriond Workshop on Antiproton–Proton Physics and the W Discovery, La Plagne, Savoie, France, 1983 (Ed. Frontières, Gif-sur-Yvette, 1983), p. 609.

[42] Arnison, G. et al. (UA1 Collab.), Phys. Lett. **139B**, 115 (1984).

[43] See the article by R.K. Ellis and W.G. Scott in this volume.

[44] Arnison, G. et al. (UA1 Collab.), Nuovo Cimento Lett. **44**, 1 (1985); Phys. Lett. **166B**, 484 (1986).
Albajar, C. et al. (UA1 Collab.), Europhys. Lett. **1**, 327 (1986).
Albajar, C. et al. (UA1 Collab.), Studies of the W and Z properties at the CERN Super Proton Synchrotron Collider (submitted to Z. Phys. C).

224

[45] Bagnaia, P. et al. (UA2 Collab.), Z. Phys. **C24**, 1 (1984).
Appel, J. et al. (UA2 Collab.), Z. Phys. **C30**, 1 (1986).
Ansari, R. et al. (UA2 Collab.), Phys. Lett. **B186**, 440 (1987); and Erratum, Phys. Lett. **B190**, 238 (1987).

[46] Collins, J.C. and Soper, D.E., Phys. Rev. **D16**, 2219 (1977).

[47] Jacob, M., Nuovo Cimento **9**, 826 (1958).

[48] Arnison, G. et al. (UA1 Collab.), Phys. Lett. **134B**, 469 (1984).

[49] Aguilar-Benitez, M. et al. (Particle Data Group), Review of Particle Properties, Phys. Lett. **170B**, 1 (1986)

[50] Albajar, C. et al. (UA1 Collab.), Phys. Lett. **185B**, 233 (1987); and Addendum, Phys. Lett. **191B**, 462 (1987).

[51] Ansari, R. et al. (UA2 Collab.), Phys. Lett. **194B**, 158 (1987).

[52] Bozzo, M. et al. (UA4 Collab.), Phys. Lett. **147B**, 392 (1984);
Bernard, D. et al. (UA4 Collab.), Phys. Lett. **198B**, 583 (1987).

[53] Duke, D.W. and Owens, J.F., Phys. Rev. **D30**, 49 (1984).

[54] Albajar, C. et al. (UA1 Collab.), Phys. Lett. **193B**, 389 (1987).

[55] Ellis, S.D. et al., Phys. Lett. **154B**, 435 (1985).

[56] Denegri, D., quoted by Altarelli, G., Proc. Workshop on Physics at Future Accelerators, La Thuile and Geneva, 1987 (CERN 87-07, Geneva, 1987), Vol. 1, p. 49.

[57] Geer, S. et al., Phys. Lett. **192B**, 223 (1987).

[58] Cabibbo, N., Proc. 3rd Topical Workshop on Proton–Antiproton Collider Physics, Rome, 1983 (CERN 83-04, Geneva, 1983), p. 567. See also:
Halzen, F. and Mursula, K., Phys. Rev. Lett. **51**, 857 (1983);
Hikasa, K., Phys. Rev. **D29**, 1939 (1984).

[59] Albajar, C. et al. (UA1 Collab.), Phys. Lett. **198B**, 271 (1987).

[60] Martin, A.D. et al., Phys. Lett. **189B**, 220 (1987).

[61] Halzen, F. et al., Phys. Rev. **D37**, 229 (1988).

[62] Colas, P., Denegri, D. and Stubenrauch, C., preprint CERN-EP/88-16 (1988) (to be published in Z. Phys. C).

[63] See for example, Physics at LEP (CERN 86-02, Geneva, 1986), Vol. 1, p. 40.

[64] Ansari, R. et al. (UA2 Collab.), Phys. Lett. **186B**, 452 (1987).

HEAVY-FLAVOUR PRODUCTION

M. Della Negra and G. Martinelli

CERN, Geneva, Switzerland

1. INTRODUCTION

The production of heavy quarks in hadron–hadron collisions is a subject of increasing experimental and theoretical interest.

Experimentally, the study of heavy flavours is more complicated and challenging in hadronic reactions than in $e^+ e^-$ annihilation because of the enormous background due to inelastic collisions containing only light particles. However, in contrast to $e^+ e^-$, the production cross-sections are rather large. Very valuable measurements of charm lifetimes have been achieved in fixed-target experiments, and several charm (c) and bottom (b) states have been identified at both fixed-target and ISR energies [1, 2].

With the increase of a factor of 10 in the available centre-of-mass energy $\sqrt{s}$ with respect to the CERN Intersecting Storage Rings, the $p\bar{p}$ Collider at CERN offers many advantages for a systematic study of heavy-flavour production. For example, the b production cross-section is expected to increase from $\simeq 0.10\ \mu b$ at the ISR to about 15 μb at $\sqrt{s} = 630$ GeV at Collider energy. This means that in a typical run of 700 nb^{-1}, about 10.5 million $b\bar{b}$ pairs are produced. Thanks to this large rate, colliders can compete favourably with $e^+ e^-$ machines on specific subjects such as B^0–$\bar{B}^0$ mixing. Finally, intermediate vector boson (IVB) decays represent an equally important source of heavy-quark jets.

At the CERN Collider, evidence of b production has been reported [3–6]. Bottom quarks are not directly identified but they can be tagged by the presence of one or more muons embedded inside a hadronic jet. Muons are preferred to electrons because they can be easily identified inside jets. Moreover, muons can be detected down to small transverse momenta ($p_T \geq 3$ GeV).

An analysis of isolated prompt-muon events has also allowed a limit to be put on the mass of the top (t) quark [7]. The t-quark is expected to be produced at a substantial rate at the Collider if its mass is not too large. For example, if $m_t = 40$ GeV, ≈ 450 $t\bar{t}$ pairs and ≈ 700 W $\rightarrow$ t$\bar{b}$ (or $\bar{t}$b) would be produced for an integrated luminosity of 700 nb^{-1}.

From the theoretical point of view — and provided the quark is heavy enough — there are arguments which suggest that the cross-section can be reliably computed in QCD by using the factorization theorem as for W/Z or jet production [8–10]. These arguments have been supported by the full calculation of the $O(\alpha_s^3)$ radiative corrections to the production cross-section which has been recently completed [11, 12].

Assuming that the factorization theorem is valid for heavy-quark production, the cross-section is calculable as a perturbative series in the QCD running coupling constant α_s, evaluated at the mass of the heavy quark.

This is strictly true when the quarks are really 'heavy', for two reasons. First, one neglects contributions which are suppressed by powers of $\bar{m}/m$, where $\bar{m}$ is a typical hadron mass scale (≈ 1 GeV) and m is the heavy-quark mass. Secondly, since the expansion is in $\alpha_s(m)$, m must be large for perturbation theory to be valid. It is commonly believed that the charm production cannot be realistically described in perturbation theory because of the small value of the c-quark mass.

On the other hand, for really 'heavy' quarks the only non-perturbative input which has to be added is the knowledge of Λ_{QCD} and of the parton densities, which must have been measured at some reference scale.

Accurate predictions for the production cross-section would make it possible to test the mechanism by which heavy quarks are produced, by doing a comparison with the experimental results. They would also provide a solid basis for the search for new flavours and for new particle states such as gluinos, scalar quarks, or technicolour particles at supercollider energies, e.g. at the Large Hadron Collider (LHC) and the Superconducting Supercollider (SSC).

Once the full-order α_s^3 radiative corrections have been computed, the main sources of uncertainty (as will be discussed below) at Collider and Tevatron energies will originate from imprecise knowledge of the gluon distribution function, from the overall uncertainty on the value of the coupling constant α_s, and from the choice of the scale at which the parton cross-section must be evaluated.

This chapter is divided as follows. In Section 2 we review the status of the QCD calculations for heavy-quark production in hadron–hadron collisions. Section 3 describes briefly the Monte Carlo programs that are necessary for a comparison of the theory with the experimental data. In Section 4 we discuss the experimental aspects of heavy-flavour detection. In particular, since only UA1 has muon detection and triggering capabilities, the UA1 muon detector is described, and muon data samples and background estimates are presented. Experimental results for c and b production in dimuon and single-muon events are given in Section 5. We also give results on large-p_T J/ψ production which indicate that a substantial fraction of J/ψ at large p_T comes from B → J/ψ + X decays [5]. A comparison with QCD theoretical predictions is also made. Section 6 presents the results of a search for the t-quark in both the muon channel and the electron channel. For the most conservative estimate [13] there is obtained a limit of m_t > 41 GeV from the 95% confidence level experimental upper-bound on the production cross-section. In Section 7 we review the evidence of B^0–$\bar{B}^0$ mixing coming from the same-sign dimuon events [3].

We note that, for simplicity, in this chapter we use units of GeV for energy, momentum, and mass.

2. HEAVY-QUARK PRODUCTION IN QCD

In the QCD-improved parton model, the production cross-section for the inclusive production of a heavy quark Q,

$$H_1(P_1) + H_2(P_2) \to Q(p_3) + X , \tag{2.1}$$

is given by

$$\frac{E_3 d^3\sigma}{d^3p_3} = \sum_{i,j} \int dx_1\, dx_2 \left[\frac{E_3 d^3\hat{\sigma}_{i,j}(x_1 P_1, x_2 P_2, p_3, \mu)}{d^3p_3} \right] F_i^1(x_1, \mu^2) F_j^2(x_2, \mu^2) , \tag{2.2}$$

where E_3 and p_3 are the heavy-quark energy and momentum, respectively; $F_{i,j}^{1,2}$ are the distribution functions of the light partons i, j (gluons, and light quarks and antiquarks) inside the hadrons 1, 2 evaluated at the scale μ (this scale is expected to be of the order of the heavy-quark mass m); $\hat{\sigma}_{i,j}$ is the short-distance parton cross-section from which all mass singularities have been factored out [14]; $\hat{\sigma}$ is calculated as a perturbative series in $\alpha_s(\mu^2)$. The lowest order which contributes in Eq. (2.2) is $O(\alpha_s^2)$.

At this order there are two relevant parton processes:

$$q(p_1) + \bar{q}(p_2) \rightarrow Q(p_3) + \bar{Q}(p_4): \text{quark–antiquark annihilation}, \tag{2.3}$$

$$g(p_1) + g(p_2) \rightarrow Q(p_3) + \bar{Q}(p_4): \text{gluon–gluon fusion}. \tag{2.4}$$

The corresponding Feynman diagrams are shown in Figs. 1 and 2a–c.

The only potentially singular diagrams are those in Figs. 2b, c. However, the internal quark line is always off-shell by at least of $O(m^2)$, i.e. in the centre-of-mass (c.m.) system of the heavy quarks the lifetime of the intermediate virtual quark is of $O(1/m)$. Thus the process is a short-distance process and can be safely computed in perturbation theory. At $O(\alpha_s^2)$, we can write Eq. (2.2) in the form

$$\frac{E_3 d^3\sigma}{d^3 p_3} = \frac{\alpha_s^2(\mu^2)}{s^2} \sum_{i,j} \int \frac{dx_1}{x_1^2} \frac{dx_2}{x_2^2} T_{i,j}(\tau_1, \tau_2, \varrho) \delta(1 - \tau_1 - \tau_2) F_i^1(x_1, \mu^2) F_j^2(x_2, \mu^2), \tag{2.5}$$

where

$$\tau_1 = \frac{P_1 \cdot p_3}{(P_1 \cdot P_2) x_2}, \tag{2.6a}$$

$$\tau_2 = \frac{P_2 \cdot p_3}{(P_1 \cdot P_2) x_1}, \tag{2.6b}$$

$$\varrho = \frac{4m^2}{x_1 x_2 s}, \qquad s = (P_1 + P_2)^2, \tag{2.6c}$$

and the transition probabilities $T_{i,j}$ are given by [15–17]

$$T_{q\bar{q}} = \frac{N^2 - 1}{2N^2} \left(\tau_1^2 + \tau_2^2 + \frac{1}{2} \varrho \right), \tag{2.7a}$$

$$T_{gg} = \frac{1}{2(N^2 - 1)N} \left(\frac{N^2 - 1}{\tau_1 \tau_2} - 2N^2 \right) \left(\tau_1^2 + \tau_2^2 + \varrho - \frac{\varrho^2}{4\tau_1\tau_2} \right). \tag{2.7b}$$

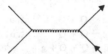

Figure 1 Lowest-order diagram for the process $q + \bar{q} \rightarrow Q + \bar{Q}$. The lines with the arrows denote the heavy-flavour quarks.

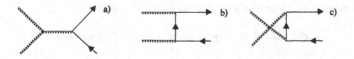

Figure 2 Lowest-order diagrams for the process $g + g \rightarrow Q + \bar{Q}$.

228

Figure 3   Flavour excitation diagram.

The phenomenological implications of the lowest-order QCD formulae can be summarized as follows:
  i) the cross-sections are predominantly central;
 ii) the average transverse momentum of the heavy quarks is of the order of their mass. The p_T distribution falls rapidly to zero when p_T becomes larger than m.

It has been suggested in the literature that besides the q–$\bar{q}$ annihilation and gg fusion, other mechanisms may play an important role in heavy-quark production; among them are
  i) flavour excitation, which contributes because of the presence of a non-zero amount of heavy quarks in the hadronic wave function (due to the evolution equations of the parton densities [8,15,18] or to some other non perturbative mechanism [19], see Fig. 3);
 ii) gluon fragmentation processes [20]:

$$g + g \rightarrow g + g$$
$$ \hookrightarrow Q + \bar{Q}. \qquad (2.8)$$

The first of the two mechanisms has been advocated to explain the large rates and the significant fraction of production in the forward direction which have been found experimentally in c production at fixed-target and ISR energies. The second mechanism, although of $O(\alpha_s^3)$, is expected to be important since, at $O(\alpha_s^2)$, the gg $\rightarrow$ q$\bar{q}$ cross-section is predicted to be about a hundred times smaller than gg $\rightarrow$ gg.

It was subsequently realized in Ref. [10] that the flavour excitation and gluon-splitting mechanisms can be correctly included only by considering the full set of $O(\alpha_s^3)$ processes. By now the full computation at $O(\alpha_s^3)$ exists for the inclusive production of one heavy quark [see Eq. (2.1)] [12].

The relevant parton subprocesses which contribute to the cross-section are

$$
\begin{aligned}
q + \bar{q} &\rightarrow Q + \bar{Q} & \alpha_s^2, \alpha_s^3, \\
g + g &\rightarrow Q + \bar{Q} & \alpha_s^2, \alpha_s^3, \\
q + \bar{q} &\rightarrow Q + \bar{Q} + g & \alpha_s^3, \\
g + g &\rightarrow Q + \bar{Q} + g & \alpha_s^3, \\
g + q &\rightarrow Q + \bar{Q} + q & \alpha_s^3, \\
g + \bar{q} &\rightarrow Q + \bar{Q} + \bar{q} & \alpha_s^3,
\end{aligned}
\qquad (2.9)
$$

Up to and including $O(\alpha_s^3)$ terms, the results can be summarized as follows. Let us write the total production cross-section, obtained by integrating Eq. (2.2) over p_3, in the form

$$\sigma(s) = \sum_{i,j} \int dx_1\, dx_2\, \hat{\sigma}_{i,j}(x_1 x_2 s, m^2, \mu^2)\, F_i^1(x_1, \mu^2) F_j^2(x_2, m^2) , \qquad (2.10)$$

where s is the square of the centre-of-mass energy of the colliding hadrons. The total short-distance cross-section $\hat{\sigma}_{i,j}$ can be written as

$$\hat{\sigma}_{i,j}(x_1 x_2 s, m^2, \mu^2) = \frac{\alpha_s^2(\mu^2)}{m^2} f_{i,j}\left(\varrho, \frac{\mu^2}{m^2}\right) . \tag{2.11}$$

The dimensionless functions $f_{i,j}$ have the following perturbative expansion:

$$f_{i,j}\left(\varrho, \frac{\mu^2}{m^2}\right) = f_{i,j}^{(0)}(\varrho) + \alpha_s(\mu^2)\left[f_{i,j}^{(1)}(\varrho) + \bar{f}_{i,j}^{(1)}(\varrho) \ln\left(\frac{\mu^2}{m^2}\right)\right] + O(\alpha_s^2) , \tag{2.12}$$

where $\alpha_s(\mu^2)$ is the running coupling constant which obeys the renormalization group equation

$$\frac{d\alpha_s(\mu^2)}{d \ln (\mu^2)} = -\beta_0 \alpha_s^2 - \beta_1 \alpha_s^3 + O(\alpha_s^4) ,$$

where

$$\beta_0 = \frac{33 - 2n_{lf}}{12\pi} , \qquad \beta_1 = \frac{153 - 19 n_{lf}}{24\pi^2} , \tag{2.13}$$

n_{lf} being the number of light flavours. Renormalization and factorization of mass singularities are done at the scale μ.

The functions $f_{i,j}^{(0)}$ defined in Eq. (2.11) are

$$f_{q\bar{q}}^{(0)}(\varrho) = \frac{\pi\beta\varrho}{27} [2 + \varrho] , \tag{2.14a}$$

$$f_{gg}^{(0)}(\varrho) = \frac{\pi\beta\varrho}{192}\left[\frac{1}{\beta}(\varrho^2 + 16\varrho + 16) \ln\left(\frac{1 + \beta}{1 - \beta}\right) - 28 - 31\varrho\right] , \tag{2.14b}$$

$$f_{gq}^{(0)}(\varrho) = f_{g\bar{q}}^{(0)}(\varrho) = 0 , \tag{2.14c}$$

where $\beta = \sqrt{1 - \varrho}$.

The $\bar{f}_{i,j}^{(1)}$ terms are determined by renormalization group arguments from the lowest-order cross-sections:

$$\bar{f}_{i,j}^{(1)}(\varrho) = \frac{1}{2\pi}\left[4\pi\beta_0 f_{i,j}^{(0)}(\varrho) - \int_{\varrho}^{1} dz_1 \, f_{k,j}^{(0)}\left(\frac{\varrho}{z_1}\right) P_{k,i}(z_1) - \int_{\varrho}^{1} dz_2 \, f_{i,k}^{(0)}\left(\frac{\varrho}{z_2}\right) P_{k,j}(z_2)\right], \tag{2.15}$$

where $P_{i,j}$ are the Altarelli–Parisi kernels.

The quantities $f^{(1)}$ depend on the scheme used for renormalization and factorization. Their explicit expression in an extension of the $\overline{MS}$ scheme has been given in Ref. [12].

The inclusion of next-to-leading corrections to the total production cross-section allows us to make an analysis of heavy-flavour production on more solid quantitative grounds. As mentioned before, the main sources of theoretical uncertainties come from our ignorance of some basic input quantities such as Λ_{QCD}, the parton densities, and the mass of the heavy quark, and from some theoretical ambiguities such as the choice of the renormalization/factorization scale [μ in Eq. (2.10)].

A recent detailed phenomenological analysis of the production of c-, b- and t-quarks at a variety of c.m. energies $\sqrt{s}$ has recently been presented in Ref. [13]. From this study it appears that not only are large values of m necessary to get accurate results, but also there must be a certain balance between m and $\sqrt{s}$. In fact, when $2m/\sqrt{s}$ goes to zero, one finds very large α_s^3 corrections coming from the t-channel exchange of gluons in gg and gq subprocesses (Figs. 4a,b). As a consequence, for small

230

Figure 4   Diagrams with gluon exchange in the t-channel.

values of $2m/\sqrt{s}$ the theoretical predictions for the total cross-sections are affected by large errors. For this reason, for example, the $O(\alpha_s^3)$ corrections to b production at $\sqrt{s} = 630$ GeV have been found to be particularly large, contrary to the t-quark case (for masses ≥ 25 GeV). In addition, large errors are expected at these energies, because the b is produced mainly by the gg subprocess in a range of small values of x where the gluon densities are poorly known. The situation worsens in the c case where, in addition to all other effects, the quark mass is probably too low to justify the use of perturbative QCD.

The production of a t-quark with a mass ≥ 25 GeV at the $p\bar{p}$ Collider and Tevatron energies is particularly favourable for relatively precise theoretical predictions. In Table 1 a compilation of cross-sections for t-quark masses in the range 20–200 GeV/c^2 at $\sqrt{s} = 0.63$, 1.8, and 2 TeV is presented. To estimate the dependence of the results on the choice of the various parameters, μ has been varied between m/2 and 2m, and Λ_{QCD} for five flavours, Λ_5, has been varied between 90 and 250 MeV. When varying Λ_{QCD}, the parton densities have been modified accordingly by reanalysing the data from deep-inelastic scattering and by performing the QCD evolution with the appropriate value of Λ_{QCD} [21]. It is quite reassuring that the error on the estimate of the cross-section is in all cases $\approx 30\%$. This is to be contrasted with the case of the b-quark where the error is much larger (typically $\geq 50\%$ at $\sqrt{s} = 630$ GeV), as can be seen from Table 2. This table also shows the results for b production for different reactions (pp, $p\bar{p}$, and $\pi^- N$) at several energies at which experimental data are available. The cross-sections for $\pi^- N$ given in the table compare well with the result of experiment WA78 [22] at CERN. This experiment finds

$$\sigma(\pi^- N \to b\bar{b}X) = (2.0 \pm 0.3 \pm 0.9) \text{ nb at } \sqrt{s} = 24.5 \text{ GeV} .$$

A much larger value had previously been reported by the NA10 Collaboration at CERN [23]:

$$\sigma(\pi^- N \to b\bar{b}X) = 14^{+7}_{-6} \text{ nb at } \sqrt{s} = 23 \text{ GeV} .$$

A comparison of the theoretical predictions for the $b\bar{b}$ production cross-section (Table 2) with the results of the UA1 Collaboration at $\sqrt{s} = 630$ GeV will be given in Section 5.

In spite of the large theoretical errors, it is interesting to note that in all cases the computed corrections of $O(\alpha_s^3)$ lead to a better agreement between the theoretical calculation and the data. With all the caveats mentioned before, this remains true even in the case of c production. In Fig. 5 we plot the c cross-section in pp collisions for energies in the range $\sqrt{s} = 10$–62 GeV. The results are given separately for $m_c = 1.5$ and 1.2 GeV, and the error band includes a variation of μ between 1 GeV and $2m_c$ and of Λ_{QCD} between 90 and 250 MeV. The predicted cross-section is compared with the data reported in Ref. [1].

We finally recall that when m $\gg \Lambda_{QCD}$, quark multiplicities are fully calculable in perturbation theory. The partonic process gg $\to q\bar{q}$ is much smaller than gg $\to$ gg in the high-energy limit. Thus

Table 1

Heavy-quark (mass m) production cross-section in $p\bar{p}$ collisions.
The cross-sections (in nb) in the second column are expressed by a 'central' value,
obtained for $\mu = m$ and $\Lambda_5 = 170$ MeV, and errors derived from varying μ and Λ_5
in the ranges $m/2 < \mu < 2m$ and 90 MeV $< \Lambda_5 <$ 250 GeV (columns 3–6)
by adding the corresponding errors in quadrature.

		$\sqrt{s} = 0.63$ TeV			
m (GeV)	σ (nb)	$\mu = m/2$ $\Lambda_5 = 170$ MeV	$\mu = 2m$ $\Lambda_5 = 170$ MeV	$\mu = m$ $\Lambda_5 = 90$ MeV	$\mu = m$ $\Lambda_5 = 250$ MeV
20	$25.6^{+5.6}_{-7.9}$	30.9	20.2	19.8	27.3
30	$3.04^{+0.62}_{-0.94}$	3.62	2.47	2.29	3.26
40	$0.643^{+0.11}_{-0.20}$	0.738	0.532	0.481	0.696
50	$0.188^{+0.025}_{-0.056}$	0.208	0.158	0.142	0.204
60	$\left(0.669^{+0.067}_{-0.190}\right) \times 10^{-1}$	0.718×10^{-1}	0.569×10^{-1}	0.508×10^{-1}	0.716×10^{-1}
70	$\left(0.267^{+0.022}_{-0.074}\right) \times 10^{-1}$	0.284×10^{-1}	0.229×10^{-1}	0.204×10^{-1}	0.281×10^{-1}
80	$\left(0.114^{+8.009}_{-0.031}\right) \times 10^{-1}$	0.122×10^{-1}	0.989×10^{-2}	0.880×10^{-2}	0.118×10^{-1}
90	$\left(0.511^{+0.037}_{-0.136}\right) \times 10^{-2}$	0.548×10^{-2}	0.443×10^{-2}	0.394×10^{-2}	0.517×10^{-2}
100	$\left(0.234^{+0.019}_{-0.062}\right) \times 10^{-2}$	0.254×10^{-2}	0.203×10^{-2}	0.181×10^{-2}	0.232×10^{-2}
110	$\left(0.110^{+0.009}_{-0.029}\right) \times 10^{-2}$	0.119×10^{-2}	0.946×10^{-3}	0.847×10^{-3}	0.107×10^{-2}
120	$\left(0.517^{+0.044}_{-0.142}\right) \times 10^{-3}$	0.562×10^{-3}	0.441×10^{-3}	0.398×10^{-3}	0.494×10^{-3}
		$\sqrt{s} = 1.8$ TeV			
m (GeV)	σ (nb)	$\mu = m/2$ $\Lambda_5 = 170$ MeV	$\mu = 2m$ $\Lambda_5 = 170$ MeV	$\mu = m$ $\Lambda_5 = 90$ MeV	$\mu = m$ $\Lambda_5 = 250$ MeV
20	228^{+55}_{-60}	277	190	182	252
40	$9.63^{+1.7}_{-2.5}$	11.4	7.97	7.74	9.97
60	$1.27^{+0.19}_{-0.35}$	1.46	1.06	1.01	1.31
80	$0.285^{+0.037}_{-0.077}$	0.322	0.241	0.222	0.296
100	$\left(0.873^{+0.109}_{-0.230}\right) \times 10^{-1}$	0.974×10^{-1}	0.755×10^{-1}	0.675×10^{-1}	0.910×10^{-1}
120	$\left(0.331^{+0.034}_{-0.086}\right) \times 10^{-1}$	0.362×10^{-1}	0.289×10^{-1}	0.257×10^{-1}	0.346×10^{-1}
140	$\left(0.144^{+0.013}_{-0.037}\right) \times 10^{-1}$	0.155×10^{-1}	0.127×10^{-1}	0.112×10^{-1}	0.150×10^{-1}
160	$\left(0.691^{+0.047}_{-0.173}\right) \times 10^{-2}$	0.732×10^{-2}	0.607×10^{-2}	0.540×10^{-2}	0.712×10^{-2}
180	$\left(0.352^{+0.019}_{-0.086}\right) \times 10^{-2}$	0.369×10^{-2}	0.311×10^{-2}	0.276×10^{-2}	0.359×10^{-2}
200	$\left(0.187^{+0.008}_{-0.045}\right) \times 10^{-2}$	0.195×10^{-2}	0.166×10^{-2}	0.147×10^{-2}	0.189×10^{-2}

Table 1 (contd.)

m (GeV)	σ (nb)	$\mu = m/2$ $\Lambda_5 = 170$ MeV	$\mu = 2m$ $\Lambda_5 = 170$ MeV	$\mu = m$ $\Lambda_5 = 90$ MeV	$\mu = m$ $\Lambda_5 = 250$ MeV
		$\sqrt{s} = 2.0$ TeV			
20	275^{+67}_{-71}	334	231	219	307
40	$12.2^{+2.3}_{-3.1}$	14.4	10.1	9.86	12.7
60	$1.66^{+0.25}_{-0.44}$	1.91	1.38	1.32	1.71
80	$0.378^{+0.051}_{-0.101}$	0.429	0.319	0.296	0.390
100	$0.117^{+0.015}_{-0.031}$	0.131	0.100	0.091	0.121
120	$\left(0.446^{+0.048}_{-0.116}\right) \times 10^{-1}$	0.491×10^{-1}	0.387×10^{-1}	0.346×10^{-1}	0.464×10^{-1}
140	$\left(0.196^{+0.018}_{-0.051}\right) \times 10^{-1}$	0.212×10^{-1}	0.172×10^{-1}	0.153×10^{-1}	0.204×10^{-1}
160	$\left(0.952^{+0.071}_{-0.240}\right) \times 10^{-2}$	0.101×10^{-1}	0.836×10^{-2}	0.743×10^{-2}	0.986×10^{-2}
180	$\left(0.496^{+0.029}_{-0.122}\right) \times 10^{-2}$	0.522×10^{-2}	0.438×10^{-2}	0.388×10^{-2}	0.509×10^{-2}
200	$\left(0.271^{+0.013}_{-0.065}\right) \times 10^{-2}$	0.283×10^{-2}	0.240×10^{-2}	0.213×10^{-2}	0.275×10^{-1}

Table 2

Same as Table 1, for b production in pp, p$\bar{\text{p}}$, and π^- N collisions

m_b (GeV)	σ	$\mu = m/2$ $\Lambda_5 = 170$ MeV	$\mu = 2m$ $\Lambda_5 = 170$ MeV	$\mu = m$ $\Lambda_5 = 90$ MeV	$\mu = m$ $\Lambda_5 = 250$ MeV
		pp, $\sqrt{s} = 41$ GeV			
4.5	23^{+21}_{-15} nb	40	12	13	34
5	$9.0^{+8.4}_{-5.9}$ nb	16	4.7	4.9	14
		pp, $\sqrt{s} = 62$ GeV			
4.5	142^{+98}_{-80} nb	231	81	91	182
5	66^{+47}_{-38} nb	109	37	41	86
		p$\bar{\text{p}}$, $\sqrt{s} = 630$ GeV			
4.5	$19^{+10}_{-8}\,\mu$b	27	15	12.5	25
5	$12^{+7}_{-4}\,\mu$b	18	10	9	16
		π^- N, $\sqrt{s} = 24.5$ GeV			
4.5	$7.6^{+4.7}_{-3.8}$ nb	12	4.6	5.2	10
5	3.1 ± 1.5 nb	4.4	1.9	2.2	3.9

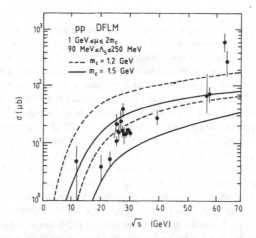

Figure 5 Total cross-section for charm production in pp collisions. The data compilation is from Ref. [1]. The solid (dashed) curves determine the bend, obtained for m_c = 1.5 GeV (1.2 GeV) by combining the theoretical uncertainties derived from independent variations of μ and Λ_5 added in quadrature.

heavy-quark production at very high energies is essentially due to gluon fragmentation into $Q\bar{Q}$ pairs. The leading logarithmic QCD series predicts the average number of $Q\bar{Q}$ pairs in a gluon jet [24]:

$$\varrho = \frac{\text{\# of } Q\bar{Q} \text{ pairs}}{\text{\# of gluon jets}} = \frac{1}{3\pi} \int_{4m^2}^{Q^2} \frac{dK^2}{K^2} \alpha_s(K^2) \left(1 + \frac{2m^2}{K^2}\right) \left(\frac{1}{4} - \frac{m^2}{K^2}\right)^{1/2} n_g(Q^2, K^2), \qquad (2.16)$$

where

$$n_g(Q^2, K^2) = [\ln(Q^2/\Lambda_{QCD}^2)/\ln(K^2/\Lambda_{QCD}^2)]^a$$

$$\times \exp[6/(\pi\beta_0) \ln(Q^2/\Lambda_{QCD}^2]^{1/2} / \exp[6/(\pi\beta_0) \ln(K^2/\Lambda_{QCD}^2)]^{1/2}$$

is the gluon multiplicity and

$$a = -\frac{1}{4}\left[1 + \frac{10n_f}{27\pi\beta_0}\right].$$

We finally mention that the rapidity and transverse-momentum distributions for the inclusive production of a heavy quark $O(\alpha_s^3)$ are to be published shortly [25]. This will allow a quite precise reconstruction of the total production cross-section using the rapidity and p_T distribution of the muon which comes from the semileptonic decay of the heavy quark. In this case, the only extra ingredient to be added is a model for the hadronization of the heavy quark in an open-flavour meson (baryon). The subsequent semileptonic decay of the meson is then described by the decay of the heavy quark, treated as in the naïve parton model. This approximation is expected to be accurate enough for masses $\geq m_b$.

234

3. MONTE CARLOS

Monte Carlo programs are based on a combination of perturbative QCD and phenomenological models for jet and beam–jet fragmentation. They are a necessary tool for confronting experimental results with QCD. Because of the importance of higher-order QCD effects in the strong production of heavy flavours, one has to choose Monte Carlos containing approximations for these effects. Two approaches exist: i) to implement only lowest-order QCD matrix elements and use a parton shower model to simulate initial- and final-state QCD radiations (ISAJET, COJETS, PYTHIA [26]); or ii) simply to add contributions from $2 \rightarrow 2$ and $2 \rightarrow 3$ processes for which explicit matrix elements exist [27], with some cut-off prescription to avoid divergences (EUROJET [28]). It is not clear *a priori* how good or bad these approximations are. In the above approaches, Monte Carlos are tuned to reproduce in the data the effects that are sensitive to higher-order corrections, such as jet multiplicities and topologies. We now discuss briefly the phenomenological assumptions used in ISAJET.

Hard-scattering cross-sections are calculated using QCD matrix elements to lowest order in α_s, the strong running coupling constant. A convolution is made with structure functions evaluated at the hard scattering $Q^2 \equiv 2stu/(s^2 + t^2 + u^2)$, where s, t, and u are the Mandelstam variables of the parton–parton scatter. ISAJET uses the structure functions of Eichten et al. (EHLQ, set I) [29] and $\Lambda = 0.2$ GeV. The heavy-quark sea at high Q^2 is obtained by evolving initial parton distributions at $Q_0^2 = 5$ GeV^2, without any intrinsic heavy-quark component. Hard partons participating in the high-Q^2 subprocesses are evolved using the basic QCD branching processes $q \rightarrow qg$, $g \rightarrow gg$, and $g \rightarrow q\bar{q}$. The scattered partons are given a virtual mass with a maximum of order Q. The partons generated during the evolution acquire successively lower virtual masses, and the evolution stops when the virtual mass falls below a fixed cut-off $\sqrt{|p^2|} = m_p + 6$ GeV, where m_p is the on-shell mass of the parton. This branching approximation [30] describes correctly the leading-log scaling violations of the structure functions and of the jet fragmentation, and also reproduces the observed jet multiplicity in jet data at SPS collider energies [31].

For heavy-quark production, three independent production mechanisms are considered:
i) heavy-flavour creation: $q\bar{q} \rightarrow Q\bar{Q}$, $gg \rightarrow Q\bar{Q}$,
ii) flavour excitation: $qQ \rightarrow qQ$, $gQ \rightarrow gQ$,
iii) final-state gluon splitting: $g \rightarrow Q\bar{Q}$.

The last two production mechanisms correspond in reality to higher-order QCD processes for which ISAJET gives an approximation valid only at large p_T. At small p_T the flavour excitation mechanism is divergent. The cut-off applied in ISAJET arises in the backward evolution [32] of the incoming heavy quark. In order to conserve heavy flavour, it is required that during this evolution the branching $g \rightarrow Q\bar{Q}$ should always occur. Hence an effective cut-off $Q > \max(6\text{ GeV}, 2m)$ is applied.

Heavy-quark production in gluon jets is a consequence of the evolution of final-state gluons. This is an abundant source of heavy quarks at large p_T [20, 24, 33], since the gluon jet cross-section is large. For gluon splitting into $b\bar{b}$ and $c\bar{c}$, the cut-off mass of 6 GeV below which the gluon is hadronized is between the $c\bar{c}$ and the $b\bar{b}$ thresholds. Consequently, the cross-section for the reaction $p\bar{p} \rightarrow gX(g \rightarrow c\bar{c})$ is underestimated. The number of heavy-quark pairs to be found in the fragmentation of a massive gluon has been calculated by Mueller and Nason, using perturbative QCD [24]. Figure 6 shows their estimate of the number of $Q\bar{Q}$ pairs in gluon jets as a function of Q, the mass of the virtual gluon. Seven per cent of the gluon jets with $Q = 30$ GeV are expected to contain a $c\bar{c}$ pair, whilst only 2% contain a $b\bar{b}$ pair. The $b\bar{b}$ yield is smaller because the higher threshold for $b\bar{b}$ production will stop the QCD evolution of the gluon shower earlier. The ISAJET gluon-shower model is in agreement with the calculation of Ref. [24] for $g \rightarrow b\bar{b}$ and $g \rightarrow t\bar{t}$ but, as expected, underestimates $g \rightarrow c\bar{c}$ because of the 6 GeV cut-off.

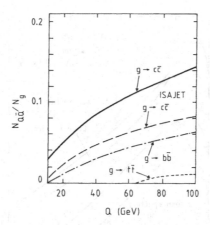

Figure 6   Fraction of gluon jets splitting into heavy quark pairs as a function of Q, the mass of the virtual gluon.

The ISAJET estimate of the differential cross-section $d\sigma/dp_T$ for the inclusive reaction $p\bar{p} \rightarrow$ b (or $\bar{b}$) + X at $\sqrt{s} = 0.63$ TeV in the central rapidity region $|y| < 1.5$ is shown in Fig. 7. The contributions from the three production mechanisms are shown as separate curves. The flavour excitation and the gluon splitting processes are suppressed in the small-p_T region ($p_T < 10$ GeV) owing to the cut-off Q > $2m_b$. As a consequence, at small p_T flavour creation dominates, and the shape is given by the 'scaling' spectrum for massive quarks [34]: $d\sigma/dp_T^2 = f(m_T/\sqrt{s})/m_T^4$, where $m_T = (p_T^2 + m_b^2)^{1/2}$. This form implies that $d\sigma/dp_T$ is maximum for $p_T \approx m_b/2$. At large p_T the three subprocesses have approximately the same p_T dependence, as expected from the scaling properties of the hard cross-section, the structure functions, and the gluon splitting function. They also contribute roughly the same amount of about 1/3 each.

As a result of the QCD evolution, the final-state partons are all on-mass-shell. They are then hadronized according to the independent fragmentation model of Field and Feynman [35]. This fragmentation model reproduces the measured properties of low-energy ($\sim$ 6 GeV) jets and justifies the use of a 6 GeV cut-off on the parton shower evolution.

Charm- and bottom-quark fragmentation has been measured extensively with e^+e^- colliders [36]. ISAJET has been tuned to reproduce these results. As for ordinary quarks, the heavy quarks of large virtual masses initiate a parton shower until a mass cut-off is reached at $\sqrt{|p^2|} = m + 6$ GeV. The resulting heavy quark is then put on the mass shell and fragmented into a leading heavy-flavour hadron in the following proportions: $Q\bar{u} : Q\bar{d} : Q\bar{s} : Qqq = 38.7\% : 38.7\% : 12.6\% : 10\%$.

The ratio of spin-1 to spin-0 (spin-3/2 to spin-1/2 for baryons) is assumed to be: $B^*/B = D^*/D \approx 3$.

The z distribution of the leading heavy hadron is parametrized with the Peterson et al. form [37]:

$$D(z)\,dz = \frac{N}{z\left[1 - \frac{1}{z} - \frac{\epsilon_Q}{(1 - z)}\right]^2}\,dz\,,$$

236

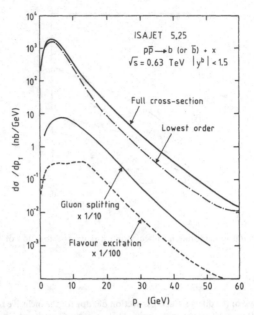

Figure 7   The inclusive differential bottom cross-section as estimated by the ISAJET Monte Carlo (version 5.25) in the central rapidity region $|y^b| < 1.5$ for antiproton–proton collisions at $\sqrt{s} = 0.63$ TeV (full line). The contributions of the different subprocesses are also indicated: lowest-order flavour creation (dash-dotted line); gluon splitting scaled down by 1/10 (second full line); flavour excitation scaled down by 1/100 (dashed line).

where

    N is a normalization factor;

    ϵ_Q is a parameter which depends on the quark mass;

    z is the fragmentation variable defined as: $z = (E + p_L)_{had}/(E + p_L)_{quark}$, where E is the energy evaluated in the same Lorentz frame as p_L, so that z is invariant with respect to Lorentz boosts along the quark axis, and p_L is the longitudinal component of the momentum with respect to the quark axis.

For ISAJET 5.23, ϵ_c was tuned to reproduce the e^+e^- results for $c \to D^*$ [36], giving $\epsilon_c = 0.30$; for b the value $\epsilon_b = 0.02$ is used. The uncertainty on ϵ_Q is the main source of error in the relation between the muon cross-section and the quark cross-section.

The weak decays of heavy hadrons are simulated using $V - A$ matrix elements. The most recent measurements or estimates of the decay branching ratios of charm and bottom hadrons [28, 38] have been incorporated. For B-hadrons an average semileptonic branching ratio is used:

$$BR(B \to e) = BR(B \to \mu) = 12\% \, .$$

The difference between ϵ_b and ϵ_c will cause the p_T spectrum for muons coming from bottom semileptonic decays to be harder than from charm. This is illustrated in Fig. 8 in the case of lowest-order flavour creation. At large p_T the b-quark and c-quark cross-sections are equal, but the

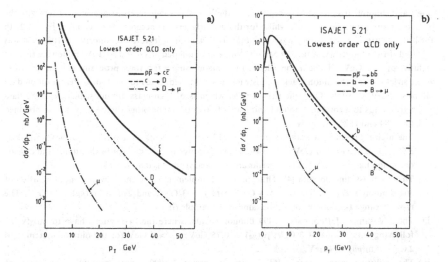

Figure 8 $d\sigma/dp_T$ for heavy quarks, heavy hadrons, and muons: a) $c\bar{c}$, b) $b\bar{b}$ (ISAJET lowest order).

charm-hadron spectrum and hence the muon spectrum is softer because of the softer Peterson fragmentation. As a consequence, large-p_T lepton triggers at hadron colliders will strongly favour bottom production over charm production.

The muon isolation is a crucial parameter in the study of heavy quarks, including the search for the t-quark. In particular, a muon from a heavy t-quark or from a $W^\pm$ or a Z decay is often isolated. In these processes, the energy flow around the muon reflects the level of activity from the spectator jets. It is therefore important to have a good description of the spectator jets in the Monte Carlo.

The remaining spectator quarks from the proton and the antiproton beams are hadronized in ISAJET, so that long-range correlations similar to those observed in minimum-bias events at CERN $p\bar{p}$ Collider energies [39] are generated. The global multiplicity distribution was adjusted so that it reproduces the height of the multiplicity plateau in high-E_T jet events far away from the jet axes. Similarly, the average transverse momentum of primary mesons is adjusted using the same jet data and fixed at $\langle p_T \rangle = 0.45$ GeV.

4.   HEAVY-FLAVOUR DETECTION

The detection of exclusive heavy-flavour states at hadron colliders is a very difficult task, given the huge combinatorial background from light particles. Fortunately, it is possible to tag heavy-flavour quarks by their semileptonic decays into large-p_T muons. The semileptonic branching ratios are large ($\approx$ 10%), and large-p_T muons can easily be identified inside jets. In practice, as will be apparent from the data, a jet containing a large-p_T muon is a clean signature for heavy-flavour production. Furthermore, these muon-jets are expected to be predominantly b-jets rather than c-jets, because of the harder fragmentation of the b-quark as explained in the Monte Carlo section. Only the UA1 detector [40] at the CERN $p\bar{p}$ Collider has muon detection and triggering capability.

Muon candidates are selected with a fast hardware trigger that requires a track pointing towards the interaction region within a cone of aperture $\pm$ 150 mrad. Muon data are recorded using inclusive single and dimuon triggers. The acceptance of the inclusive single-muon trigger covers typically the pseudorapidity range $|\eta| < 1.5$, whilst for dimuon triggers the acceptance extends to $|\eta| < 2$. In 1983, the centre-of-mass energy was $\sqrt{s} = 0.54$ TeV, and an integrated luminosity of 108 nb^{-1} was achieved. In 1984 and 1985 a further 556 nb^{-1} were accumulated at $\sqrt{s} = 0.63$ TeV. Following this, an inclusive $p_T > 3$ GeV off-line muon selection was made using a fast filter program. A final selection, where strict cuts on the muon track reconstruction are applied, leads to the four different muon data samples described below. Only the first sample, which is of high-mass dimuons, includes 1983 data and corresponds to a total integrated luminosity of 664 nb^{-1}. The other three samples were obtained with the 1984 and 1985 data only.

i) The high-mass dimuon data [3] consist of 512 dimuon events with $p_T(\mu_1) > 3$ GeV, $p_T(\mu_2) > 3$ GeV, and $m(\mu\mu) > 6$ GeV. The pseudorapidity range is $|\eta| < 2$ for both muons, and at least one muon has $|\eta| < 1.5$. The Z events are removed from this data sample.

ii) The low-mass dimuon data [4]. This sample is complementary to the previous one. It consists of 304 dimuon events with $p_T(\mu_1) > 3$ GeV, $p_T(\mu_2) > 3$ GeV, and $2m_\mu < m(\mu\mu) < 6$ GeV. The rapidity range for both muons is the same as for sample (i).

iii) The J/ψ sample [5] consists of 494 dimuon events, where the two muons have to satisfy the following cuts: $p_T(\mu_1) > 3$ GeV, $p_T(\mu_2) > 0.75$ GeV, $p_T > 4$ GeV for the dimuon system, and 2 GeV $< m(\mu\mu) < 4$ GeV.

iv) The inclusive muon sample [6] consists of about 20,000 events containing at least one reconstructed muon with $p_T(\mu) > 6$ GeV in the central rapidity range $|y| < 1.5$.

The main sources of background to the prompt muon signature in UA1 are: decays of kaons and pions in flight; non-interacting hadrons; shower leakage; leakage through cracks; cosmic rays; missassociation of the central detector and the muon chamber tracks. The dominant background contribution comes from π/K decays in flight. Details concerning the estimate of the decay background can be found in Refs. [3] and [6].

5. CHARM AND BOTTOM PRODUCTION

At moderate p_T values ($p_T < 15$ GeV), where W and Z decays do not contribute, prompt muon events can come from a variety of sources: the Drell–Yan mechanism, J/ψ and Υ decays, and heavy-flavour processes. The first three processes are in fact the background, which has to be subtracted from the different muon data samples in order to study heavy-flavour production. This background is characterized by muons which are produced isolated, in contrast to heavy-flavour events (bottom and charm) where the muon is embedded in a jet. Thus the main tool for separating the background from the heavy-flavour component will be the muon isolation.

5.1 The High-Mass Dimuons

Since both muons have $p_T > 3$ GeV, the 512 dimuon events with $m(\mu\mu) > 6$ GeV correspond to a topology where both muons are produced in opposite azimuthal hemispheres, and therefore the dimuon system has a small transverse momentum. An isolation variable is defined as $S = [\Sigma E_T(\mu_1)]^2 + [\Sigma E_T(\mu_2)]^2$, where $\Sigma E_T(\mu)$ is the scalar sum of the transverse energy E_T measured in calorimeter cells in a cone of $\Delta R = (\Delta\eta^2 + \Delta\phi^2)^{1/2} < 0.7$ around the muon, excluding the expected energy deposited by the muons; η is the pseudorapidity and ϕ is the azimuthal angle measured in radians. The distribution of S for the unlike-sign pairs (Fig. 9a, 355 events) shows a clear enhancement at $S < 9$ GeV^2 due to Drell–Yan and Υ decays, which is not observed for the like-sign events (Fig. 9b, 157 events). The dimuon events with $S < 9$ GeV^2 are therefore classified as isolated.

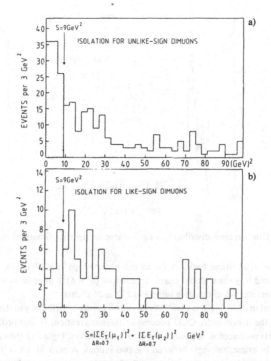

Figure 9   Distribution of the isolation variable S for a) 355 unlike-sign events and b) 157 like-sign events. The shaded regions, $S < 9$ GeV/c^2, define the isolated events samples.

According to the charge and to the isolation requirement of $S < 9$ GeV^2, the 512 dimuon events are divided into four categories: 98 isolated unlike-sign events (8 background), 15 isolated like-sign events (8 background), 257 non-isolated unlike-sign (58 background), and 142 non-isolated like-sign (58 background). The background figures indicated correspond to the total estimated background from non-prompt muons dominated by decay background.

The 98 isolated unlike-sign events are used to estimate the background from Drell–Yan and Υ decays. The dimuon mass distribution for these events is shown in Fig. 10. A clear peak due to Υ production is visible, and the continuum above 15 GeV is dominated by Drell–Yan production. The contributions from the different processes were determined by fitting the sum of their individual mass distributions to the data after background subtraction. Allowance must be made for the presence of isolated dimuons from semileptonic heavy-flavour decays. Having fixed the background contribution to 8 ± 1.5 events, the fit results in 40 ± 7 events from Υ decays, 29 ± 9 events from the Drell–Yan process, and 22 ± 9 events from heavy-flavour decays. Correcting for the efficiency of the isolation cut (82%), the total contribution of Drell–Yan and Υ is estimated to be 84 events. In summary, after subtraction of the 134 non-prompt muon background and of the 84 Drell–Yan and Υ events, the 512 events result in 294 events due to heavy-flavour production.

According to ISAJET, 90% of these events are due to $b\bar{b}$ and only 10% come from $c\bar{c}$. The relative transverse momentum p_T^{rel} between the muon and its accompanying jet can be used to separate

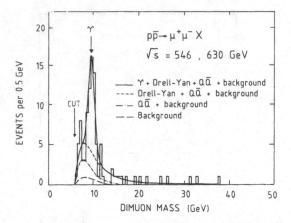

Figure 10   Dimuon mass distributions for isolated unlike-sign dimuon events (98 events).

b-jets and c-jets on a statistical basis. A fit to the observed p_T^{rel} distributions, using ISAJET p_T^{rel} distributions for c and b, gives a global charm fraction of $(8 \pm 6)\%$ in agreement with expectations. Therefore the $b\bar{b}$ contribution to this data sample is 272 ± 36 events.

Another prediction of ISAJET is that 75% of the $b\bar{b}$ events contributing to the high-mass dimuon sample come from the lowest-order QCD process of flavour creation. In this process the two muons come from two b-jets produced in opposite azimuthal hemispheres. Figure 11 shows the distribution of $\Delta\phi$, the difference in azimuthal angle between the two muons. A peak at $\Delta\phi = 180°$ is observed, in agreement with lowest-order dominance, but a tail of acoplanar events with low values of $\Delta\phi$ is also present. The curves shown are the ISAJET lowest-order and higher-order predictions, the

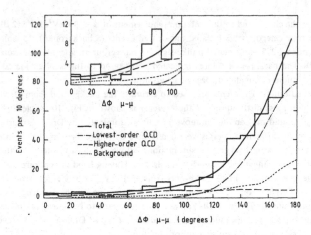

Figure 11   Azimuthal angle difference between the muons for the non-isolated pairs. The curves are predictions from ISAJET normalized to the 399 events observed.

background, and the sum of all contributions. The total ISAJET prediction has been normalized to the 399 non-isolated events after subtraction of 116 background events. As expected, the lowest-order contribution peaks at 180° and disappears for $\Delta\phi < 100°$. It is clear that the data are in agreement with the ISAJET $b\bar{b}$ production model. In particular, higher-order processes are also needed at low $\Delta\phi$, and ISAJET estimates their relative contribution (21%) correctly.

The ISAJET absolute prediction for the number of $b\bar{b}$ events contributing to the high-mass dimuon sample is 392 events ($\pm 50\%$). Hence the ratio N = DATA/ISAJET = 0.69 ($\pm 51\%$). Using this normalization factor, one can extract from the data a measurement of the b cross-section in the kinematical range relevant to the high-mass dimuon sample. This will be done at the end of this section for the four muon data samples together, and the result will be compared with recent full QCD calculations at $O(\alpha_s^3)$ [41].

5.2 The Low-Mass Dimuons

The sample of low-mass dimuons consists of 304 events. The simultaneous requirements of large transverse momentum ($p_T > 3$ GeV) for each muon and a low dimuon mass $2m_\mu < m(\mu\mu) < 6$ GeV imply that the dimuon system has a large transverse momentum, typically $p_T > 6$ GeV. This transverse momentum has to be balanced by hadronic activity opposite to the dimuon system in the transverse plane, irrespective of the production mechanism. On the other hand, the hadronic activity in the immediate vicinity of the muons depends on the production mechanism. Low-mass high-p_T dimuons are expected to receive contributions from Drell–Yan processes, heavy-flavour production, J/ψ decays, and leptonic decays of light mesons (ϱ, ω, ϕ, η, η'). Again the isolation of the muons is used to separate the heavy-flavour component from the Drell–Yan processes. A muon pair is classified as isolated if $\Sigma E_T(\mu) < 3$ GeV for each muon and $\Sigma E_T(\mu\mu) < 3$ GeV for the dimuon system, where ΣE_T is defined as the scalar sum of E_T in a cone of $\Delta R < 0.7$ around the direction of the muon or dimuon momentum vector.

The data sample of 304 events is divided into 93 isolated unlike-sign events, 174 non-isolated unlike-sign events, and a total of 37 like-sign events. Figure 12a shows the dimuon mass spectrum for the 174 non-isolated unlike-sign events and Fig. 12b for the 93 isolated unlike-sign events. A clear J/ψ peak is visible on both spectra, indicating that different mechanisms are at work for J/ψ production. It is found that in fact non-isolated J/ψ's come from B-hadron decays [5]. The J/ψ events will be

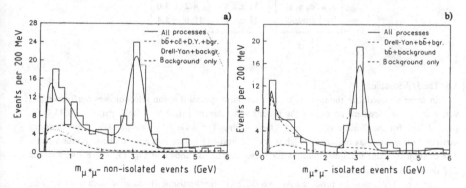

Figure 12 Dimuon mass distribution: a) 174 non-isolated unlike-sign muon pairs, b) 93 isolated unlike-sign muon pairs. The curves correspond to the contributions listed in Table 3.

discussed separately in the next subsection. Outside the J/ψ peak, the continuum shows, on the left, a broad enhancement extending down to the threshold of two muon masses ($2m_\mu$). In the case of isolated events (Fig. 12b), this enhancement is mainly due to the Drell–Yan production mechanism in an unusual kinematical regime: $m^2 \ll p_T^2$ [4]. In the case of non-isolated events (Fig. 12a), the low-mass enhancement is mainly due to leptonic decays of light mesons (ϱ, ω, ϕ, η, η'). The dominant contribution from heavy-flavour decays originates from chain decays of single b-quarks: $b \rightarrow c\mu\nu(c \rightarrow s\mu\nu)$, resulting in unlike-sign non-isolated dimuons with masses around 2 GeV. As in the case of high-mass dimuons, the contribution from $c\bar{c}$ processes is suppressed owing to the softer fragmentation for c-quarks. Comparing the non-isolated mass spectrum (Fig. 12a) with the isolated one (Fig. 12b), an excess of non-isolated events is clearly visible near 2 GeV. This is interpreted as evidence for the observation of b-chain decays. The ISAJET Monte Carlo was used to generate the shapes of the mass distributions for all these different processes and to estimate the efficiency of the isolation cut.

A fit of these shapes to the experimental mass spectra was then performed, fixing the non-prompt muon background contribution to its estimated value [4]. The results of these fits are given in Table 3 and the corresponding fitted contributions are shown as separated curves in Fig. 12. A contribution of 48 ± 11 events due to heavy-flavour production is obtained for the non-isolated events. Comparing the fitted number of events with the absolute ISAJET prediction, a scale factor of N = DATA/ISAJET = 0.58 ($\pm 44\%$) is needed to explain the data. This scale factor will be used to extract an independent measurement of the inclusive b cross-section.

Table 3
Contributions to the unlike-sign dimuon sample

Process	Non-isolated dimuons	Isolated dimuons
J/ψ	58 ± 12	36.8 ± 5.8
Heavy flavour	48 ± 11	5.3 ± 2.7
Drell–Yan	9 ± 4	36.2 ± 5.8
$\varrho, \omega, \phi, \eta, \eta'$	31 ± 8	4.2 ± 3.0
Background	27 ± 7	10.6 ± 5.4

5.3 The J/ψ Sample

In order to maximize the number of J/ψ events, a special dimuon selection was performed [5], where the p_T requirement on one of the muons was relaxed: $p_T(\mu_1) > 3$ GeV, $p_T(\mu_2) > 0.75$ GeV, $p_T > 4$ GeV for the dimuon system, and 2 GeV $< m(\mu\mu) < 4$ GeV. A total number of 494 dimuons satisfying these cuts was obtained, which is divided into 434 unlike-sign and 60 like-sign muon pairs. The dimuon effective mass distributions for these samples are shown in Fig. 13, where a very strong J/ψ signal can be seen.

Large-p_T J/ψ's may be produced by two different mechanisms: i) radiative decays of large-p_T χ-states [42] produced by gluon fusion [43]: $g + g \rightarrow \chi + g(\chi \rightarrow J/\psi + \gamma)$; ii) production and decay of large-p_T B-hadrons [44]: $B \rightarrow J/\psi + X$. ISAJET was used to simulate the two production processes.

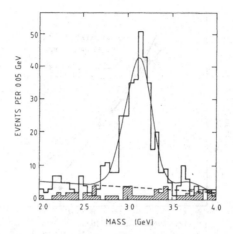

Figure 13   The effective mass distribution for $\mu^+\mu^-$ pairs (open histogram) and for $\mu^+\mu^+$ plus $\mu^-\mu^-$ pairs (shaded histogram). The full curve is the result of a Gaussian fit to the J/ψ mass peak, giving m(J/ψ) = 3.110 ± 0.011 GeV. The dashed line represents the fitted non-resonant background.

For the first process the differential cross-sections of Humpert [45] are used. For the second process, bottom production was simulated as already described. The branching fraction for B-hadron decay into J/ψ was taken to be 1.1%, and the fractions of B-hadrons decaying through two-body and quasi-two-body channels were taken to be 10% and 40%, respectively, based on the measurements of Ref. [46].

The measured differential cross-sections times muonic branching ratio of the J/ψ, BR(dσ/dp$_T$), is shown in Fig. 14, corrected for the experimental acceptance. The error bars do not include the overall normalization error of 15% arising from the uncertainty in the total integrated luminosity. Taking the whole region p$_T$ > 5 GeV and $|y|$ < 2, one obtains an overall cross-section times branching ratio of: BR(J/$\psi \rightarrow \mu^+\mu^-$)·$\sigma$(p$\bar{\text{p}} \rightarrow$ J/ψ) = 7.5 ± 0.7 ± 1.2 nb, where the first error is statistical and the second is systematic. The systematic error comes mainly from the uncertainty in the integrated luminosity, but also contains small contributions from the residual model dependence of the acceptance and from uncertainties in the simulation of the ranging-out of muons in the detector. The shapes are compared with Monte Carlo calculations, normalized to the observed cross-section for $|y|$ < 2 and p$_T$ > 5 GeV. The steep slope of the J/ψ p$_T$ spectrum is better reproduced by the χ production process than by the bottom production process. It is possible to determine the relative contributions of the two processes by looking at the J/ψ isolation. The J/ψ's from B-hadron decays will normally be accompanied both by the other decay products of the B-hadron and by the other fragmentation products of the parent b-quark. In contrast, J/ψ's from the decay of χ's are expected to be isolated, accompanied by only the photon from the χ decay, with no fragmentation products. A fraction, 0.76 ± 0.08 ± 0.12, of the J/ψ's with p$_T$ > 5 GeV is estimated to come from χ processes. Using the measured overall J/ψ cross-section for $|y|$ < 2, p$_T$ > 5 GeV, one obtains a cross-section times muonic branching ratio for J/ψ production through χ processes of 5.7 ± 0.8 ± 1.3 nb, and 1.8 ± 0.6 ± 0.9 nb for J/ψ production through B-hadrons.

The ISAJET Monte Carlo calculation of bottom production predicts a value of 1.8 nb (±35%) for the J/ψ production cross-section times branching ratio, where most of the cross-section comes

244

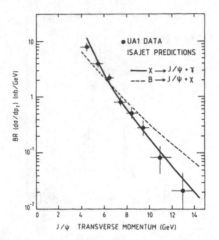

Figure 14   The differential cross-section $BR(d\sigma/dp_T)$ for $p + \bar{p} \rightarrow J/\psi \rightarrow \mu^+\mu^-$ in the region $|y| <$ 2, corrected for acceptance. The dashed lines show the Monte Carlo calculations described in the text, normalized to the data for $p_T > 5$ GeV.

from lowest-order QCD processes. Hence, for the measurement of the bottom cross-section through large-p_T J/ψ's, the normalization factor N = DATA/ISAJET = 1 ($\pm 69\%$) will be used.

5.4   The Inclusive Muon Sample

The inclusive muon cross-section $d\sigma/dp_T$ is shown in Fig. 15 for $p_T(\mu) > 6$ GeV and $|\eta(\mu)| < 1.5$, after subtraction of the decay background. In this figure and in the following analysis $p_T(\mu)$ has not been corrected for smearing due to momentum measurement errors in the central detector, which typically exceed 10% for $p_T(\mu) > 20$ GeV. The dashed curve shows the equivalent prediction for the muon cross-section obtained by summing all the physics processes considered in ISAJET, namely: c and b production, W and Z decay, and Drell-Yan, J/ψ, and Υ production. Except for the heavy-flavour contribution, all the other processes in ISAJET are fixed from their observation in other channels: the W, Z contribution is fixed from the electron channel, and the Drell-Yan, J/ψ, and Υ production is fixed from the dimuon channel. The dashed-dotted curve shows the predicted contribution from $W \rightarrow \mu\nu$ alone; the Jacobian peak is spread out owing to the limited momentum resolution at 40 GeV ($\sim 20\%$). The W contribution increases with p_T from $\approx 10\%$ at 15 GeV and exceeds the $b\bar{b}/c\bar{c}$ contribution for $p_T(\mu) > 25$ GeV.

Below 10 GeV the errors on the measured cross-section are dominated by the uncertainties on the decay background subtraction. Table 4 gives a summary of the contributions to the inclusive cross-section in the region 10 GeV < $p_T(\mu)$ < 15 GeV, $|\eta(\mu)| < 1.5$. After subtraction of $\sim 36\%$ decay background, the measured cross-section is

$$\sigma_{\text{data}}(10\,\text{GeV} < p_T(\mu) < 15\,\text{GeV}, \ |\eta(\mu)| < 1.5) = 4.33 \pm 0.20^{+0.83}_{-0.56}\,\text{nb},$$

where the first error is statistical and the second is systematic, dominated by the error on the background subtraction. The main contribution comes from semileptonic decays of charm and bottom particles.

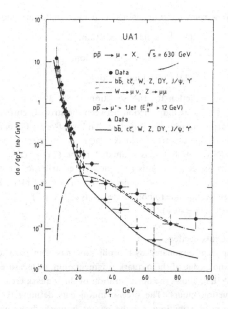

Figure 15   The inclusive muon p_T spectrum (corrected for decay background and acceptance) for all events and for events with at least one jet with $E_T > 12$ GeV. The curves show the sum of the ISAJET predictions. The $W \to \mu\nu$ contribution is shown as a separate curve.

Table 4

Cross-sections for physics processes
contributing to the inclusive muon spectrum in the range
$10\,\text{GeV} < p_T(\mu) < 15\,\text{GeV},\ |\eta^\mu| < 1.5$

Process	Cross-section (nb)
$b \to \mu$	2.45 ± 0.03
$c \to \mu$	0.83 ± 0.02
Drell–Yan, J/ψ, Υ	0.19 ± 0.009
$W, Z \to \mu$	0.07 ± 0.005
Total Monte Carlo	3.54 ± 0.04
Data (background subtracted)	$4.33 \pm 0.20\,^{+0.83}_{-0.56}$

ISAJET predicts that a substantial fraction ($\approx 2/3$) of heavy-flavour production at large p_T comes from the flavour-excitation and gluon-splitting mechanisms. The main topological difference between these processes and the lowest-order flavour-creation process is the angular separation of the two heavy quarks. This implies a different distribution in $\Delta\phi$, the azimuthal angle difference between the two muons. A three-component fit to the $\Delta\phi$ distribution has been made of the different shapes expected for the different processes.

The results of the fit as percentages of the data after background subtraction are:

$$(45^{\pm}\,{}^{18}_{9})\% \text{ flavour creation,}$$

$$(33^{\pm}\,{}^{31}_{17})\% \text{ flavour excitation,}$$

$$(22^{\pm}\,{}^{29}_{19})\% \text{ gluon splitting,}$$

compared with the ISAJET prediction for this selection: 33%, 39%, and 28%, respectively. The statistical errors are large, with no real distinction between the strongly correlated flavour-excitation and gluon-splitting components; nevertheless it is clear that lowest-order flavour creation alone cannot explain the $\Delta\phi$ distribution. Using the distribution of the muon momentum perpendicular to the jet axis p_T^{rel}, the ratio $N_\mu(b)/[N_\mu(b) + N_\mu(c)]$ is estimated to be $(76 \pm 12)\%$. Subtracting from the measured inclusive muon cross-section in this p_T region all contributions not arising from bottom production, a muon cross-section due to bottom decays, $\sigma(\text{Data}) = 3.1$ nb $(\pm 25\%)$, is obtained. The corresponding ISAJET cross-section due to bottom decays is $\sigma(\text{ISAJET}) = 2.5$ nb $(\pm 31\%)$. Hence the ratio $N = \text{Data}/\text{ISAJET} = 1.27\,(\pm 40\%)$.

5.5 The Bottom Inclusive Cross-Section

A heavy-flavour component dominated by b production has been extracted from four different muon data samples. The knowledge of the b-quark fragmentation, as measured in e^+e^- experiments and as implemented in ISAJET, allows the normalization of the b-quark cross-section to be inferred, within the ISAJET production model. The precise muon cuts defining the different muon data samples translate into the p_T distributions for the parent b-quark shown in Fig. 16. The arrows indicate the value p_T^{min}, such that 90% of the parent b-quarks have $p_T > p_T^{min}$. One can therefore extract an inclusive cross-section for b or $\bar{b}$ production with $p_T > p_T^{min}$ in the rapidity range $|y^b| < 1.5$, corresponding to the acceptance of the muon trigger. In Table 5 the values of p_T^{min} for the different muon data samples are listed together with the corresponding ISAJET cross-sections. The two highest-p_T measurements $p_T > 23$ GeV and $p_T > 32$ GeV (rows 5 and 6) have been obtained from the inclusive muon cross-section measured in two more p_T intervals: 15 GeV $< p_T(\mu) < 20$ GeV and 20 GeV $< p_T(\mu) < 25$ GeV. The previously obtained normalization ratios $N = \text{Data}/\text{ISAJET}$ (column 4, Table 5) lead to the b cross-section measurements given in this table (last column). The total error affecting these measurements varies from 45% (single muons) to 72% (J/ψ sample) and is dominated by the uncertainty on the b-quark fragmentation and that on the shape of the bottom production spectrum $(\pm 20\%)$. Figure 17 shows the six measurements of the cross-section $\sigma(p_T > p_T^{min})$ obtained with the UA1 data for the inclusive reaction $p\bar{p} \to b$ (or $\bar{b}$) + X at $\sqrt{s} = 0.63$ TeV in the central rapidity region $|y^b| < 1.5$. Three curves are shown for comparison with the data: the $O(\alpha_s^2)$ and $O(\alpha_s^3)$ QCD calculations of Nason et al. [41] obtained with a scale $\mu = m = 5$ GeV and the DFLM structure functions [47], and the full ISAJET calculation. The $O(\alpha_s^2)$ and $O(\alpha_s^3)$ spectra have similar shapes. The ISAJET spectrum used to estimate the acceptance of the muon cuts does not agree with the $O(\alpha_s^3)$ shape in the region 5 GeV $< p_T < 15$ GeV. An acceptance correction factor resulting from changing the shape of the p_T spectrum from the full ISAJET calculation to the lowest-order one was therefore applied (column 5, Table 5). For p_T values large compared to the mass of the bottom quark, the $O(\alpha_s^3)$ QCD calculation is unreliable. For this reason the curve of Nason et al. is shown only up to 15 GeV; in this region it agrees well with the data. At higher p_T, an extrapolation of the $O(\alpha_s^3)$ curve seems to underestimate the data. The parton shower model of ISAJET is compatible with the data over the entire p_T range measured, including the region $p_T > 15$ GeV.

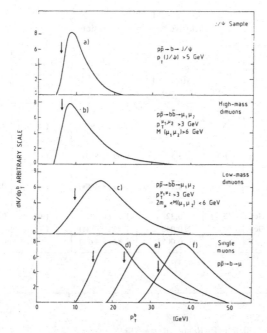

Figure 16 Parent b-quark p_T distribution for the six muon samples: a) $J/\psi \rightarrow \mu^+\mu^-$, b) high-mass dimuons, c) low-mass dimuons, d) inclusive muon sample 10 GeV $< p_T^\mu < 15$ GeV, e) 15 GeV $< p_T^\mu <$ 20 GeV, f) 20 GeV $< p_T^\mu < 25$ GeV. The arrows indicate p_T^{min} such that 90% of the muon events have $p_T^b > p_T^{min}$.

It is possible to estimate the total cross-section for $b\bar{b}$ pair production by using the data of Fig. 17 and the $O(\alpha_s^3)$ calculation of Nason et al. to extrapolate the data down to $p_T = 0$ GeV and to the full rapidity range. This estimation assumes in particular that, in the forward direction, bottom production is entirely described by QCD.

For this estimation, the point from the high-mass dimuon sample, which is not from a truly inclusive measurement (the two muons come from two different b's) has been excluded. Also the points above 15 GeV, where the $O(\alpha_s^3)$ QCD calculation is unreliable, have not been used.

This gives $\sigma(b\bar{b}) = 7.3 \pm 2.3 \, \mu b$ for $b\bar{b}$ pair production in the central rapidity region $|y| < 1.5$ and

$$\sigma(b\bar{b}) = 10.2 \pm 3 \, \mu b \,,$$

for the full rapidity range.

The quoted errors are just the propagation of the three experimental errors, assuming no further error coming from the uncertainty on the theoretical shape used for the extrapolation. The $b\bar{b}$ pair production cross-section estimated at $O(\alpha_s^3)$ by Altarelli et al. [13] (Table 2) is $12^{+7}_{-4} \, \mu b$ for $m_b = 5$ GeV and $19^{+10}_{-8} \, \mu b$ for $m_b = 4.5$ GeV. Within the uncertainties affecting both the measurements and the theory, reasonable agreement is observed between the data and QCD.

Table 5

Inclusive bottom cross-sections

Muon sample	p_T^{min} (GeV)	σ(ISAJET) $p_T > p_T^{min}$ $|y(b)| < 1.5$ (μb)	$N = \dfrac{\text{Data}}{\text{ISAJET}}$	Acceptance correction factor	σ(b) $p_T > p_T^{min}$ $|y(b)| < 1.5$ (μb)
J/ψ sample	6	4.1	1.0 ($\pm 69\%$)	1.1	4.5 ($\pm 72\%$)
High-mass dimuons	6.5	3.5	0.69 ($\pm 51\%$)	1.0	2.4 ($\pm 55\%$)
Low-mass dimuons	10	1.2	0.58 ($\pm 44\%$)	1.2	0.83 ($\pm 48\%$)
Single muons					
$10 < p_T(\mu) < 15$ GeV	15	0.32	1.27 ($\pm 40\%$)	1.05	0.42 ($\pm 45\%$)
$15 < p_T(\mu) < 20$ GeV	23	0.049	1.56 ($\pm 41\%$)	1.0	0.076 ($\pm 46\%$)
$20 < p_T(\mu) < 25$ GeV	32	0.0083	2.83 ($\pm 44\%$)	1.0	0.023 ($\pm 48\%$)

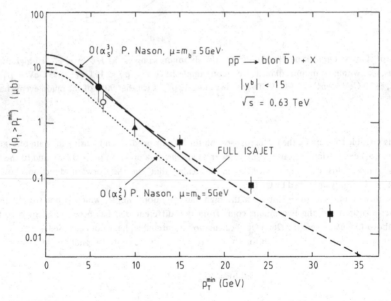

Figure 17 The cross-section for b or $\bar{b}$ production with $p_T > p_T^{min}$ and $|y^b| < 1.5$ in p$\bar{p}$ collisions at $\sqrt{s} = 0.63$ TeV as a function of p_T^{min}. The six experimental points come from the independent measurements discussed in the text: J/$\psi \to \mu^+\mu^-$ (solid circle), high-mass dimuons (open circle), low-mass dimuons (triangle), inclusive muon samples (squares). The curves are the absolutely normalized QCD predictions discussed in the text.

6. SEARCH FOR THE TOP-QUARK

The existence of the t-quark is an important open question. In the Standard Model with six quarks (u, d, s, c, b, and t), flavour-changing neutral currents are forbidden. On the other hand, if the t-quark does not exist, the b-quark must be a left-handed isosinglet, and processes such as b → (s or d) $\mu^+\mu^-$ cannot be avoided. The experimental limit [(b → $\mu^+\mu^-$ X)/(b → all) ≤ 10^{-3}] [48] implies either that the t-quark exists or that the Standard Model is not correct.

In the framework of the Standard Model, theoretical upper limits can be placed on the t-quark mass [49] from the measurement of ϱ, the parameter that specifies the relative strengths of the neutral and charged weak currents, and from a comparison of the measurements of ($\sin^2 \theta_w$) from UA1 and UA2 [50] with those at low Q^2. Thus, taking into account the recent lower limit from TRISTAN [51, 52], the available evidence suggests that the t-quark exists and that its mass lies in the range 27.7 GeV < m_t < 200 GeV. An indirect indication of a lower mass limit of about 50 GeV has recently been obtained from analyses of B^0-$\bar{B}^0$ mixing [53]. For the b′, a hypothetical d-quark of the fourth generation, a lower mass limit of 26.3 GeV at 95% confidence level has been measured at TRISTAN [52].

6.1 Top-Quark Production Mechanisms

In proton–antiproton collisions at $\sqrt{s}$ = 0.63 TeV, the t-quark can be produced by two different mechanisms:

$$p\bar{p} \rightarrow W + X(W \rightarrow t\bar{b}), \tag{6.1}$$

$$p\bar{p} \rightarrow t\bar{t} + X. \tag{6.2}$$

The production through the decay Z → $t\bar{t}$ has a comparatively negligible cross-section and is not considered. For process (6.1) the production cross-section is derived from the measurement of σ(W → $\ell\nu$) in the same experiment [54]. The branching ratio BR(W → $t\bar{b}$) depends on m_t and on QCD corrections to the partial decay widths [55]. The estimated cross-section for process (6.1), $\sigma(t\bar{b})$, is given in Table 6 (column 2) as a function of m_t.

The cross-section for process (6.2) can now be estimated, using perturbative QCD, at the next-to-leading order, as discussed in Section 2. The best estimate by Altarelli et al. [13] of the

Table 6
Top-quark cross-section

Quark mass (GeV)	$\sigma(t\bar{b})$ (nb)	$\sigma(t\bar{t})$ ADMN [13] (nb)	$\sigma(t\bar{t})$ EUROJET (nb)
25	1.77	8.5	12.8
30	1.68	3.04	5.1
40	1.41	0.64	1.1
45	1.26	0.33	0.57
50	1.09	0.188	0.31
55	0.89	0.11	0.18
60	0.68	0.0669	0.11

cross-section as a function of m_t, resulting from these calculations, is given in Table 6 (column 3). When the UA1 analysis was performed, the $O(\alpha_s^3)$ calculation did not exist, and the reference value for $\sigma(t\bar{t})$ was taken from the EUROJET Monte Carlo program [28]. In the EUROJET calculation, terms of $O(\alpha_s^2)$ and $O(\alpha_s^3)$ are included using exact QCD matrix elements, but not including virtual gluon contributions. The calculation depends on an arbitrary cut-off on the p_T of the soft gluon in the $O(\alpha_s^3)$ terms. The chosen value of this cut-off (5 GeV) gives the cross-section estimate shown in Table 6 (column 4). The higher values of $\sigma(t\bar{t})$ obtained with EUROJET are not due to an overestimation of the (arbitrary) contribution from $O(\alpha_s^3)$ terms, but are essentially due to the use of a different value for α_s and to a different choice of gluon structure functions: EHLQ set 1 instead of DFLM [47].

As discussed in Section 2 and Ref. [13], given the uncertainty on the gluon parametrization and on the precise value of α_s, $\sigma(t\bar{t})$ is less well known than $\sigma(t\bar{b})$. The absence of a t-quark signal in UA1 will be presented as an upper bound for $\sigma(t\bar{t})$ as a function of m_t, after subtraction of the $t\bar{b}$ contribution. This upper bound depends on the assumed shape for the t-quark differential cross-section $d\sigma/dydp_T$, and on the t-quark fragmentation model. The differential cross-section is taken from the lowest-order QCD calculation of ISAJET: a heavy t-quark is predicted to be produced centrally with a broad p_T spectrum having $\langle p_T \rangle \approx m_t/2$.

In ISAJET, when a t-quark fragments, it is assumed that, because of its large mass, almost all the original t-quark energy is transferred to the t-hadron. In fact the parametrization of Peterson et al. [37] for the z-distribution of the t-hadron results in a fragmentation function which is almost a δ-function at $z = 1$, since the ϵ parameter becomes very small: $\epsilon_t = 0.5/m_t^2$.

Once a t-hadron is formed, the constituent t-quark is decayed into a lepton, a neutrino, and a b-quark jet (semileptonic decay) or into three quark jets (hadronic decay). For each lepton channel, $V - A$ matrix elements are used and a branching ratio of 11% is assumed. For hadronic decays (branching ratio 66%), three-body phase space and standard jet fragmentation are used.

Figure 18 shows the expected $[\sigma(t\bar{t})$ from EUROJET] number of t-quark semileptonic decays (electrons and muons) at the CERN $p\bar{p}$ Collider for an integrated luminosity of 700 nb^{-1}. No selection

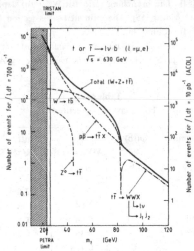

Figure 18   The expected number of events with one semileptonic decay of a t-quark $(e + \mu)$ for an integrated luminosity equivalent to the total UA1 data sample of 700 nb^{-1}.

cuts have been applied. For a t-quark mass of 40 GeV, about 300 such semileptonic decays are expected. The contribution from Z decays never exceeds 20% of the contribution for W decays and disappears for $m_t > m_Z/2$. For m_t between about 40 and 80 GeV the production via W decays is predicted to dominate.

6.2 Top-Quark Signature

Independently of the production mechanism, an event resulting from a t-quark decaying semileptonically will have a characteristic topology, namely an isolated lepton accompanied by two or more jets and a neutrino, all concentrated in the central pseudorapidity region ($|\eta| < 1.5$). These properties make it possible to discriminate other sources of charged leptons—W, Z, Drell-Yan, J/ψ, Υ, $b\bar{b}$, anc $c\bar{c}$, accompanied by gluons—that constitute the main physics backgrounds to a new heavy-quark signal.

Jets are defined in the calorimeters using the UA1 jet algorithm [56]. A jet is counted if its axis lies outside a cone of radius $\Delta R \equiv (\Delta\eta^2 + \Delta\phi^2)^{1/2} = 1$ around the lepton direction, where $\Delta\phi$ ($\Delta\eta$) is the difference in azimuthal angle (pseudorapidity) between the lepton and the jet axis, and $|\eta_{jet}| < 2.5$. Thus a jet containing the charged lepton, called the lepton jet, is not counted. Events are considered if they have at least one jet with $E_T > 12$ GeV, where E_T is the uncorrected transverse energy of the jet as defined by the algorithm. Jet 1 is defined as the highest-E_T jet in the event. The other jets (if any), with $E_T > 7$ GeV, are numbered in order of decreasing E_T and must be validated by the presence of a charged track with $p_T > 0.5$ GeV and within $\Delta R = 0.4$ from the direction of the calorimeter jet axis. With this jet definition, an event coming from the semileptonic decay of a b-quark or c-quark will generally result in a topology with fewer jets than events coming from a t-quark decay.

In the semileptonic decay of a new massive quark, the charged lepton will usually be emitted at a large angle with respect to the directions of the other decay products and will therefore tend to be isolated. Since the decays of b-quarks and c-quarks produce mainly non-isolated leptons, lepton isolation is a good way to discriminate between b-quarks, c-quarks, and heavier quarks. As muons can be identified even if they are not isolated, they can be used to study the isolation properties of heavy-flavour events. In the UA1 detector, the activity around the charged lepton can be measured in two independent ways: i) in the calorimeter (ΣE_T); ii) in the central track detector (Σp_T), where either calorimeter cells or charged tracks within a cone of radius $\Delta R = 0.7$ around the lepton directions are considered. The transverse energy E_T and transverse momentum p_T are measured with respect to the beam axis. The two measurements are combined into a single isolation variable: $I = [(\Sigma E_T/3)^2 + (\Sigma p_T/2)^2]^{1/2}$, where the relative weights are chosen to reflect the average values of these quantities for the underlying event, and ΣE_T and Σp_T are expressed in GeV.

The neutrino transverse energy $E_T(\nu)$ is estimated by the missing transverse energy in an event. Events in which the charged lepton–neutrino transverse mass $m_T(\ell,\nu)$ exceeds 45 GeV (electrons) or 40 GeV (muons) are classified as W candidates. The transverse mass is defined as $m_T(\ell,\nu) \equiv [2E_T(\ell)E_T(\nu)(1 - \cos\Delta\phi_{\ell\nu})]^{1/2}$, where $E_T(\ell)$ and $E_T(\nu)$ are the transverse energies of the charged lepton and of the neutrino, respectively, and $\Delta\phi_{\ell\nu}$ is the azimuthal angle difference between the charged lepton and the neutrino directions. The slightly different cuts reflect the different energy resolutions for electrons and muons.

6.3 The $b\bar{b}g$ Background

Isolated lepton + jet events can be generated by W, Z, Drell-Yan pairs, J/ψ, and Υ produced at large p_T in association with jets. The t-quark background due to these sources has already been discussed.

A more difficult background comes from $b\bar{b}g$ ($c\bar{c}g$) events, where one of the b (c)-quarks fragments in an almost isolated lepton. To estimate this background, the UA1 analysis relies on experimental measurements of this reaction, when the lepton is not isolated. This can only be done with muons, since, in UA1, electrons cannot be identified inside jets.

A reference sample of muon–jet events dominated by $b\bar{b}$ ($c\bar{c}$) production is used to check in detail how well ISAJET reproduces their properties. The muon is required to have a transverse momentum in the interval: $10\ \text{GeV} < p_T(\mu) < 15\ \text{GeV}$. In this range $b\bar{b}$ ($c\bar{c}$) production dominates (subsection 5.4). The jet requirement is that at least one calorimeter jet in the event has $E_T > 12\ \text{GeV}$, outside a cone of $\Delta R = 1$ around the muon direction. With these cuts, 658 muon–jet events are obtained.

The predicted contributions to this sample from decay background, $b\bar{b}$ and $c\bar{c}$ production, and residual processes are given in Table 7. The quoted errors for the Monte Carlo predictions are statistical; the first error on the decay background is statistical, the second reflects the systematic uncertainty in the decay tables, and the third represents the $\pm 15\%$ uncertainty in the luminosity of the jet data sample used to estimate the background. Additional jets in each event are considered if they have $E_T > 7\ \text{GeV}$, as explained in subsection 6.2. The measured and predicted jet multiplicity distributions are shown in Fig. 19 and Table 7. The absolute numbers of muon events with one, two, three, four and even five jets are reproduced to $\approx 20\%$. A variable which allows the $b\bar{b}g$ background to be differentiated from the t is $|\cos \theta^*_{\text{jet2}}|$, where θ^*_{jet2} is the angle between jet 2 and the beam direction in the rest system of the muon, jet 1, jet 2, and the missing transverse energy. Gluon jets coming from initial-state bremsstrahlung in $b\bar{b}g$ events tend to have large values of $|\cos \theta^*_{\text{jet2}}|$, whilst for the t-quark a flat distribution is expected. A detailed study shows that ISAJET reproduces the distributions of various kinematical quantities equally well, including the $|\cos \theta^*_{\text{jet2}}|$ distribution and the distribution of the isolation variable $I = [(\Sigma E_T/3)^2 + (\Sigma p_T/2)^2]^{1/2}$ shown in Fig. 20. The good agreement observed between the data and ISAJET demonstrates that the shape of the isolation distribution is well understood, and that the number of isolated leptons from $b\bar{b}g$ events can be safely estimated.

Table 7
Classification of muon + jet events in the range
$10\ \text{GeV} < p_T(\mu) < 15\ \text{GeV}$; $E_T(\text{jet 1}) > 12\ \text{GeV}$

	Monte Carlo					Data
	K/π decays	W/Z	Drell–Yan J/ψ Υ	$b\bar{b}$ $c\bar{c}$	Total	
μ + 1 jet	$147 \pm 18 \,^{+22}_{-44} \pm 22$	3.6 ± 0.4	17 ± 1	176 ± 4	344 ± 18	375
μ + 2 jets	$65 \pm 12 \,^{+13}_{-19} \pm 10$	1.1 ± 0.3	2.1 ± 0.3	145 ± 4	213 ± 12	204
μ + $\geq$ 3 jets	$23 \pm 7 \,^{+5}_{-7} \pm 3$	0.2 ± 0.1	0.2 ± 0.1	59 ± 2	82 ± 7	79
Total	$235 \pm 23 \,^{+40}_{-70} \pm 35$	4.9 ± 0.5	19 ± 1	380 ± 6	639 ± 23	658

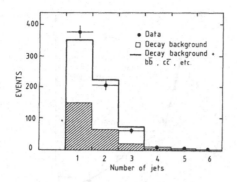

Figure 19 Jet multiplicity distribution for events with at least one jet $E_T > 12$ GeV and one muon with $10\,\text{GeV} < p_T(\mu) < 15$ GeV.

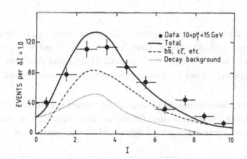

Figure 20 The isolation variable $I \equiv [(\Sigma E_T/3)^2 + (\Sigma p_T/2)^2]^{1/2}$ for events with at least one jet $E_T > 12$ GeV and one muon with $10\,\text{GeV} < p_T(\mu) < 15$ GeV.

6.4 Search for the Top-Quark in the Muon Channel

Figure 21 shows the p_T spectrum for the initial inclusive muon sample with $p_T(\mu) > 6$ GeV after subtraction of decay background. The Monte Carlo prediction for W, Z, Drell–Yan, J/ψ, Υ, $b\bar{b}$, and $c\bar{c}$ reproduces both the shape and the normalization of the distribution. The predicted t-quark contribution, shown for masses between 25 and 50 GeV, is everywhere small.

Selective cuts have been applied to increase the sensitivity to a possible t-quark signal. It is instructive to illustrate the way in which the expected signal-to-noise ratio changes as the cuts are tightened on the muon and jets, as follows:

a) $p_T(\mu) > 10$ GeV, $E_T^{jet1} > 12$ GeV, $m_T(\mu\nu) < 40$ GeV, no jet-2 requirement;

b) $p_T(\mu) > 12$ GeV, $E_T^{jet1} > 15$ GeV, $m_T(\mu\nu) < 40$ GeV, no jet-2 requirement;

c) $p_T(\mu) > 12$ GeV, $E_T^{jet1} > 15$ GeV, $m_T(\mu\nu) < 40$ GeV, $E_T^{jet2} > 7$ GeV.

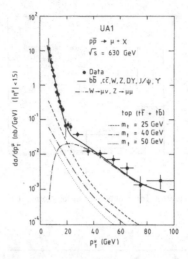

Figure 21 The inclusive muon momentum spectrum $d\sigma/dp_T(\mu)$ versus $p_T(\mu)$. The data are compared with Monte Carlo predictions including: W, Z, Drell–Yan, J/ψ, Υ, $b\bar{b}$ and $c\bar{c}$. The t-quark production for three different masses is shown (m_t = 25, 40, and 50 GeV). The data have been corrected for decay background and for acceptance, but not for muon momentum measurement errors.

Lepton isolation is the principal tool for distinguishing t-quark events from others. Figures 22a–c show the data and the Monte Carlo predictions for the isolation variable I for the three sets of cuts. Also shown are the expected contributions for a t-quark of 30 GeV mass. It can be seen that in each

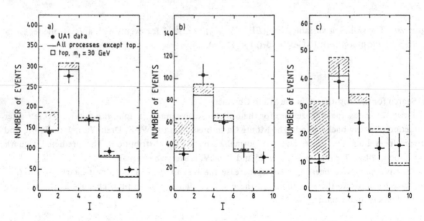

Figure 22 The isolation variable distribution {$I = [(\Sigma E_T/3)^2 + (\Sigma p_T/2)^2]^{1/2}$ in a cone $\Delta R = 0.7$ around the muon} for the three sets of cuts described in the text. The solid histogram is the Monte Carlo prediction of the background processes without top. The hatched area represents the contribution for a t-quark of 30 GeV mass. The black points with error bars are the UA1 data.

case the whole distribution is well reproduced without the need for a contribution from a new heavy quark. In the region I < 2 (isolated muons), for m_t = 30 GeV, the expected signal-to-noise ratio is 0.3, 0.8, and 1.8 for cuts (a), (b), and (c), respectively. The third set of cuts (c), together with I < 2, is used to enrich the data with possible t-quark candidates. After these cuts, 10 events remain with an isolated muon and two or more jets. Table 8 gives the expected contributions from the previously discussed background sources to the 10 isolated μ + $\geq$ 2 jets, together with the results for one-jet events. In Table 8 and in subsequent tables, when two errors are quoted the first error is statistical and the second is systematic. In the μ + $\geq$ 2 jets channel the expected contributions from W, Z, Drell–Yan, J/ψ, Υ, and $b\bar{b}$, $c\bar{c}$, amount to 9.2 ± 0.8 ± 1.0 events, with another 2.3 ± 0.4 ± 0.7 events from K/π decays. This accounts well for the observed 10 events, without the need for an additional contribution from a t-quark. The number of t-quark events expected for different masses in the isolated muon + jet events is given in Table 9. The absence of a t-quark signal enables an upper bound to be extracted for the

Table 8
Sources of isolated muon + jet events with $p_T(\mu)$ > 12 GeV

	Monte Carlo					Data
	K/π decays	W/Z	Drell–Yan J/ψ Υ	$b\bar{b}$ $c\bar{c}$	Total	
μ + 1 jet	7.2 ± 1.7 ± 2.2	2.5 ± 0.5	7.3 ± 0.7 ± 3.6	6.3 ± 0.6	23.3 ± 2.0 ± 4.2	22
μ + $\geq$ 2 jets	2.3 ± 0.4 ± 0.7	0.6 ± 0.2	2.0 ± 0.4 ± 1.0	6.6 ± 0.7	11.4 ± 0.9 ± 1.2	10

Table 9
Expected t-quark event rates for isolated muons
with $p_T(\mu)$ > 12 GeV

Number of jets = 1:

	t-quark mass (GeV)			
	25	30	40	50
$t\bar{b}$	2.3 ± 0.3	3.3 ± 0.5	2.8 ± 0.4	1.7 ± 0.3
$t\bar{t}$	7.1 ± 0.6	5.0 ± 0.5	1.4 ± 0.2	0.5 ± 0.1
Total	9.4 ± 0.7	8.3 ± 0.7	4.2 ± 0.4	2.2 ± 0.3

Table 9 (contd.)

Number of jets ≥ 2:

	t-quark mass (GeV)			
	25	30	40	50
t$\bar{\text{b}}$	2.5 ± 0.3	3.9 ± 0.6	3.6 ± 0.5	3.1 ± 0.5
t$\bar{\text{t}}$	21.8 ± 1.1	16.6 ± 0.9	6.0 ± 0.5	2.2 ± 0.3
Total	24.3 ± 1.1	20.5 ± 1.1	9.6 ± 0.7	5.3 ± 0.6

t-quark cross-section based on event rates in the muon channel only. A better limit is obtained by combining results from the muon and the electron channels and by making use of distributions of event properties, which differentiate top from background.

6.5 Search for the Top-Quark in the Electron Channel

For the t-quark search, UA1 uses a sample of inclusive electrons with $E_T > 15$ GeV. The rather high value of 15 GeV for the threshold is dictated by the background to the electron signature. The total integrated luminosity for this sample is 689 nb^{-1}.

An electron is identified by the presence of a high-p_T ($p_T > 10$ GeV) charged track associated with an electromagnetic cluster ($E_T > 15$ GeV). Owing to the large cell size of the UA1 electromagnetic calorimeter, strong isolation cuts are applied around the electron candidate in order to allow identification of the electron. These isolation cuts reject electrons coming from b-quark and c-quark decay, and therefore already enhance the t-quark signal. A sample of 205 electrons with $E_T(e) > 15$ GeV is obtained.

The 205 electron events are divided into three categories: 119 events have $m_T(e, \nu) > 45$ GeV and are classified as W candidates; 26 events have a second electron candidate, defined by a charged track in the central detector with $p_T > 5$ GeV pointing to an isolated electromagnetic cluster with $E_T > 6$ GeV in the calorimeter, and are classified as dielectron events. These events are background-free and are well understood in terms of W, Z, Drell–Yan, J/ψ, and Υ production. Finally, the remaining 60 events are classified according to their topology: 34 events have no jet, 19 have one jet, and 7 are produced with two jets, where the jets are defined as in subsection 6.2.

The two main background sources are $\pi^\pm \pi^0$ overlaps and unidentified converted photons, mainly from π^0's. Details regarding the estimate of the electron background can be found in Ref. [7].

Table 10 summarizes all contributions to the isolated electron events, according to event topology.

All physics contributions—W, Z, Drell–Yan, J/ψ, Υ, b$\bar{\text{b}}$ and c$\bar{\text{c}}$ are normalized from the data [3–5]. In particular, the b$\bar{\text{b}}$ and c$\bar{\text{c}}$ background is normalized to the non-isolated muon + jet events [5].

With the electron sources summarized in Table 10, the observed event rate is explained for all topologies, without the need for a new heavy-quark contribution. The expected rates from t-quark sources are summarized in Table 11 as a function of mass.

Table 10

Sources of isolated electron + jet events with $E_T(e) > 15$ GeV, $m_T(e, \nu) < 45$ GeV

	Monte Carlo					Data
	Overlap + conversions	W/Z	Drell–Yan J/ψ Υ	$b\bar{b}$ $c\bar{c}$	Total	
e + 0 jet	3 ± 0.2 ± 0.5	26.9 ± 2 ± 1.73	2.1 ± 0.4 ± 0.72	1.8 ± 0.4 ± 0.45	33.8 ± 2.1 ± 2.0	34
e + 1 jet	5.5 ± 0.3 ± 1.0	5.3 ± 0.3 ± 0.34	4.3 ± 0.5 ± 1.5	1.6 ± 0.35 ± 0.4	16.7 ± 0.7 ± 1.9	19
e + ≥ 2 jets	2.6 ± 0.3 ± 0.5	0.8 ± 0.2 ± 0.06	1.1 ± 0.3 ± 0.38	2.2 ± 0.45 ± 0.55	6.7 ± 0.6 ± 0.7	7
e + ≥ 1 jet	8.1 ± 0.5 ± 1.5	6.1 ± 0.4 ± 0.4	5.4 ± 0.6 ± 1.9	3.8 ± 0.6 ± 0.95	23.4 ± 0.9 ± 2.7	26

Table 11

Expected t-quark event rates for isolated electrons
with $E_T(e) > 15$ GeV and accompanied by at least one jet

	t-quark mass (GeV)					
	25	30	40	45	50	55
t$\bar{b}$	3.3 ± 0.3	5.2 ± 0.5	6.2 ± 0.6	6.3 ± 0.6	6.0 ± 0.6	6.4 ± 0.6
t$\bar{t}$	20.7 ± 0.9	14.1 ± 0.7	7.2 ± 0.5	3.2 ± 0.3	3.3 ± 0.3	2.0 ± 0.2
Total	24.0 ± 0.9	19.4 ± 0.9	13.4 ± 0.8	9.5 ± 0.7	9.3 ± 0.7	8.4 ± 0.6

6.6 Upper Bound on t and b′ Cross-Sections

To derive a limit on the t-quark cross-section as a function of mass, UA1 uses the shapes of distributions of event properties that are able to differentiate between t-quark and background processes. For the muon channel the following variables are selected: the muon isolation variable I, $E_T(\nu)$, E_T^{jet1}, and $|\cos \theta^*_{jet2}|$. The distributions for the four variables are shown in Fig. 23 for non-t processes, for a t-quark with a mass of 40 GeV, and for the data. The shapes of t and non-t distributions are different enough to improve the sensitivity to a possible t-quark signal. In addition, the normalization of the non-t contribution is strongly constrained from the I-distribution (Fig. 23a) in the region I > 2, where t-quark events do not contribute.

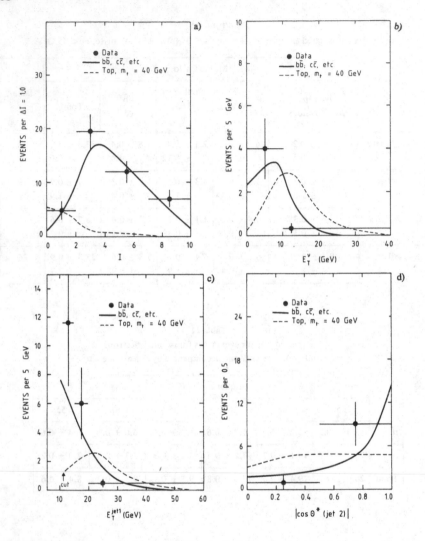

Figure 23 Variables used to derive the t-quark mass limit from the muon-jet data, shown for events with $p_T(\mu) > 12$ GeV and ≥ 2 jets:

a) The isolation variable I (104 events);

b) The missing transverse energy: $E_T(\nu)$ (17 events);

c) The transverse energy of jet 1: E_T^{jet1} (10 events);

d) $|\cos(\theta^*)|$ for jet 2 (10 events).

The variables (b), (c), and (d) are given for events with $I < 2$, and (a), (c), and (d) for events with $E_T^{jet1} > 15$ GeV. The points with error bars represent the data, the solid line all processes except top, and the broken line the expected t-quark contribution for a mass of 40 GeV.

For the electron channel the following two distributions are chosen:

i) The $E_T(\nu)$ spectrum displayed in Fig. 24a for the data, for 'background events' (defined as the list of contributions given in Table 10), for W's, and for t-decay electrons (from a t-quark of 40 GeV mass). The curves are normalized to the expected event rate of each contribution, for a total integrated luminosity of 689 nb^{-1}.

ii) The number of jets in the event: t-quarks contribute more to events with ≥ 2 jets; this effect is more striking for increasing m_t, but the expected rates are lower. The experimental measurements together with the expected rates for top and background processes are shown in Fig. 24b.

The four muon variables [I, $E_T(\nu)$, E_T^{jet1}, $|\cos \theta^*_{jet2}|$] and the two electron variables [$E_T(\nu)$, N^{jets}] are used, in a global fit, to get an upper limit on the t-quark cross-section, which is shown as a function of mass in Fig. 25. Systematic errors have been included for the energy scale, the selection efficiency, the integrated luminosity, the cross-section $\sigma(t\bar{b})$, and the background calculation. The dashed curve represents the lower limit coming from $\sigma(t\bar{b})$ only. It can be seen that the experiment is not sensitive to the process $W \rightarrow t\bar{b}$ alone. A limit on m_t can be obtained only if the contribution from $p\bar{p} \rightarrow t\bar{t}X$ is taken into account. The full curve is the predicted cross-section using the EUROJET calculation for $t\bar{t}$ production. From the intersection with the 95% CL contour, one would obtain $m_t > 56$ GeV at 95% CL. Taking into account the uncertainties arising when the $t\bar{t}$ cross-section is computed at lowest order only, UA1 quotes a mass limit $m_t > 44$ GeV at 95% CL. It should be noted that this mass limit does not take into account uncertainties on α_s.

The method used for the b'-quark is the same as that used for the t-quark, except that in this case there is assumed to be no contribution from W decays. A 95% CL limit on the b' cross-section as a function of the b' mass is obtained, using the combined electron and muon information. The most pessimistic lowest-order cross-section that was used for $t\bar{t}$ gives $m_{b'} > 32$ GeV (95% CL).

ISOLATED ELECTRONS + >1 JET

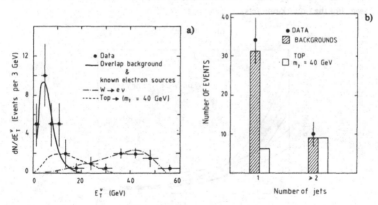

Figure 24   Variables used for t-quark mass limits from electron-jet data.
a) The missing transverse energy $E_T(\nu)$, for events with at least one jet.
b) The jet multiplicity N^{jet}.
The points with error bars represent the data, the solid line all processes except top, the broken line the expected t-quark contribution for a mass of 40 GeV.

260

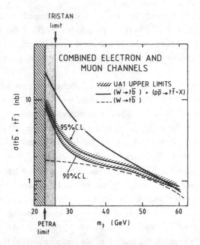

Figure 25 Confidence level contours in the t-quark cross-section versus m_t plane from electron and muon information combined. The regions above the curves are excluded at the 90% and 95% confidence levels, respectively. The TRISTAN limit ($m_t > 26$ GeV) is indicated; ($p\bar{p} \rightarrow t\bar{t}X$) refers to the EUROJET calculation.

6.7 Final Mass Limits using $O(\alpha_s^3)$ Cross-Sections

Figure 26 shows the best estimate by Altarelli et al. [13] of the heavy-quark cross-section and its error band as a function of the heavy-quark mass m, using the next-to-leading calculation described in Section 2. The 95% CL upper limit for the t and b' cross-sections from the UA1 analysis are superimposed. For the t-quark one gets

$$m_t > 41 \text{ GeV} .$$

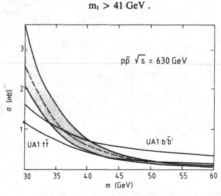

Figure 26 A comparison of the predicted cross-section for heavy-quark production with the UA1 95% CL upper bound on $t\bar{t}$ and $b'\bar{b}'$ pair production cross-sections. The shaded area is obtained by combining in quadrature the uncertainties from $m/2 \leq \mu \leq 2m$ and 90 MeV $\leq \Lambda_5 \leq 250$ MeV. One obtains $m_t > 41$ GeV, $m_{b'} > 34$ GeV. The dashed line corresponds to the lower edge of the cross-section band, if the EHLQ parton densities are used.

The value of the lower limit varies between 41 and 48 GeV if the theoretical prediction is moved from the lower to the upper edge of the error band. On the b'-quark, one finds

$$m_{b'} > 32 \, \text{GeV} ,$$

or a limit between 34 and 40 GeV. Thus this analysis confirms the UA1 values [7] stated previously. However, this limit is now much better justified. The reason why a better limit is not found [although the theoretical error band is smaller for the cross-section computed at $O(\alpha_s^3)$ than it was for the lowest-order case] is that a softer gluon density (DFLM) is preferred [47]. For example, the dashed line in Fig. 26 is the lower limit obtained with the EHLQ parton density (harder glue), which would lead to larger values for m_t and $m_{b'}$. Also, for a given Λ computed at two-loop accuracy, α_s is smaller than what is usually used in lowest-order calculations.

7. B^0-$\bar{B}^0$ MIXING

Weak interactions allow transitions between B^0 and $\bar{B}^0$ states through the exchange of two W's (box diagrams) [57]. In the same way as for the K^0-$\bar{K}^0$ system, and neglecting CP violation, the mass eigenstates are the linear combinations $B_H = (B^0 + \bar{B}^0)/\sqrt{2}$ and $B_L = (B^0 - \bar{B}^0)\sqrt{2}$. The mass difference $\Delta m = m(B_H) - m(B_L)$ results in a time-dependent phase difference between the B_H and B_L wave functions. If the decay time $\tau_{decay} = 1/\Gamma$ is sufficiently long with respect to the oscillation time $\tau_{osc} = 1/\Delta m$, oscillations may be observed. The degree of mixing can be related to measurable quantities, such as the probability that a B^0 decays into a 'wrong'-sign muon ($\Gamma_L = \Gamma_H = \Gamma$):

$$\chi = P(B^0 \to \bar{B}^0 \to \mu^-)/P(B^0 \to \mu^{\pm}) = (1/2)(\Delta m/\Gamma)^2/[1 + (\Delta m/\Gamma)^2] .$$

In the case of maximal mixing ($\tau_{decay} \gg \tau_{osc}$ or $\Delta m/\Gamma \gg 1$), χ approaches 0.5. Oscillations can occur for the two neutral-meson states $B_d^0 = (\bar{b}d)$ and $B_s^0 = (\bar{b}s)$. The mass difference Δm can be calculated from the box diagrams [57] for the two states, but many parameters are unknown (in particular the t-quark mass); hence precise predictions of the strength of B^0-$\bar{B}^0$ oscillations are not possible. Theoretical estimates of χ are in the range 0.1 to 0.48 for B_s^0 and less than 0.2 for B_d^0. In the Standard Model with three families, existing limits on Kobayashi–Maskawa (KM) matrix elements imply [58, 59] that

$$\chi_d < 0.21^2 \chi_s/[1 - 2\chi_s(1 - 0.21^2)] .$$

7.1 Evidence for B^0-$\bar{B}^0$ Mixing

Non-isolated dimuons at the CERN p$\bar{p}$ Collider come essentially from b$\bar{b}$ production. Therefore the dimuon system is a good laboratory for the study of B^0-$\bar{B}^0$ oscillations [3]. In the absence of B^0-$\bar{B}^0$ oscillations, dimuons coming from first-generation decays of b$\bar{b}$ pairs must have opposite signs. Like-sign dimuons are expected from second-generation decays, where one muon arises from b decay and the other from the c decay of the associated bottom–charm cascade. The signature for B^0-$\bar{B}^0$ oscillations is a yield of like-sign dimuon events in excess of that expected for second-generation decays.

In the non-isolated dimuon sample of the UA1 experiment (399 events), there are 142 like-sign events (background 58) and 257 unlike-sign events (background 58). The contributions from Drell–Yan and Υ decays are negligible. The ratio of like-sign to unlike-sign dimuons is therefore

$$R = N(\pm \pm)/N(+ -) = (142 - 58)/(257 - 58) = 0.42 \pm 0.07 \pm 0.03 ,$$

where the first error is statistical and the second error is the systematic error coming from the background subtraction. In the absence of oscillations, the expected value is $R = N_s/(N_f + N_c)$, where

Table 12

Predictions for R without oscillations ($\chi = 0$) and
for three values of the mixing parameter χ

Reference	$\chi = 0$	$\chi = 0.05$	$\chi = 0.10$	$\chi = 0.15$	Charm content
Barger and Phillips [60]	0.25	0.31	0.36	0.42	0.23
Halzen and Martin [61]	0.25	0.33	0.41	0.48	0.11
ISAJET	0.26	0.34	0.42	0.50	0.10
EUROJET	0.21	0.28	0.36	0.43	0.15

N_f is the number of dimuons from first-generation $b\bar{b}$ decays, N_s is the number due to one first-generation plus one second-generation decay from $b\bar{b}$, and N_c is the number due to $c\bar{c}$ decays. Using ISAJET to estimate N_s (0.75 nb), N_f (2.52 nb), and N_c (0.35 nb), the expected value for no mixing is 0.26 ± 0.03. The error is due to uncertainties from the b and c fragmentation and the semileptonic branching ratios. The ISAJET value agrees well with independent calculations [60, 61] (Table 12). The measured value of R is larger than that expected in the absence of mixing. When the mixing is possible, the value of R is related to χ—the probability of a b (or $\bar{b}$) decaying into a 'wrong'-sign muon—by

$$R = \{2\chi(1-\chi)N_f + [(1-\chi^2) + \chi^2]N_s\}/\{[(1-\chi^2) + \chi^2]N_f + 2\chi(1-\chi)N_s + N_c\}.$$

Table 12 shows the expected value of R for different values of χ and for different calculations. The preferred value for χ is between 0.10 and 0.15. To determine χ more precisely, a maximum likelihood fit to the two-dimensional p_T distribution of the two muons is performed. Second-generation muons

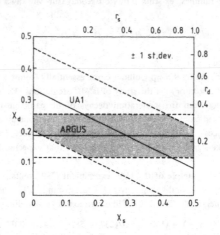

Figure 27 The ± 1 standard deviation bands from UA1 and ARGUS: $f_s = 0.18$ for UA1, whilst $f_s = 0.15$ was used for MARK II. This is justified by the lower Q^2 of the $e^+ e^-$ data.

have a softer p_T spectrum than have primary muons. The likelihood fit takes into account the following errors: ±50% error on the c contribution; ±25% error on the background normalization; error on average semileptonic branching ratios, BR(B → e) = (12 ± 0.7)%, BR(D → e) = (13 ± 1.3)%. The fit results in $\chi = 0.121 ± 0.047$ or $\chi > 0.065$ at 90% CL.

The mixing parameter χ is related to χ_s and χ_d by $\chi = f_s\chi_s + f_d\chi_d$, where f_s and f_d are the probabilities for a b to pick up an $\bar{s}$ or a $\bar{d}$ from the sea to form a $\bar{B}^0_s$ or a $\bar{B}^0_d$. From the ratio $K^+/\pi^+ = 0.5$ measured at large p_T at the CERN ISR, and allowing for 10% B-baryons, UA1 uses $f_s = 0.18$ and $f_d = 0.36$. The non-zero value of χ, measured by UA1, results in the ±1 st. dev. band in the χ_s–χ_d plane shown in Fig. 27.

7.2 Comparison with other Experiments

Searches for B^0–$\bar{B}^0$ oscillations were made in e^+e^- experiments by CLEO [62], ARGUS [63], MARK II [64], and JADE [65]. Their results are given in Table 13.

Table 13

Results on B^0–$\bar{B}^0$ oscillations from e^+e^- experiments

Reference	$b\bar{b}$ source	$\sqrt{s}$ (GeV)	Result
CLEO	$\Upsilon(4s)$	10.6	$\chi_d < 0.19$ (90% CL)
ARGUS	$\Upsilon(4s)$	10.6	$\chi_d = 0.19 ± 0.07$
MARK II	$e^+e^- \to b\bar{b}$	29	$\chi < 0.12$ (90% CL)
JADE	$e^+e^- \to b\bar{b}$	34	$\chi < 0.13$ (90% CL)

CLEO and ARGUS use the $\Upsilon(4s)$ resonance, which decays into $B^\pm$ and B^0_d; thus they are sensitive to B^0_d mixing only. MARK II and JADE use continuum data well above the B^0_s threshold and are therefore, like UA1, sensitive to a linear contribution of B^0_d and B^0_s mixing, depending on f_s and f_d. The only positive result coming from ARGUS, $\chi_d = 0.187 ± 0.069$, is shown in Fig. 27, together with the UA1 result. The two measurements agree over a large area of the χ_s–χ_d plane.

Figure 28 shows the allowed region at 90% CL in the χ_s–χ_d plane, resulting from all existing measurements; χ_d is restricted to between 0.08 and 0.20 mainly due to ARGUS, whilst there is no

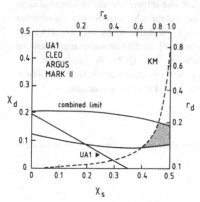

Figure 28   Combined limit on χ_d and χ_s from UA1, MARK II, CLEO, and ARGUS (solid lines 90% CL). The dashed line indicates the allowed region coming from present limits on KM matrix elements. The dotted region is allowed by both constraints. The lower limit coming from UA1 alone is also indicated: $\chi > 0.065$ at 90% CL.

restriction on χ_s. All values between 0 (no mixing) and 0.5 (maximal mixing) are possible. The dashed line in Fig. 28 indicates the allowed region in the χ_s–χ_d plane from the existing limits on the KM matrix elements discussed previously [58]. If this constraint is combined with the experimental limits, only a small allowed region remains, implying almost maximal mixing in the B_s^0 channel: $\chi_s > 0.40$ at 90% CL ($\chi_s > 0.30$ is from UA1 alone).

8. CONCLUSIONS

Detailed measurements of the b production cross-section have been achieved at the CERN $p\bar{p}$ Collider, using muons to tag the b jet. The main limitation of this technique comes from the uncertainty on the fragmentation $b \to \mu$, which is not measured in the same experiment. Within the uncertainties affecting both the measurements and the theory, good agreement is observed between the data and QCD expectations for b production at $\sqrt{s} = 0.63$ TeV.

The ratio of like-sign to unlike-sign dimuons is not affected by the above uncertainties and is a good tool for studying B^0–$\bar{B}^0$ mixing. The first evidence for this mixing came from UA1 [3] and was later confirmed by ARGUS [63]. So far, bottom physics could not be studied in the electron channel. The new generation of detectors at the upgraded CERN $p\bar{p}$ Collider using ACOL and at the Fermilab Tevatron ($\sqrt{s} = 1.8$ TeV), with better electron identification, may extend the feasibility of bottom physics to electrons. More precise measurements of χ are therefore expected in both the muon and the electron channel, which will put more constraint on the Standard Model in the χ_s–χ_d plane.

The limit $m_t > 41$ GeV at 95% confidence level is now firmly established. Two elements were necessary to achieve this: the good experimental control of the $b\bar{b}g$ background in the muon channel, and the reduced theoretical error on $\sigma(t\bar{t})$ now computed at the next-to-leading order. Future experiments with ACOL and at the Tevatron will be in severe competition to find the t-quark. If $m_t < m_W$, ACOL will compensate for its smaller centre-of-mass energy by having better luminosity. With an expected integrated luminosity of 10 pb^{-1} and provided $m_t < m_W$, ACOL can find the t-quark from the dominant channel $W \to t\bar{b}$ only. If, on the other hand, $m_t > m_W$, the Tevatron will benefit from its larger centre-of-mass energy: for $m_t = 100$ GeV, for example, the $t\bar{t}$ cross-section increases by a factor of 40 between $\sqrt{s} = 0.63$ TeV and $\sqrt{s} = 1.8$ TeV (see Table 1). This large factor is only marginally compensated by the expected higher luminosity with ACOL (about a factor of 10). If the t-quark is heavier than the W, an interesting situation occurs. The t-quark can decay into a real W, so that the process $p\bar{p} \to t\bar{t}X$ leads to a final state with two real large-p_T W's. The average p_T of each t-quark with respect to the beam axis is $\langle p_T(t) \rangle \approx m_t/2$. In the decay $t \to Wb$, the W — because of its mass — will carry most of the transverse momentum of the parent t-quark and is therefore also produced at large p_T: $\langle p_T(W) \rangle \approx 45$ GeV. If one of the W's decays into $e\nu$ (or $\mu\nu$) and the other into two jets, the resulting $\ell\nu$jj topology is similar to the QCD production of a large-p_T W balanced by two jets. UA1 has observed two Wjj events with $p_T(W) > 80$ GeV and $m_{jj} \approx m_W$ [66]. In the kinematic range where the events are observed, 0.05 ± 0.03 events are predicted for QCD W production and $0.02^{+0.02}_{-0.01}$ events for $t\bar{t}$ production with $m_t = 90$ GeV. No conclusion can be drawn from the present statistics. Future experiments with ACOL and at the Tevatron should clarify the situation.

Acknowledgement

We wish to thank the Scientific Reports Editing and Text Processing Sections at CERN for their dedicated help in preparing this paper.

REFERENCES

[1] Tavernier, S.P.K., Progr. in Physics **50**, 1439 (1987).

[2] Schubert, K.R., Proc. Int. Europhysics Conf. on High-Energy Physics, Uppsala, 1987, ed. O. Botner (Uppsala Univ., 1987), vol. II, p. 791.

[3] Albajar, C. et al. (UA1 Collab.), Phys. Lett. **186B**, 237 and 247 (1987).

[4] Albajar, C. et al. (UA1 Collab.), preprint CERN-EP/88-46 (1988), to appear in Phys. Lett. B.

[5] Albajar, C. et al. (UA1 Collab.) Phys. Lett. **200B**, 380 (1988).

[6] Albajar, C. et al. (UA1 Collab.), Z. Phys. **C37**, 489 (1988).

[7] Albajar, C. et al. (UA1 Collab.), Z. Phys. **C37**, 505 (1988).

[8] Mazzanti, P. and Wada, S., Phys.Rev. **D26**, 602 (1982).

[9] Brodsky, S.J., Collins, J.C., Ellis, S.D., Gunion, J.F. and Mueller, A.H., Proc. 1984 Summer Study on the Design and Utilization of the SSC, Snowmass, Colo., 1984, eds. R. Donaldson and J. Morfin (AIP, New York, 1985), p. 221.

[10] Collins, J.C., Soper, D.E. and Sternan, G., Nucl. Phys. **B263**, 37 (1986).

[11] Ellis, R.K., Proc. 21st Rencontre de Moriond: Strong Interactions and Gauge Theories, Les Arcs, 1986 (Éd. Frontières, Gif-sur-Yvette, 1987), p. 339.

[12] Dawson, S., Ellis, R.K. and Nason, P., preprint Fermilab-Pub-87/222-T (1987).

[13] Altarelli, G., Diemoz, M., Martinelli, G. and Nason, P., preprint CERN–TH.4978 (1988).

[14] Ellis, R.K., Georgi, H., Machacek, M., Politzer, H.D. and Ross, G.G., Nucl. Phys. **B152**, 285 (1979), and references therein.

[15] Combridge, B.L., Nucl. Phys. **B151**, 429 (1979).

[16] Babcock, J., Sivers, D. and Wolfram, S., Phys. Rev. **D18**, 162 (1978).

[17] Hagiwara, K. and Yoshino, T., Phys. Lett. **80B**, 282 (1979).
Jones, L.M. and Wild, H., Phys. Rev. **D17**, 782 (1978).
Georgi, H. et al., Ann. Phys. **114**, 273 (1978).

[18] Barger, V., Halzen, F. and Keung, W.Y., Phys. Rev. **D25**, 112 (1979).

[19] Brodsky, S.J., Hoyer, P., Peterson, C. and Sakai, N., Phys. Lett. **93B**, 451 (1980).
Brodsky, S.J., Peterson, C. and Sakai, N., Phys. Rev. **D23**, 2745 (1981).

[20] Halzen, F. and Hoyer, P., Phys. Lett. **154B**, 324 (1985).

[21] Diemoz, M., Ferroni, F., Longo, E. and Martinelli, G., in preparation.
See also Ref. [1].

[22] Catanesi, M.G. et al. (WA78 Collab.), Phys. Lett. **202B**, 453 (1988).

[23] Bordalo, P. et al. (NA10 Collab.), preprint CERN-EP/88-39 (1988), to appear in Z. Phys. C.

[24] Mueller, A.H. and Nason, P., Phys. Lett. **157B**, 226 (1985); Nucl. Phys. **B266**, 265 (1986).

[25] Dawson, S., Ellis, R.K. and Nason, P., in preparation.

[26] Paige, F. and Protopopescu, S.D., ISAJET, BNL 38034 (1986).
Odorico, R., COJETS, Nucl. Phys. **B228**, 381 (1983).
Bengtsson, H.-U. and Sjöstrand, T., PYTHIA, Comput. Phys. Commun. **46**, 43 (1987).

[27] Kunszt, Z. and Pietarinen, E., Nucl. Phys. **B164**, 45 (1980).
Ellis, R.K. and Sexton, J.C., Nucl. Phys. **B269**, 445 (1986).
Kunszt, Z. and Gunion, J.F., Phys. Lett. **178B**, 296 (1986).

[28] Ali, A. et al., Nucl. Phys. **B292**, 1 (1987).
van Eijk, B., Ph.D. thesis, University of Amsterdam (1987).

[29] Eichten, E. et al., Rev. Mod. Phys. **56**, 579 (1984) and **58**, 1065 (1986).

[30] Fox, G.C. and Wolfram, S., Nucl. Phys. **B168**, 285 (1980).

[31] Albajar, C. et al. (UA1 Collab.), Z. Phys. **C36**, 33 (1987).

[32] Gottschalk, T.D., Caltech preprint CALT–68–1241 (1985).
Sjöstrand, T., Phys. Lett. **157B**, 321 (1985).

[33] Köpp, G. et al., Phys. Lett. **153B**, 315 (1985).
Ali, A. and Ingelman, G., Phys. Lett. **156B**, 111 (1985).

[34] Berger, E.L., Proc. Topical Seminar on Heavy Flavours, San Miniato, 1987 [Nucl. Phys. B (Proc. Suppl.) **1B** (1988)], p. 425.

[35] Field, R.D. and Feynman, R.P., Nucl. Phys. **B136**, 1 (1978).

[36] Chen, A. et al., Phys. Rev. Lett. **52**, 1084 (1984).
Green, J. et al., Phys. Rev. Lett. **51**, 347 (1983).
Csorna, S.E. et al., Phys. Rev. Lett. **54**, 1894 (1985).
Bacino, W. et al., Phys. Rev. Lett. **43**, 1073 (1979).
Schindler, R. et al., Phys. Rev. **D24**, 78 (1981).
Baltrusaitis, R. et al., Phys. Rev. Lett. **54**, 1976 (1985).
Bethke, S., Z. Phys. **C29**, 175 (1985), and references therein.

[37] Peterson, C. et al., Phys. Rev. **D27**, 105 (1983).

[38] Particle Data Group, Review of Particle Properties, Phys. Lett. **170B** (1986).

[39] Rushbrooke, J.G., Proc. 16th Int. Symp. on Multiparticle Dynamics, Kiryat-Anavim (Israel), 1985 (Éd. Frontières, Gif-sur-Yvette, 1985).

[40] Eggert, K. et al., Nucl. Instrum. Methods **176**, 217 and 223 (1987).

[41] Nason, P., talk presented at Les Rencontres de Physique de la Vallée d'Aoste, La Thuile, Italy (1988).

[42] Baier, R. and Rückl, R., Phys. Lett. **102B**, 364 (1981); Z. Phys. **C19**, 251 (1983).

[43] Baier, R. and Rückl, R., Nucl. Phys. **B208**, 381 (1982).
Halzen, F. et al., Phys. Rev. **D30**, 700 (1984).
Glover, E.W.N., Halzen, F. and Martin, A.D., Phys. Lett. **185B**, 441 (1987).

[44] Fritzsch, H., Phys. Lett. **86B**, 164 and 343 (1979).

[45] Humpert, B., preprint CERN. TH–4551/86 (1986).

[46] Alam, M. et al., Phys. Rev. **D34**, 3279 (1986).
Albrecht, H. et al., Phys. Lett. **162B**, 395 (1985); also Heidelberg preprint IHEP–HD/86–3 (1986).

[47] Diemoz, M., Ferroni, F., Longo, E. and Martinelli, G., Z. Phys. **C39**, 21 (1988).

[48] Adeva, B. et al. Phys. Rev. Lett. **50**, 799 (1983).
Bartel, W. et al., Phys. Lett. **132B**, 241 (1983).
Avery, P. et al., Phys. Rev. Lett. **53**, 1309 (1984).
Bean, A. et al. (CLEO Collaboration), Cornell Univ. preprint CLNS 87/73 (1987).

[49] Amaldi, U. et al., Phys. Rev. **D36**, 1385 (1987), and references therein.
Altarelli, G., Proc. Int. Europhysics Conf. on High-Energy Physics, Uppsala, 1987, ed. O. Botner (Uppsala Univ., 1987), vol. I, p. 372.
Sirlin, A., talk given at the Int. Symp. on Lepton and Photon Interactions at High Energies, Hamburg, 1987.

[50] Albajar, C. et al. (UA1 Collab.), Europhys. Lett. **1**, 327 (1986).
Ansari, R. et al. (UA2 Collab.), Phys. Lett. **186B**, 440 (1987); Erratum, Phys. Lett. **190B**, 238 (1987).

[51] Sugahara, R. (TOPAZ Collab.), talk given at the 22nd Rencontre de Moriond on Current Issues in Hadron Physics, Les Arcs, 1988.

[52] Sagawa, H. (AMY Collab.), ibid.

[53] Ellis, J., Hagelin, J.S. and Rudaz, S., Phys. Lett. **192B**, 201 (1987).
Barger, V., Hau, T. and Nanopoulos, D.V., Phys. Lett. **194B**, 312 (1987).
Bigi, I.I. and Sanda, A.I., Phys. Lett. **194B**, 307 (1987).
Chau, L.L. and Keung, W.Y., UC Davis preprint, UCD-87-02 (1987).
Harari, H. and Nir, Y., Stanford preprint SLAC-PUB-4341 (1987).
Altarelli, G. and Franzini, P.J., preprint CERN-TH.4745/87 (1987).
Albrecht, H. et al. (ARGUS Collaboration), Phys. Lett. **192B**, 245 (1987).

[54] Arnison, G. et al. (UA1 Collab.), Lett. Nuovo Cimento **44**, 1 (1985).
Albajar, C. et al. (UA1 Collab.), Studies of the W and Z^0 properties at the CERN Proton-Antiproton Collider, in preparation.

[55] Chang, T.H. et al., Amsterdam preprint NIKHEF-H 81/34 (1981).
Gusken, S. et al., Phys. Lett. **155B**, 185 (1985).
Kühn, J.H. et al., Nucl. Phys. **B272**, 560 (1986).
Tholl, H.D., Diploma thesis, Aachen (1985).

[56] Arnison, G. et al. (UA1 Collab.), Phys. Lett. **132B**, 214 (1983).

[57] Chau, L.L., Phys. Rep. **95**, 1 (1983).
Buras, A.J. et al., Nucl. Phys. **B245**, 3691 (1984).
Ali, A. and Jarlskog, C., Phys. Lett. **144B**, 266 (1984).

[58] Eggert, K. and Moser, H.-G., Aachen report PITHA 87-10 (1987).

[59] Kleinknecht, K. and Renk, B., Z. Phys. **C34**, 209 (1987).

[60] Barger, V. and Phillips, R.J.N., Wisconsin preprints MAD/PH/155, 239, and 266 (1984).

[61] Halzen, F. and Martin, A., Wisconsin preprint DTP/84/14 (1984).

[62] Bean, A. et al. (CLEO Collab.), Phys. Rev. Lett. **58**, 183 (1987).

[63] Albrecht, H. et al. (ARGUS Collab.), Phys. Lett. **192B**, 245 (1987).

[64] Schaad, T. et al. (MARK II Collab.), Phys. Lett. **160B**, 188 (1985).

[65] Bartel, W. et al. (JADE Collab.), Phys. Lett. **146B**, 437 (1984).

[66] Arnison, G. et al. (UA1 Collab.), Phys. Lett. **193B**, 389 (1987).

SEARCHES FOR NEW PHYSICS

John Ellis and Felicitas Pauss

CERN, Geneva, Switzerland

1. INTRODUCTION

The CERN $p\bar{p}$ Collider has been the first accelerator to operate in a completely new energy domain, reaching centre-of-mass energies an order of magnitude larger than those previously available with the Intersecting Storage Rings (ISR) at CERN, or with the Positron–Electron Tandem Ring Accelerator (PETRA) at DESY and the Positron–Electron Project (PEP) at SLAC. Naturally there has been great interest in the searches for new physics in this virgin territory. Theorists have approached these searches from either or both of two rival points of view. Either they have had an *a priori* prejudice as to what new physics should be searched for, and what its signatures should be, or they have tried to interpret *a posteriori* some experimental observations. Whilst some experimentalists' searches have been moulded by such *a priori* prejudices, many have emerged from systematic studies of all the measurable parameters of the events. The basic building-blocks of new physics in the 100 GeV energy domain are jets j, charged leptons ℓ, photons γ, and missing transverse energy E_T. Therefore searches have been conducted in channels which are combinations of these elements. Table 1.1 lists the various combinations of j, ℓ, γ, and E_T which have been explored by the UA1 and UA2 experiments. It also shows some of the main *a priori* theoretical prejudices which can be explored in each of these channels. Entries in Table 1.1 indicate the major searches which have been made, and which are reviewed here.

The layout of the rest of this paper is as follows. There are sections discussing each of the major prejudices: the Standard Model in Section 2; supersymmetry in Section 3; extra gauge degrees of freedom in Section 4; composite models in Section 5; and other possibilities in Section 6. Each of these sections contains a description of the motivations and characteristics of the new physics to be searched for, followed by a review of the searches made up to now at the CERN $p\bar{p}$ Collider. Finally, Section 7 summarizes the lessons to be learnt so far from searches for new physics at the CERN $p\bar{p}$ Collider, and previews some of the prospects for the next rounds of collider searches at CERN and FNAL.

Table 1.1
Possible new physics and signatures

	Standard Model	Supersymmetry	Extra gauge bosons	Composite models etc.
j–j			$Z' \to q\bar{q}$ W'	$\Lambda_{contact}$ q^*
j–ℓ	t, b'			Leptoquark
j $\not{E}_T$	N_ν $W \to L\nu_L$	$\tilde{g}, \tilde{q}$		$X \to ZW$ ZZ Leptoquark
$\ell\ell$			$Z' \to \ell\ell$	
$\ell \not{E}_T$	$W \to L\nu_L$	$W \to \tilde{\ell}^* \bar{\nu}$	$W' \to \ell\nu$	
jℓ $\not{E}_T$	t, b'	$\tilde{W}, \tilde{Z}$	$Z' \to WW$ $W' \to ZW$	$X \to WW$ ZW Leptoquark
j$\ell\ell$			$W' \to WZ$ $Z' \to ZZ$	$X \to WZ$ ZZ Leptoquark
$\ell\ell \not{E}_T$	$Z \to L^+L^-$	$Z \to \tilde{\ell}\tilde{\ell}$ $Z \to \tilde{W}\tilde{W}$		
γX				$Z \to \ell\ell\gamma$ $W \to \ell\nu\gamma$

2. THE STANDARD MODEL

Within the minimal version of the Standard Model [2.1] with just three generations of quarks and leptons and just one Higgs doublet, there are just three more elementary particles waiting to be discovered, namely the ν_τ, the t-quark, and the neutral Higgs boson. Going beyond the minimal version, the Standard Model could easily accommodate more quark and lepton generations — implying more neutrinos, new charged heavy leptons L, and quarks, and/or more Higgs doublets — in turn implying more neutral Higgs bosons and also charged ones. Searches for more quarks at the CERN $p\bar{p}$ Collider are discussed elsewhere [2.2], as are constraints on the number N_ν of neutrino species inferred from the production rates for $W^\pm$ and Z^0 bosons [2.3]. In this section we concentrate on neutrino counting using events with large $\not{E}_T$, and on the searches for heavy leptons and the Higgs boson.

2.1 Neutrino Counting

Even though the ν_τ has not been observed directly, there are two indirect arguments for its existence. One is the agreement of the τ lifetime and leptonic branching ratio with the Standard Model prediction [2.4], and the other is the observation of $W \to \tau\nu$ decay at a rate consistent with e-μ-τ universality of the weak charged-current coupling at $Q^2 = m_W^2$ [2.3]. Defining the ratio of coupling strengths $(g_1/g_2)^2 \equiv \Gamma(W \to \ell_1\nu)/\Gamma(W \to \ell_2\nu)$, and combining the $\sqrt{s} = 546$ GeV and 630 GeV $W \to \ell\nu_e$ ($\ell = e, \mu, \tau$) data samples, UA1 obtained [2.5]

$$g_\mu/g_e = 1.00 \pm 0.07 \pm 0.04, \quad g_\tau/g_e = 1.01 \pm 0.10 \pm 0.06, \qquad (2.1)$$

where the first error is statistical and the second one is due to systematic uncertainties. All the neutrinos are expected to be massless in the minimal version of the Standard Model, but could in principle acquire masses from extensions of it. Direct measurements of τ decay tell us that $m_{\nu_\tau} < 35$ MeV/c^2 [2.6], and cosmological considerations using reasonable hypotheses about possible ν_τ decay modes then suggest that $m_{\nu_\tau} \lesssim 100$ eV/c^2 [2.7]. Therefore any neutrino counting experiment with a kinematic range $\gtrsim 1$ keV should find $N_\nu \geq 3$.

An important historical role in bounding N_ν has been played by cosmology through the successful confrontation of primordial nucleosynthesis calculations with observation [2.8]. The best fit is obtained with $N_\nu = 3$, but the upper bound on N_ν depends on other inputs. If we assume [2.9] (more conservatively [2.10]) that the primordial ^{4}He abundance $Y(^4\text{He}) < 0.254$ (0.26), that the baryon-to-photon ratio $N_B/N_\gamma \gtrsim 3 \times 10^{-10}$ (2×10^{-10}) as suggested by the abundance of D + ^{3}He (^{7}Li), and that the neutron half-life $t_{1/2}(n) > 10.4$ (10.2) minutes, we find for the two assumptions

$$N_\nu < 4.0 \ (5.2). \qquad (2.2)$$

(Keeping track of the decimal is not useful for counting conventional left-handed neutrinos, but it is useful when extending the analysis to other light neutral particle species, such as right-handed neutrinos.) More recently, astrophysics has provided an interesting upper bound on N_ν, thanks to the observation of neutrinos from the supernova SN 1987a [2.11]. The events seen are presumably due to $\bar{\nu}_e$, and can be used to infer the energy output $E_{\bar{\nu}_e}$ of the supernova through $\bar{\nu}_e$. As a first approximation, this can be multiplied by $2N_\nu$ to infer the total energy output through all neutrinos. This must be less than the binding energy of a neutron star, which is $< 4 \times 10^{53}$ erg. Thus the data can be used to bound N_ν [2.12]:

$$N_\nu \leq 6. \qquad (2.3)$$

(We have rounded down to the nearest integer: any extension to right-handed neutrinos requires more input on the core of the neutron star.) The best fit to the supernova data is again with $N_\nu = 3$.

The best particle physics experimental limit, apart from those obtained at the CERN $p\bar{p}$ Collider, is provided by searches for the reaction $e^+e^- \to \gamma$ + nothing. At the present PEP and PETRA energies, the cross-section for this reaction is $\propto (N_\nu + 4)$. A compilation of the world's data (ASP, MAC, CELLO) gives [2.13]

$$N_\nu < 4.9 \qquad (90\% \text{ CL}). \qquad (2.4)$$

This upper bound is increased to 7.3 if one assumes that there are at least three neutrino species [2.14]. At the SLAC Linear Collider (SLC) and at the CERN Large Electron–Positron storage ring (LEP) the cross-section for this process just above the Z peak will be $\propto N_\nu$, and it should be possible to obtain high enough statistics to measure N_ν with a 10% error [2.15].

One of the ways to obtain N_ν at the CERN $p\bar{p}$ Collider is to measure the ratio $\sigma \cdot \text{BR}(W^\pm \to e^\pm \nu)/\sigma \cdot \text{BR}(Z \to e^+e^-)$. Measurements of $W^\pm$ and Z production and their properties are discussed elsewhere in this volume [2.3]. Here we just note that the numerator in this ratio is approximately $\propto (N_\nu + 12)$, with an uncertainty depending on other possible unknown decay modes of the Z and/or $W^\pm$. Ignoring all such possibilities except $W \to t\bar{b}$ and $Z \to t\bar{t}$, a compilation of UA1 and UA2 data yields

$$N_\nu < 5.9 \qquad (90\% \text{ CL}) \qquad (2.5)$$

for $m_t > 44$ GeV/c^2 if one imposes the bound $N_\nu \geqslant 3$. Details can be found in Ref. [2.16].

The other way to measure N_ν at the CERN $p\bar{p}$ Collider is via the contribution to the missing-energy cross-section due to the process $p\bar{p} \to$ gluon or quark + $(Z \to \nu\bar{\nu})$ + X, i.e. Z production at large p_T. In this case the p_T^Z is transformed into large $E\!\!\!/_T$ since the neutrinos go undetected. We then observe the recoil gluon (or quark) which appears as a jet in the detector. The cross-section is $\propto N_\nu$, and can be not only calculated using QCD but also calibrated by using gluon + $(W \to e\nu)$ events, i.e. W's produced at large p_T [2.17]. However, there are significant backgrounds to high-p_T Z production: for example, from the production of heavy quarks (with subsequent semileptonic decay of the heavy quark, where the charged lepton goes undetected) and $W \to \tau\nu$ (where the τ decays hadronically), which produce genuine $E\!\!\!/_T$; and from instrumental effects due to mismeasurements of jet energies in QCD events with no ν emission, which produce fake $E\!\!\!/_T$. Since missing-energy events play key roles in the searches for other new types of physics, e.g. heavy charged leptons L, squarks $\tilde{q}$, and gluinos $\tilde{g}$, we discuss here in more detail the UA1 1983 + 1984 + 1985 $E\!\!\!/_T$ event sample and possible background sources.

The aim was to define an event sample in which the background due to fluctuations in the detector response was small. The significance of the measured $E\!\!\!/_T$ was defined by

$$N_\sigma = E\!\!\!/_T / 0.7 \sqrt{\Sigma E_T}, \qquad E\!\!\!/_T \equiv |\sum_i \mathbf{E}_T^i|, \qquad (2.6)$$

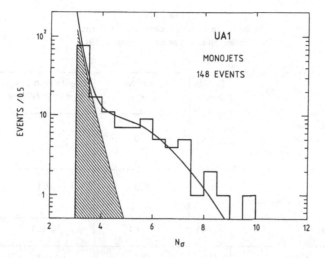

Fig. 2.1 Distribution of N_σ [Eq. (2.6)] for $> 3\sigma$ monojet events (only one jet with $E_T^j > 12\,\text{GeV}$) shown as histogram, compared with all expected contributions (solid line) and jet fluctuation contributions only (shaded area) [2.19].

where all energies are in units of GeV, and E_T^i is a vector in the direction of the calorimeter cell i with magnitude E_T^i and ΣE_T is the scalar sum of transverse energy observed in all calorimeter cells. The background due to fluctuations of the detector response was evaluated with a Monte Carlo technique using real jet data [2.5]. Figure 2.1 shows the distribution of N_σ for events with $N_\sigma > 3$ compared with all expected contributions (solid line) and jet fluctuation contributions only (shaded area). For $N_\sigma < 4$ the observed event rate is dominated by contributions from jet fluctuations. Accordingly, an inclusive selection was made of events with isolated $\not{E}_T$ with $N_\sigma > 4$, fulfilling the following conditions: i) $\not{E}_T > 15\,\text{GeV}$ and $N_\sigma > 4$; ii) at least one jet with $E_T^j > 12\,\text{GeV}$ in $|\eta| < 2.5$ and a matching central detector (CD) track of $p_T > 1\,\text{GeV/c}$; (iii) no e or μ candidates; (iv) no back-to-back jets, i.e. remove events with a second jet[*] in an azimuthal angle region of $\pm 30°$ opposite to the most energetic jet and around the $\not{E}_T$ direction; and (v) additional $\not{E}_T$ validation cuts were applied. A total of 56 events passed the selection cuts. The observed rate is in very good agreement with the contributions expected from Standard Model physics, as seen in the first column of Table 2.1. The calculations for all known physics processes were done with the ISAJET Monte Carlo program [2.18] including i) simulation of the full proton–antiproton collision to take into account the

---

[*] Jet observed in the calorimeter with $E_T > 8\,\text{GeV}$, or jet observed in the CD drift chambers with $p_T > 5\,\text{GeV/c}$.

Table 2.1

Predicted rates for processes giving large $\not{E}_T$
using all event selection cuts [2.19]

Process	Events (total)	Events with $L_\tau < 0$	Events with $L_\tau < 0$ and $E_T^j < 40\,\text{GeV}$
$W \to e\nu$ $\to \mu\nu$ $\to \tau\nu \to$ leptons	3.6	2.0	1.4
$W \to \tau\nu$ $\to \nu\bar{\nu} +$ hadrons	36.7	8.0	7.1
$W \to c\bar{s}$	< 0.1	< 0.1	< 0.1
$Z \to \tau^+\tau^-$	0.5	0.1	0.1
$Z \to \nu\bar{\nu}$ (3 neutrino species)	7.4	7.1	5.6
$Z \to c\bar{c}$ and $b\bar{b}$	< 0.1	< 0.1	< 0.1
$c\bar{c}$ and $b\bar{b}$ (direct production)	0.2	0.2	0.2
Jet fluctuations (fake missing energy)	3.8	3.4	3.4
Total	52.2	20.8 ± 5.1 ± 1.0	17.8 ± 3.7 ± 1.0

effects of the spectator particles, and ii) simulation of the UA1 detector including the hardware triggers. The accuracy of these Monte Carlo predictions has been checked by loosening the cuts, e.g. by relaxing to $N_\sigma > 3$, and by removing the back-to-back cut for events with $N_\sigma > 4$.

A scatter-plot of the 56 $\not{E}_T$ events remaining after the primary selection is shown in Fig. 2.2. The horizontal axis is the E_T^j of the highest-E_T jet, whilst the vertical axis is a measure of the τ-likelihood L_τ of the event, defined by

$$L_\tau = \ln\left(P_F P_R P_N\right). \tag{2.7}$$

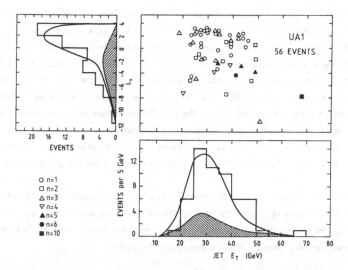

Fig. 2.2 Scatter plot of L_T [Eq. (2.7)] versus the transverse energy of the highest-E_T jet in the event. The projections for L_T and E_T^j are plotted for the data (histogram), for all predicted contributions (solid line), and for non-τ contributions (shaded area). The different symbols indicate the charged multiplicity N [Eq. (2.10)] of the highest-E_T jet for the 56 events [2.19].

Here P_F, P_R, and P_N are the relative probabilities for values in the distributions of F, R, and N:
— F is the fraction of the jet energy, as measured in the calorimetry

$$F \equiv \frac{\Sigma\, E_T \text{ in } \Delta R < 0.4}{\Sigma\, E_T \text{ in } \Delta R < 1}, \qquad (2.8)$$

where $\Delta R = (\Delta\theta^2 + \Delta\eta^2)^{1/2}$; so that F is a measure of the collimation of the jet;
— R is the angular separation in η–ϕ space,

$$R \equiv [(\phi_{T_1} - \phi_J)^2 + (\eta_{T_1} - \eta_J)^2]^{1/2}, \qquad (2.9)$$

where the subscript T_1 refers to the CD track with the largest p_T, and J refers to the calorimeter jet axis; and
— N is the number of charged tracks in a cone of $\Delta R = 0.4$ about the calorimeter jet axis,

$$N \equiv \text{No. of charged tracks } (p_T > 1 \text{ GeV/c}). \qquad (2.10)$$

Tau leptons produced in W decay have a large Lorentz boost, therefore the hadrons coming from the τ decay form a narrow, high-p_T hadronic jet. In addition, ~ 99% of the hadronic decays of the τ have charged multiplicities of one or three. These τ characteristics are combined in L_τ [Eq. (2.7)], which is used to select a low background sample of W $\to \tau\nu$ candidates. Events with $L_\tau > 0$ (32 events) can mainly be understood as W $\to \tau\nu_\tau$ decays with $\tau \to$ hadrons $+ \nu$, and were used to extract the lepton universality result (2.1) [2.5]. The remaining 24 events with $L_\tau < 0$, which are less likely to be W $\to \tau\nu_\tau$ decays, have been used to search for new physics [2.19]. The second column of Table 2.1 shows Monte Carlo predictions for expected contributions to the $L_\tau < 0$ sample from known sources. The overall number agrees well with the observed number of 24 events in this category. Note that the contribution of jet $+$ (Z $\to \nu\bar{\nu}$) events in Table 2.1 is estimated using the experimental distribution of large-p_T W events [2.20] for normalization, and assuming just $N_\nu = 3$ neutrino flavours.

To establish an upper limit on N_ν, the number of additional events expected with $L_\tau < 0$ and $E_T^j < 40$ GeV was calculated for each additional neutrino species: 1.8 event per species. Comparing this prediction with the numbers in the third column of Table 2.1, which already include three neutrino species, it was concluded [2.19] that

$$N_\nu \leqslant 10 \quad (90\% \text{ CL}). \tag{2.11}$$

This result can alternatively be expressed as an upper limit on all Z decays into any light neutral non-interacting particles (assuming a partial width of 180 MeV for each Z $\to \nu\bar{\nu}$ decay mode):

$$\Gamma(Z \to \text{nothing}) \leqslant 1.8 \text{ GeV} \quad (90\% \text{ CL}). \tag{2.12}$$

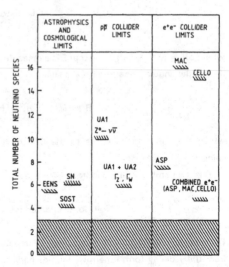

Fig. 2.3 Summary of current limits on the number of neutrino species.

Note that in the calculation for the total expected background, no contribution from the top quark was included (i.e. $W \to t\bar{b}$, $Z \to t\bar{t}$, and $pp \to t\bar{t} + X$). Adding these contributions would improve the upper bound on the total number of neutrino species by an amount depending on m_t.

The limit (2.11) is not yet competitive with the most stringent of the other limits shown in Fig. 2.3, but it is more direct than all except the limits from $e^+e^- \to \gamma +$ nothing. We can expect that the 'gluon-tagging' method for neutrino counting will be improved by data taken with the Antiproton Collector (ACOL) and the upgraded detectors, which should significantly reduce the statistical and systematic errors.

2.2 Fourth-Generation Charged Leptons

These are not included in the minimal three-generation version of the Standard Model, but we have no model-independent reason to exclude them theoretically. The pair-production of heavy leptons has been sought in e^+e^- collisions at centre-of-mass energies $\sqrt{s} \leqslant 46.6\,\text{GeV}$ with negative results, allowing us to conclude [2.13] that

$$m_L > 22.7\,\text{GeV}/c^2 \quad (90\%\,\text{CL}), \quad\quad (2.13)$$

and improved upper limits are becoming available now from TRISTAN [2.21]. In principle, we could exclude indirectly the existence of any heavy lepton by determining that $N_\nu < 4$, but the bounds (2.2) to (2.5) and (2.11) on N_ν are not yet sufficiently precise. There is a weak upper bound on the possible mass of any heavy lepton from a global analysis of electroweak neutral- and charged-current data, which bounds radiative corrections due to a heavy lepton, telling us that

$$m_L < 290\,\text{GeV}/c^2 \quad (90\%\,\text{CL}), \quad\quad (2.14)$$

if any such particle exists [2.22].

There are at least three ways of producing a heavy lepton at the CERN $p\bar{p}$ Collider: through $W \to L\nu_L$, through γ^*, $Z \to L^+L^-$, and through heavy-quark decays such as $t \to L\nu_L b$. In practice, only the first channel has been studied [2.23], since it has a cross-section much larger than that of the second, and the t-quark has not yet been observed. A conventional sequential heavy lepton is expected to have the branching ratio

$$\Gamma(W \to L\nu_L)/\Gamma(W \to e\nu_e) = 1[1 - 1.5\varrho + 0.5\varrho^3], \quad \varrho \equiv (m_L/m_W)^2, \quad\quad (2.15)$$

for $m_L < m_W$, assuming $m_{\nu_L} = 0$. Its total decay rate is expected to be

$$\Gamma(L \to \text{all}) = (G_F^2 m_L^5/192\pi^3)[3 + 6(1 + \alpha_s/\pi)], \quad\quad (2.16)$$

278

where the first term corresponds to the almost equally probable $L \to \nu_L e \bar{\nu}_e$, $\nu_L \mu \bar{\nu}_\mu$, and $\nu_L \tau \bar{\nu}_\tau$ decay modes, and the second term corresponds to the hadronic decay modes $L \to \nu_L(u\bar{d}, c\bar{s})$. For $m_L = 40$ GeV/c^2, the expression (2.16) gives a lifetime $t_L \approx 10^{-19}$ s, which is too short to leave a detectable track. The leptonic decay modes of the L ($L \to \ell \bar{\nu}_\ell \nu_L; \ell = $ e, μ, τ) would be difficult to disentangle from the $W \to e\bar{\nu}_e$, $\mu \bar{\nu}_\mu$ and $W \to \tau \bar{\nu}_\tau (\tau \to \ell \bar{\nu}_\ell \nu_\tau)$ backgrounds, which have larger rates. Accordingly the search performed by UA1 has focused on the $L \to \nu_L q\bar{q}$ decays [2.19] with a total branching ratio of 67% and with an expected event topology of jet(s) + $\not{E}_T$.

This decay pattern gives events which are qualitatively similar to $W \to \tau \bar{\nu}_\tau (\tau \to$ hadrons + $\nu_\tau)$ except for the phase-space factor (2.15) and the typically larger invariant mass of the $q\bar{q}$ system. The fragmentation into hadrons has been modelled using a modified version of ISAJET where spin effects have been included [2.24]. Since heavy leptons come from W decays, such $W \to L\bar{\nu}_L$ events would mostly populate the $E_T^j < 40$ GeV region of Fig. 2.2. Also, the wider jets fall mainly into the $L_\tau < 0$ region, as can be seen in Fig. 2.4 for heavy-lepton Monte Carlo events ($m_L = 55$ GeV/c^2). Figure 2.5 shows the number of events expected from $W \to L\bar{\nu}_L \to q\bar{q}\nu_L\bar{\nu}_L$ in the $L_\tau < 0$, $E_T^j < 40$ GeV domain of Fig. 2.4, as a function of m_L.

The net acceptance for a heavy lepton of mass 40 GeV/c^2, decaying semihadronically, is 6.2%. This low efficiency comes mainly from the trigger efficiency and the $N_\sigma > 4$ requirement. With the jet definition used, most of the heavy-lepton decays are predicted to give monojet events. The fraction of dijet events ($E_T^j > 12$ GeV) ranges from $\sim$ 18% ($m_L = 25$ GeV/c^2) to $\sim$ 27% ($m_L = 65$ GeV/c^2).

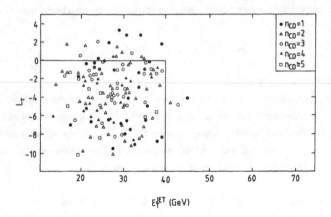

Fig. 2.4 Scatter plot of L_τ versus E_T^j of the highest-E_T jet for the heavy-lepton Monte Carlo events with $m_L = 55$ GeV/c^2. The symbols indicate the charged track multiplicity as defined in Eq. (2.10).

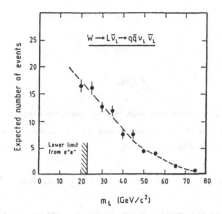

Fig. 2.5 The rate of heavy-lepton events passing the isolated 4σ $\not{E}_T$ selection and the cuts $L_\tau < 0$ and $E^j_T < 40$ GeV as a function of the heavy-lepton mass (solid points). Also indicated is the lower limit on m_L from e^+e^- experiments.

Using the calculated heavy-lepton contributions and including the contribution of one additional neutrino ν_L to missing-energy events from jet $+$ $(Z \to \bar{\nu}_L\nu_L)$, it was found [2.19] that

$$m_L > 41 \text{ GeV/c}^2 \qquad (90\% \text{ CL}). \qquad (2.17)$$

Note that in computing this limit, no t-quark contributions were added to the background, and $L \to t\bar{b}\bar{\nu}$ decays were not included in the predicted event rates. These contributions, if included, would increase the limit.

Studies have shown that the fraction of dijet $+$ $\not{E}_T$ events increases as the mass of the heavy lepton increases [2.24]. These events are expected to have a more distinctive signature: two high-p_T jets balanced by $\not{E}_T$ in a 'Mercedes'-type configuration. This event topology is not expected to be dominant in dijet events coming from the standard physics processes. Therefore the improved statistics and systematics of data accumulated with ACOL (including better calorimeter granularity of the upgraded UA1 detector) should make it possible to search for a signal for high-mass heavy leptons ($m_L \gtrsim 45$ GeV/c^2) in the dijet event topology also. In the absence of a signal, the limits are expected to be improved considerably — in principle up to the phase-space limit of the W decay (2.15).

2.3 The Higgs Boson

The presence of at least one physical neutral Higgs boson H^0 is essential for the renormalizability of the Standard Model. In the minimal version of the Standard Model with just one Higgs doublet, there is a single H^0 with well-defined couplings:

$$g_{H f \bar{f}} \quad = g m_f / 2 m_W , \qquad (2.18a)$$

$$g_{H W^- W^+} = g m_W , \qquad (2.18b)$$

$$g_{H Z Z} \quad = g m_Z , \qquad (2.18c)$$

where $G_F / \sqrt{2} = g^2 / 8 m_W^2$. Since its couplings increase with the mass of the other particles, we expect the Higgs to decay into the heaviest available particles. For example, a Higgs with mass between 11 GeV/c^2 and $2 m_t$ or $2 m_W$ has

$$BR(H \to e^+ e^- : \mu^+ \mu^- : \tau^+ \tau^- : c \bar{c} : b \bar{b}) = 1 : m_\mu^2 / m_e^2 : m_\tau^2 / m_e^2 : 3 m_c^2 / m_e^2 : 3 m_b^2 / m_e^2 .$$

Whilst the couplings of this minimal Higgs boson are known, its mass is not. Radiative corrections in the minimal version of the Standard Model suggest [2.25]

$$m_H \geq [O(\alpha) m_W^2]^{1/2} \approx \text{a few GeV/c}^2 , \qquad (2.19)$$

but this can be evaded if $m_t \approx 100$ GeV/c^2. Imposing tree-level unitarity on WW scattering amplitudes suggests [2.26] that

$$m_H \leq [O(m_W^2 / \alpha)]^{1/2} \approx 1 \text{ TeV/c}^2 . \qquad (2.20)$$

However, tree-level unitarity does not necessarily hold if the Higgs sector is strongly coupled, and although it has been argued that some observable effect should show up in the 1 TeV/c^2 mass range even in this case, it would surely not take the form of an identifiable narrow particle state. Phenomenologically, one knows from the absence of Higgs-boson effects in low-energy neutron–nucleus scattering, nuclear decays, and the spectra of muonic X-rays, that [2.27]

$$m_H \geq 15 \text{ MeV/c}^2 . \qquad (2.21)$$

With a view to searching for heavier Higgs bosons, estimates have been made of the branching ratio for $\eta' \to \eta + H$ [2.28] and $\psi' \to J/\psi + H$ [2.29], and of the flavour-changing vertices for $s \to d + H$ and $b \to s + H$ [2.30] which control the branching ratios for $K \to \pi + H$ and $B \to K + H$, respectively. Of these, the branching ratio for $\eta' \to \eta + H$ is difficult to estimate reliably, and the experimental upper limit [2.31] on $\psi' \to J/\psi + H$ is not sensitive enough to be interesting. The flavour-changing vertices for $s \to d + H$ and $b \to s + H$ can, in principle, be calculated reliably, but in practice the available calculations [2.30] disagree amongst themselves and with a previous general theorem [2.32] on the form of such vertices which should be satisfied. Some of the existing calculations have been used to argue that unsuccessful searches for $K^+ \rightarrow \pi^+ + H$ and $B \rightarrow K + H$ [2.33] exclude certain ranges of Higgs masses [2.34, 2.35]. We believe these conclusions to be premature, and await the results of a new calculation of the flavour-changing Higgs vertices which is now under way [2.36]. However, we note that any vertex of the general form previously derived would probably yield branching ratios for $K^+ \rightarrow \pi^+ + H$ and $B \rightarrow K + H$ that are below present experimental sensitivities.

Although searches for a Higgs boson in $J/\psi \rightarrow \gamma + X$ have not yet reached sufficient sensitivity to rule out $m_H \lesssim m_{J/\psi}$, searches for a Higgs boson in a combined sample of $\Upsilon \rightarrow \gamma + X$ and $\Upsilon'' \rightarrow \gamma + X$ decays [2.37] have now reached sufficient sensitivity to rule out a Standard Model Higgs in the range

$$0.6 \lesssim m_H \lesssim 3.9 \, \text{GeV}/c^2, \qquad (2.22)$$

even after QCD radiative corrections [2.38] to the branching ratio are included. This is the only range of Higgs-boson masses above 15 MeV that seems to be reliably excluded.

Two ways to search for a Higgs boson at the CERN $p\bar{p}$ Collider have been considered. They are $p\bar{p} \rightarrow [W^* \rightarrow (W + H)] + X$ or $[W \rightarrow (W^* + H)] + X$, and $p\bar{p} \rightarrow H + X$. The former exploits a direct-channel W pole and the relatively large coupling (2.18b). However, even for light Higgses in the Standard Model the number of such events is expected to be $\lesssim 2 \times 10^{-3}$ [2.39]. Since only about 400 W events have been detected in UA1 and about 250 W events in UA2, neither experiment yet has sensitivity to the $(W + H)$ process. Even with more W events, it would not be easy to pick out any accompanying Higgs from the conventional hadronic background, although if $m_H > 2m_b$ this would be less difficult with a microvertex detector if it allows the b-quark to be tagged. The dominant mechanism for unaccompanied Higgs production at the CERN $p\bar{p}$ Collider is expected to be gluon–gluon fusion: $gg \rightarrow H$ via a t-quark loop. The cross-section for this reaction to leading order in α_s is [2.40]

$$\sigma [p\bar{p} \rightarrow (gg \rightarrow H) + X] = (\pi/32) \, (\alpha_s/\pi)^2 \, (G_F/\sqrt{2}) \, (N^2/9) \, \tau \int d\tau \, G(\sqrt{\tau} \, e^y) \, G(\sqrt{\tau} \, e^{-y}), \quad (2.23)$$

where $\tau = m_H^2/s$, $G(\xi)$ is the gluon distribution function, and $N = 1$ if the t-quark is the only quark much heavier than m_H. Unfortunately, the expression (2.23) gives unobservably low rates for H production in the absence of a clear decay signature, whatever the values of m_H.

Some excitement was caused by the final analysis of events in the W + jet(s) data sample of UA1 [2.20]. Two events were observed at very large transverse momentum of the W ($p_T^W \simeq 100 \, \text{GeV}/c$) in which the W candidate recoils against an energetic dijet system. The mass of the jet–jet system is compatible with the W mass. These events could be interpreted as WW, WZ or ZZ pair-production, and one might ask whether they could be due to Higgs production and decay into W^+W^- and/or ZZ. However, the cross-section for producing a

Higgs boson at the CERN p$\bar{\text{p}}$ Collider is very small [2.40]. As an example, UA1 estimated the expected number of events for $H^0 \to W^+W^-$ passing all selection cuts and being consistent with the parameters of the observed events, to be $\leqslant 2 \times 10^{-5}$. Therefore, neither in the present Collider data sample nor in the future data set accumulated with ACOL does one expect an observable event rate from Higgs boson decay into W^+W^- or ZZ.

3. SUPERSYMMETRY

There are many reasons for postulating a supersymmetric extension of the Standard Model, starting with the 'fundamental' ones of elegance or of utility in constructing a unified Theory of Everything (TOE). However, the only motivation for having light supersymmetric particles in an accessible mass range is as a solution to the naturalness or hierarchy problem [3.1]. This has its roots in the fact that radiative corrections to the masses of elementary scalar bosons, such as the Higgs boson of the Standard Model, are quadratically divergent,

$$\delta m_H^2 = O(\alpha/\pi) \, \Lambda^2, \tag{3.1}$$

where Λ is a loop momentum cut-off. It can be seen from Eq. (3.1) that in order for the physical mass of the Higgs to be natural, i.e. $\delta m_H^2 \leqslant m_H^2$, where $m_H^2 \approx m_W^2$, Λ should be $\leqslant 1$ TeV. The supersymmetric solution to this naturalness problem starts from the observation that the quadratically divergent boson and fermion loop corrections (3.1) to m_H^2 have opposite signs, and that if bosons B and fermions F occur in pairs with identical couplings as in supersymmetry, these quadratically divergent corrections will cancel, leaving behind a residual

$$\delta m_H^2 = O(\alpha/\pi) \, |m_B^2 - m_F^2|. \tag{3.2}$$

Therefore the role of the cut-off Λ in Eq. (3.1) is taken over by the boson–fermion mass difference. In this approach

$$|m_B^2 - m_F^2| \leqslant 1 \text{ TeV}/c^2, \tag{3.3}$$

hence one expects the supersymmetric partners of the known particles listed in Table 3.1 to have masses $\leqslant 1$ TeV/c^2.

The gauge and Yukawa couplings of the known particles have the following counterparts for their supersymmetric partners:

$$g\bar{f}\gamma_\mu V^\mu f \to \sqrt{2} g\bar{f}\bar{V}\tilde{f} + \text{h.c.}, \quad g\tilde{f}^* \overset{\leftrightarrow}{\partial}_\mu \tilde{f} V^\mu, \quad (g^2/2)|\tilde{f}^* \tilde{f}|^2 \tag{3.4a}$$

$$\lambda \tilde{f} f H \rightarrow \lambda \tilde{f} f \tilde{H} + \dots, \quad \lambda^2 [|\tilde{f}^c \tilde{f}|^2 + |\tilde{f} H|^2 + |\tilde{f}^c H|^2]. \tag{3.4b}$$

It is easy to see that these interactions all conserve multiplicatively a quantum number called R-parity which is defined to be $+1$ for Standard Model particles and -1 for all their supersymmetric partners [3.2]. It is possible to relate R-parity to other quantum numbers conserved in the Standard Model:

$$R = (-1)^{B + 3L + 2S}. \tag{3.5}$$

Table 3.1

Spectrum of SUSY particles

Particle	Spin	Sparticle		Spin
quark $q_{L,R}$	$\frac{1}{2}$	squark $\tilde{q}_{L,R}$		0
lepton $\ell_{L,R}$	$\frac{1}{2}$	slepton $\tilde{\ell}_{L,R}$		0
photon γ	1	photino $\tilde{\gamma}$		$\frac{1}{2}$
gluon g	1	gluino $\tilde{g}$	gauginos	$\frac{1}{2}$
W	1	wino $\tilde{W}$		$\frac{1}{2}$
Z	1	zino $\tilde{Z}$		$\frac{1}{2}$
Higgs H	0	shiggs $\tilde{H}$		$\frac{1}{2}$
graviton G	2	gravitino $\tilde{G}$		$\frac{3}{2}$

The conservation of R-parity has three important phenomenological consequences.

i) Sparticles are always produced in pairs, e.g. $e^+ e^- \rightarrow \tilde{\mu}^+ \tilde{\mu}^-$, $p\bar{p} \rightarrow (\tilde{g}\tilde{g}$ or $\tilde{g}\tilde{q}$ or $\tilde{q}\tilde{q}) + X$. This property can also be seen directly from the forms of the interactions (3.4).

ii) Every sparticle decays into another sparticle, e.g. $\tilde{e} \rightarrow e + \tilde{\gamma}$, $\tilde{g} \rightarrow q\bar{q}$ or $q\bar{q}\tilde{\gamma}$, $\tilde{q} \rightarrow q + \tilde{g}$ or $q + \tilde{\gamma}$.

iii) The lightest supersymmetric particle (LSP) is absolutely stable, since it has no legal decay mode.

Since the LSP is absolutely stable, it should be present in large numbers in the Universe today as a cosmological relic from the Big Bang. If the relic LSPs had strong and/or electromagnetic interactions, they would have bound with conventional nuclei to form exotic isotopes. These have not been seen [3.3] even at a level far below the standard predictions for LSP abundances [3.4]. Therefore we conclude that the LSP has no strong or electromagnetic interactions; thus it can escape from experimental detection in much the same way as a conventional neutrino. Just as $\not{E}_T$ was used as a signature in the search for W $\rightarrow e\nu, \mu\nu, \tau\nu$, so $\not{E}_T$ could also be a signature for sparticle pair-production.

In Table 3.1 there are many neutral weakly interacting sparticles which are candidates for being the LSP. They include the sneutrinos $\tilde{\nu}_e$, $\tilde{\nu}_\mu$, $\tilde{\nu}_\tau$ of spin 0, the shiggs $\tilde{H}$, photino $\tilde{\gamma}$ and zino $\tilde{Z}$ of spin $\frac{1}{2}$, and the gravitino $\tilde{G}$ of spin $\frac{3}{2}$. In fashionable models the sneutrinos $\tilde{\nu}$ and gravitino $\tilde{G}$ are usually heavier than the lightest spin-$\frac{1}{2}$ sparticle, and unsuccessful dark-matter searches squeeze the $\tilde{\nu}$ possibilities still further. Hence the LSP seems most likely to be some $\tilde{H}/\tilde{\gamma}/\tilde{Z}$ mixture [3.5]. Model studies easily produce examples where the LSP is an almost pure $\tilde{H}$ or $\tilde{\gamma}$ state. The cosmological relic density of an $\tilde{H}$ LSP tends to be too high unless $m_{\tilde{H}} > m_b$ or m_t. Thus there is a tendency to assume that the LSP is (at least approximately) a $\tilde{\gamma}$, although this is by no means ironclad. Searches for sparticle production have indeed generally assumed that the LSP is the $\tilde{\gamma}$ and that $m_{\tilde{\gamma}}$ is negligible compared with the other sparticle masses, which will also be our working hypothesis here. In most cases, the limits are essentially unchanged for $m_{\tilde{\gamma}}$ up to half the decaying sparticle masses [3.6].

3.1 Squarks and Gluinos

Since these sparticles have strong interactions, they have the largest cross-sections at hadron–hadron colliders. Unsuccessful searches in e^+e^- collisions [3.7] tell us that

$$m_{\tilde{q}} \geqslant 21.5 \, \text{GeV/c}^2 \quad (90\% \, \text{CL}). \tag{3.6}$$

It is normally assumed [3.8, 3.9] that the supersymmetric partners of the left- and right-handed quarks have essentially equal masses,

$$m_{\tilde{q}_L} \simeq m_{\tilde{q}_R}, \tag{3.7}$$

and that the masses of the $\tilde{u}$, $\tilde{d}$, $\tilde{s}$, $\tilde{c}$, and $\tilde{b}$ squarks are also essentially equal,

$$m_{\tilde{u}} \simeq m_{\tilde{d}} \simeq m_{\tilde{s}} \simeq m_{\tilde{c}} \simeq m_{\tilde{b}}. \tag{3.8}$$

Neither of these assumptions is above suspicion, since unequal vacuum expectation values (v.e.v.'s) for the two Higgs doublets in the minimal supersymmetric extension of the Standard Model would split the squark masses. However, if these v.e.v.'s were equal, one would expect $|m_{\tilde{q}_1}^2 - m_{\tilde{q}_2}^2| \simeq |m_{q_1}^2 - m_{q_2}^2| \ll m_{\tilde{q}}^2$, and this assumption of degeneracy would be justified. When $m_{\tilde{q}} > m_{\tilde{g}}$, we expect $\tilde{q} \rightarrow q\tilde{g}$ to dominate, whereas when $m_{\tilde{q}} < m_{\tilde{g}}$, a large branching ratio for $\tilde{q} \rightarrow q\tilde{\gamma}$ decay is expected, although $\tilde{q} \rightarrow q\tilde{W}$ or $q\tilde{Z}$ decay could also be important. The implications of non-degenerate squark masses and of $\tilde{q} \rightarrow q(\tilde{W}, \tilde{Z})$ decays for hadron–hadron searches are discussed elsewhere [3.10, 3.11]. Here we assume both degenerate squark masses and the dominance of $\tilde{q} \rightarrow q\tilde{\gamma}$ decays. We mainly consider the case where the $\tilde{\gamma}$ is stable, but the case where $\tilde{\gamma} \rightarrow \gamma + \tilde{H}$ will also be discussed.

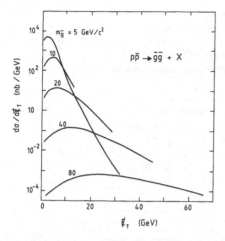

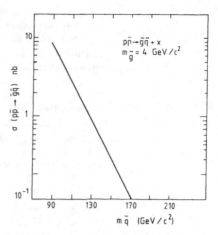

Fig. 3.1 Missing transverse energy spectrum from p$\bar{\text{p}}$ → $\tilde{\text{g}}\tilde{\text{g}}$ + X for different values of $m_{\tilde{\text{g}}}$ at $\sqrt{s}$ = 630 GeV.

Fig. 3.2 Cross-section for p$\bar{\text{p}}$ → $\tilde{\text{g}}\tilde{\text{q}}$ + X production as a function of $m_{\tilde{\text{q}}}$ and a fixed gluino mass of 4 GeV/c^2.

The lower limits on $m_{\tilde{\text{g}}}$ from fixed-target hadron–hadron collisions, from Υ decays, and from bottomonium $^3P_1(b\bar{b})$ → g + $\tilde{\text{g}}\tilde{\text{g}}$ searches are in the range [3.7, 3.12]

$$m_{\tilde{\text{g}}} \gtrsim (2 \text{ to } 5) \text{ GeV}/c^2, \tag{3.9}$$

depending on the squark mass. When $m_{\tilde{\text{g}}} > m_{\tilde{\text{q}}}$, we expect $\tilde{\text{g}}$ → $\bar{\text{q}}\tilde{\text{q}}$ or $q\tilde{\text{q}}$ decays to dominate, whereas when $m_{\tilde{\text{g}}} < m_{\tilde{\text{q}}}$, we expect $\tilde{\text{g}}$ → $q\bar{q}\tilde{\gamma}$ to dominate. Thus gluino decays always give rise to the final state $\bar{q}q\tilde{\gamma}$, whilst squark decays produce the dominant final states $q\tilde{\gamma}$ (if $m_{\tilde{\text{g}}} > m_{\tilde{\text{q}}}$) or $\bar{q}qq\tilde{\gamma}$ (if $m_{\tilde{\text{q}}} > m_{\tilde{\text{g}}}$). Therefore, we expect multi-jets and large $\not{E}_T$ as an event signature.

When $m_{\tilde{\text{g}}} \ll m_{\tilde{\text{q}}}$, the dominant production process is p$\bar{\text{p}}$ → $\tilde{\text{g}}\tilde{\text{g}}$ + X, which produces the $\not{E}_T$ spectrum shown in Fig. 3.1. However, the associated production process p$\bar{\text{p}}$ → $\tilde{\text{g}}\tilde{\text{q}}$ + X can also be important, although its cross-section falls rapidly with increasing $m_{\tilde{\text{q}}}$, as seen in Fig. 3.2. The relative importance of the different $\tilde{\text{g}}\tilde{\text{g}}$, $\tilde{\text{g}}\tilde{\text{q}}$, and $\tilde{\text{q}}\tilde{\text{q}}$ production processes is shown in Table 3.2 for different choices of $m_{\tilde{\text{g}}}$ and $m_{\tilde{\text{q}}}$.

When $m_{\tilde{\text{q}}} < m_{\tilde{\text{g}}}$, $\tilde{\text{q}}\tilde{\text{q}}$ pair-production dominates the production of strongly interacting particles, whereas $\tilde{\text{g}}\tilde{\text{g}}$ dominates if $m_{\tilde{\text{q}}} > m_{\tilde{\text{g}}}$, as illustrated in Fig. 3.3. Production of associated $\tilde{\text{g}}\tilde{\text{q}}$ pairs can also be important if $m_{\tilde{\text{q}}} \approx m_{\tilde{\text{g}}}$, as also seen in Table 3.2, and $\tilde{\text{q}}\tilde{\gamma}$ and $\tilde{\text{g}}\tilde{\gamma}$ final states are not completely negligible[*]. All these processes should be included in a

---

[*] For instance, direct photino production p$\bar{\text{p}}$ → $\tilde{\gamma}\tilde{\text{q}}$ contributes ~ 3% for $m_{\tilde{\text{g}}}$ = 90 GeV/c^2, $m_{\tilde{\text{q}}}$ = 50 GeV/c^2, increasing to ~ 20% of the total cross-section as $m_{\tilde{\text{g}}} \to \infty$.

Table 3.2

Relative cross-section contributions
from gluino and squark production
$(p\bar{p} \rightarrow \tilde{g}\tilde{g}, \tilde{g}\tilde{q}, \tilde{q}\bar{\tilde{q}} + X)$ at $\sqrt{s} = 630\,\text{GeV}$
for selected values of the gluino and squark masses

Mass (GeV/c^2)		Fraction (%)		
$m_{\tilde{g}}$	$m_{\tilde{q}}$	$\tilde{g}\tilde{g}$	$\tilde{g}\tilde{q}$	$\tilde{q}\bar{\tilde{q}}$
120	50	< 1	7	93
50	120	96	4	< 1
70	90	39	46	15

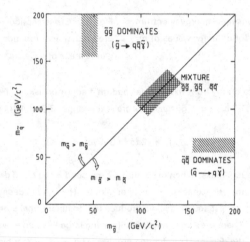

Fig. 3.3 Schematic picture of supersymmetric processes dominating different areas of the $m_{\tilde{q}}$–$m_{\tilde{g}}$ plane.

complete analysis of squark and gluino production at a hadron–hadron collider, also taking into account QCD radiative corrections if possible, and certainly in the analysis of light $\tilde{g}\tilde{g}$ pair-production, where they can be important. The results of analyses are generally published as allowed domains in the $(m_{\tilde{q}}, m_{\tilde{g}})$ plane.

We start by describing the squark and gluino search by the UA1 Collaboration [3.13]. A special selection was made to optimize the signal-to-background ratio for heavy $\tilde{g}$ and $\tilde{q}$ and for light $\tilde{g}$, taking into account the complexity of different topologies depending on the $\tilde{g}$ and $\tilde{q}$ mass combinations. The standard missing-energy event selection described in subsection 2.2

was used ($\not{E}_T$ > 15 GeV and N_σ > 4), and the subsample with L_τ < 0 was selected to reduce contamination from W → $\tau\bar{\nu}_\tau$ (τ → hadrons + ν). Two changes were made in the event selection: i) instead of at least one jet with E_T^j > 12 GeV, at least *two* jets were required; ii) instead of the stringent cut on the isolation of the $\not{E}_T$ vector, the difference $\Delta\phi$ in azimuthal angle between the two highest-E_T jets was required to satisfy $\Delta\phi$ < 140°. A total of four events passed all the above selection criteria. The expected contributions from Standard Model processes and from jet fluctuation background were evaluated using ISAJET and the jet fluctuation Monte Carlo as described in subsection 2.2. The predictions for the numbers of events from different sources are shown in Table 3.3. The total expected number of 5.2 ± 1.9 ± 1.0 events should be compared with the four events observed. Figure 3.4 shows the

Table 3.3

Predicted rates for processes giving ⩾ 2 jets + $\not{E}_T$ events
passing the SUSY event selection cuts [3.13]

Process	No. of events
W/Z decays	3.0
b$\bar{b}$/c$\bar{c}$	2.0
Jet fluctuations	0.2
Total*)	5.2 ± 1.9 ± 1.0

*) The first error is statistical, the
second is systematic.

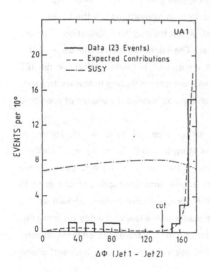

Fig. 3.4 Distribution of $\Delta\phi$, the azimuthal angle between the two highest-E_T jets in the event: for data (histogram), for the expectation from conventional processes plus background (dashed curve), and for squark and gluino production (dot–dashed curve), using $m_{\tilde{q}}$ = 60 GeV/c² and $m_{\tilde{g}}$ = 70 GeV/c². The supersymmetric particle production was multiplied by a factor of 10 [3.13].

Table 3.4

Production cross-sections, experimental acceptances, and predicted event rates from squark and gluino production ($p\bar{p} \to \tilde{g}\tilde{g}$, $\tilde{g}\tilde{q}$, $\tilde{q}\tilde{q}$, $\tilde{g}\tilde{\gamma}$, $\tilde{q}\tilde{\gamma}$ + X) for selected values of the squark and gluino mass. The errors quoted are the statistical errors resulting from the Monte Carlo generation [3.13].

$m_{\tilde{g}}$ (GeV/c^2)	$m_{\tilde{q}}$ (GeV/c^2)	Cross-section (nb)	Acceptance (%)	Events
30	200	20.20	0.2	24.0 ± 5.6
50	120	0.71	1.3	6.8 ± 0.8
50	90	0.78	1.4	7.7 ± 0.9
60	70	0.49	3.2	11.2 ± 0.8
70	90	0.11	2.5	2.0 ± 0.2
200	30	4.35	0.6	17.0 ± 2.2
120	50	0.44	3.8	12.0 ± 0.8
90	50	0.64	3.8	17.5 ± 1.2
70	60	0.49	6.0	20.9 ± 1.2
130	70	0.07	6.6	3.3 ± 0.2
90	70	0.13	7.0	6.5 ± 0.3
70	70	0.27	5.3	10.2 ± 0.8

$\Delta\phi$ distribution for the large-$\not{E}_T$ multijet sample before applying the cut $\Delta\phi < 140°$. As can be seen, most events with two jets are almost back-to-back in azimuth, as are the background contributions (i.e. heavy flavour and jet fluctuations), whilst the angular distribution expected from a typical supersymmetric scenario is quite flat. The contributions to the $\Delta\phi < 140°$ multijet sample from squark and gluino production were again evaluated using the ISAJET Monte Carlo program, with full simulation of the UA1 detector including hardware triggers. Events were generated for a wide range of squark and gluino masses. Examples of predicted event rates and efficiencies are given in Table 3.4.

The selection efficiency is relatively high for large values of $m_{\tilde{g}}$ ($\epsilon \approx 3\%$ for $m_{\tilde{g}} = 70$ GeV/c^2, $m_{\tilde{q}} \to \infty$) but quite small for small values of $m_{\tilde{g}}$ ($\epsilon \approx 0.01\%$ for $m_{\tilde{g}} = 20$ GeV/c^2, $m_{\tilde{q}} \to \infty$), if only the lowest-order 2-to-2 production processes gg $\to \tilde{g}\tilde{g}$ and $q\bar{q} \to \tilde{g}\tilde{g}$ are considered. However, for $m_{\tilde{g}} \lesssim 10$ GeV/c^2, 'indirect' higher-order production mechanisms in which a gluon splits to yield a $\tilde{g}\tilde{g}$ pair becomes the dominant contribution, shown as the dashed line in Fig. 3.5. A technical problem is that because of the low efficiency for small $m_{\tilde{g}}$, it is difficult to generate enough Monte Carlo statistics to compute the expected rates reliably. Cuts had to be placed on the Monte Carlo event generation (for 'direct' and 'indirect' gluino

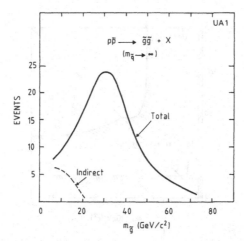

Fig. 3.5 Predicted contribution to the $\Delta\phi < 140°$ multijet + $\not{E}_T$ sample from supersymmetric particle production as a function of gluino mass, calculated for the case where the squark mass becomes infinitely large. The dashed curve shows the contribution from $p\bar{p} \to gX$, $g \to \tilde{g}\tilde{g}$ only [3.13].

production). This would therefore underestimate the true rates. Other assumptions which tended to underestimate the true rate included the disregard of any K-factor corresponding to higher-order QCD effects, and the neglect of other indirect production processes.

In the case when $m_{\tilde{q}} \ll m_{\tilde{g}}$, the acceptance for $\tilde{q}\bar{\tilde{q}}$ final states is about a factor of three higher than for direct $\tilde{g}\tilde{g}$ production in the limit considered above. However, the cross-sections are smaller by a similar factor, so that the expected event rates finish up being comparable, as seen in Fig. 3.6. Since e^+e^- data exclude $m_{\tilde{q}} \lesssim 21.5 \text{ GeV}/c^2$, only the region of Fig. 3.6 where $m_{\tilde{q}} \gtrsim 20 \text{ GeV}/c^2$ was considered.

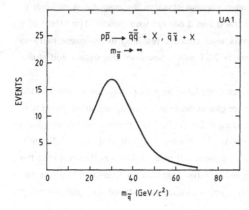

Fig. 3.6 Predicted contributions to the $\Delta\phi < 140°$ multijet + $\not{E}_T$ sample from supersymmetric particle production as a function of squark mass, calculated for the case when the gluino mass becomes infinitely large [3.13].

290

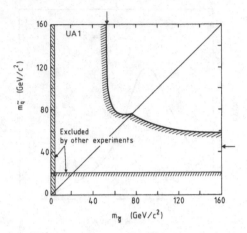

Fig. 3.7 Limits on squark and gluino masses (90% CL) obtained by UA1. The arrows indicate the asymptotic values of the 90% CL contours as the squark or gluino mass becomes infinitely large [3.13].

The domain of the $(m_{\tilde{g}}, m_{\tilde{q}})$ plane excluded by this UA1 analysis is shown in Fig. 3.7. Particular cases of interest are

$$m_{\tilde{q}} > 45 \text{ GeV}/c^2 \text{ in the limit } m_{\tilde{g}} \to \infty \quad (90\% \text{ CL}), \qquad (3.10a)$$

$$m_{\tilde{g}} > 53 \text{ GeV}/c^2 \text{ in the limit } m_{\tilde{q}} \to \infty \quad (90\% \text{ CL}), \qquad (3.10b)$$

$$\tilde{m} > 75 \text{ GeV}/c^2 \text{ if } m_{\tilde{q}} = m_{\tilde{g}} = \tilde{m} \quad (90\% \text{ CL}). \qquad (3.10c)$$

Note that the detailed shape of the exclusion contour close to the diagonal is difficult to determine because the branching ratios of the $\tilde{g}$ and $\tilde{q}$ decays vary rapidly. The effect of a non-zero photino mass on the derived limits was studied, resulting in an insensitivity to photino masses up to $\sim 20 \text{ GeV}/c^2$. For $m_{\tilde{\gamma}} > 20 \text{ GeV}/c^2$, however, the squark and gluino mass limits decrease rapidly.

The overall picture of Fig. 3.7 needs to be completed by a more detailed analysis of the light gluino case. The dominant mechanism for gluino decay is via virtual squarks, giving $t_{\tilde{g}} \approx 4 \times 10^{-8} \text{ s} \times (m_{\tilde{q}}^4/m_{\tilde{g}}^5)$, with $m_{\tilde{q}}$ in TeV/c^2 and $m_{\tilde{g}}$ in GeV/c^2 [3.14]. This means that the gluino becomes long-lived when $m_{\tilde{q}}$ is large (e.g. $t_{\tilde{g}} \gtrsim 10^{-10} \text{ s}$ for $m_{\tilde{g}} = 4 \text{ GeV}/c^2$ if $m_{\tilde{q}} \gtrsim 1 \text{ TeV}/c^2$). When $t_{\tilde{g}} \gtrsim 10^{-10} \text{ s}$, a significant fraction of the produced gluinos decay before reaching the calorimeter, so the $\not{E}_T$ signature is lost. However, other searches for long-lived hadrons can exclude gluinos with $t_{\tilde{g}} \gtrsim 10^{-10} \text{ s}$, under certain assumptions about the gluino–hadron

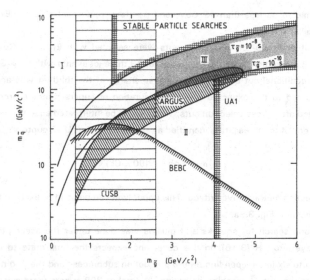

Fig. 3.8 Region of gluino and squark masses excluded by different experiments. Also shown are curves of equal gluino lifetime (see text) [3.13].

interaction cross-section [3.12]. A compilation of limits for small $m_{\tilde{g}}$ and large $m_{\tilde{q}}$ is shown in Fig. 3.8. Note that there are two regions not yet excluded: i) domain I, where $m_{\tilde{g}} <$ 0.7 GeV/c^2, and ii) domain II, where $m_{\tilde{g}} \approx$ (2.7 to 4) GeV/c^2 and $m_{\tilde{q}} \approx$ (70 to 800) GeV/c^2. Domain III (dotted area) is only excluded for certain assumptions on the gluino interaction cross-section [3.12]. Further dedicated searches to exclude these remaining domains are desirable, but the p$\bar{\text{p}}$ Collider is not the best machine for this task.

The UA2 Collaboration has also searched for squark and gluino production followed by $\tilde{q} \to q\tilde{\gamma}$ and $\tilde{g} \to q\bar{q}\tilde{\gamma}$ decays [3.15]. Two cases were considered: i) the photino is the LSP and is therefore stable and ii) the photino decays into $\gamma\tilde{H}$. We first discuss the analysis performed under the assumption that the photino is stable. In this case, as mentioned above, the expected supersymmetric signature would be multijets with significant $\not{E}_T$. The background from QCD multijet events is expected to be large, particularly since large $\not{E}_T$ can be faked in UA2 by a jet escaping at polar angles θ smaller than 20° with respect to the beams, or depositing only a small fraction of its energy in the forward calorimeters, which have only 1.5 absorption lengths. Therefore, rather stringent cuts had to be applied to remove this background. Two- or three-jet events were selected with $E_T^j >$ 12 GeV, in which at least one pair of jets had $\Delta\phi(j_1, j_2) <$ 95° (to suppress the QCD background which peaks at $\Delta\phi \approx$ 180°), and $\Delta\phi(j_1, j_2) >$ 15° to suppress beam-halo events. Additional technical cuts had to be applied for jets entering the forward calorimeter: the total energy measured in the forward calorimeter had to be less than 12 GeV, and $\Delta\phi$ (total p_T of central jets, ΣE_T of forward calorimeter cells) < 135°.

A total of 203 two- or three-jet events were selected with $\not{E}_T$ > 30 GeV. To study possible $\tilde{q}$ or $\tilde{g}$ production, only the subsample of 13 three-jet events with $\not{E}_T$ > 30 GeV were retained. An additional cut on the jet transverse energy of the third jet was applied ($E_T^{j_3}$ < 40 GeV), tailored to the expected event configuration of supersymmetric processes. One event remained after these selection cuts, which could be interpreted as a (W → eν) + 2 jets event where one of the three jets is identified as an electron. Thus UA2 quote an upper limit of

$$\sigma < 8\,\text{pb} \qquad (90\%\,\text{CL}) \qquad\qquad (3.11)$$

on events meeting all the above criteria. This upper limit can be used to exclude the domains of ($m_{\tilde{g}}$, $m_{\tilde{q}}$) shown in Fig. 3.9a.

The second search for squarks and gluinos was done under the assumption that the photino decays into $\gamma\tilde{H}$ [3.16]. In this case, one expects the final state to contain two photons, two to six jets (depending on the production subprocess and the $\tilde{g}$, $\tilde{q}$ masses), and small $\not{E}_T$ (from the two $\tilde{H}$ escaping detection). A total of 309 events were selected with at least two photons of E_T > 6 GeV and $m_{\gamma\gamma}$ > 10 GeV/c^2, most of which are single or multiple

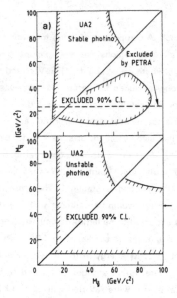

Fig. 3.9 Limits on squark and gluino masses (90% CL) obtained by UA2 [3.15]: a) in the case of a stable photino, using three-jet events with large $\not{p}_T$; b) in the case of an unstable photino, using photon pairs accompanied by jets.

π^0 events. The requirement of $m_{\gamma\gamma} > 10$ GeV/c^2 and at least two additional jets with $E_T^j >$ 10 GeV left UA2 with no events, corresponding to an observed cross-section of

$$\sigma < 12 \text{ pb} \quad (90\% \text{ CL}). \tag{3.12}$$

From a Monte Carlo simulation using the cross-sections from Ref. [3.14], UA2 inferred an overall efficiency of 10^{-5} to 10^{-1} for $m_{\tilde{g}}$, $m_{\tilde{q}}$ in the range 10 to 60 GeV/c^2. Combined with the limit (3.12), this led to the exclusion of the domain shown in Fig. 3.9b. Particular cases of interest are

$$m_{\tilde{q}} < 9 \text{ GeV/c}^2 \text{ or } > 46 \text{ GeV/c}^2 \quad \text{in the limit } m_{\tilde{g}} \rightarrow \infty, \tag{3.13a}$$

$$m_{\tilde{g}} < 15 \text{ GeV/c}^2 \text{ or } > 50 \text{ GeV/c}^2 \quad \text{in the limit } m_{\tilde{q}} \rightarrow \infty, \tag{3.13b}$$

$$\tilde{m} \geqslant 60 \text{ GeV/c}^2 \text{ if } m_{\tilde{g}} = m_{\tilde{q}} = \tilde{m}. \tag{3.13c}$$

In the region of small $\tilde{g}$ and $\tilde{q}$ masses an uncertainty of a few GeV/c^2 on the limits is expected from fragmentation effects and higher-order terms. The limits in the high-mass region are insensitive to $m_{\tilde{\gamma}} \leqslant 30$ GeV/c^2.

The quoted results do not depend on the exact value of the photino lifetime, provided $m_{\tilde{\gamma}} > 1$ GeV/c^2, corresponding to an average decay path in UA2 smaller than ~ 1 cm.

3.2 Sleptons

Unsuccessful searches for $e^+e^- \rightarrow \tilde{e}^+\tilde{e}^-$, $\tilde{\mu}^+\tilde{\mu}^-$, and $\tilde{\tau}^+\tilde{\tau}^-$ establish the lower limits [3.17]

$$m_{\tilde{e}} > 22 \text{ GeV/c}^2, \quad m_{\tilde{\mu}} > 20 \text{ GeV/c}^2, \quad m_{\tilde{\tau}} > 17 \text{ GeV/c}^2. \tag{3.14}$$

The previously mentioned searches for $e^+e^- \rightarrow \gamma + \text{nothing}$ also impose significant constraints in the $(m_{\tilde{e}}, m_{\tilde{\gamma}})$ plane, implying, for example, that

$$m_{\tilde{e}} > 67 \text{ GeV/c}^2 \quad (90\% \text{ CL}) \tag{3.15}$$

if $m_{\tilde{\gamma}} = 0$. However, they give no bound at all on $m_{\tilde{e}}$ if $m_{\tilde{\gamma}} > 13$ GeV/c^2 [3.17].

The most important slepton production mechanisms at the CERN p$\bar{\text{p}}$ Collider are $W^{\pm} \rightarrow \tilde{\ell}^{\pm}\tilde{\nu}$ decay [3.18] and $Z \rightarrow \tilde{\ell}^+\tilde{\ell}^-$ decay [3.19]. In each case, $\tilde{\ell}^{\pm} \rightarrow \ell^{\pm}\tilde{\gamma}$ is expected to dominate, whilst even if the $\tilde{\nu}$ is not the LSP and hence unstable, in many models its dominant decay is the invisible $\tilde{\nu} \rightarrow \nu\tilde{\gamma}$. Therefore we expect $\ell^{\pm} + \not{E}_T$ final states from $W^{\pm} \rightarrow \tilde{\ell}^{\pm}\tilde{\nu}$ decay, and $\ell^+\ell^- + \not{E}_T$ final states from $Z \rightarrow \tilde{\ell}^+\tilde{\ell}^-$ decay. The former may be distinguishable from

294

conventional $W^{\pm} \rightarrow \ell^{\pm}\nu$ decay by having a softer $\ell^{\pm}$ spectrum with a different angular distribution. The $Z \rightarrow \tilde{\ell}^{+}\tilde{\ell}^{-}$ decay can be distinguished from the conventional Drell–Yan background by its $\not{E}_T$ signature.

The UA1 Collaboration has searched for $p\bar{p} \rightarrow W + X$ followed by $W \rightarrow \tilde{e}\tilde{\nu}$, $\tilde{e} \rightarrow e\tilde{\gamma}$, $\tilde{\nu} \rightarrow \nu\tilde{\gamma}$ decays [3.20]. It was assumed in the analysis that the $\tilde{\gamma}$ is stable and $m_{\tilde{\gamma}} \ll m_{\tilde{\ell}}$. Left- and right-handed sleptons are mass-degenerate and the $\widetilde{W}$ and $\widetilde{Z}$ are sufficiently heavy to be ignored. The signature sought was an excess of events with a transverse mass of e and $\not{E}_T$ lower than those for the standard $W \rightarrow e\nu$ decay, and with the isotropic angular distribution in the W centre of mass characteristic of spin-0 particles, to be distinguished from the sharply forward-peaked distribution characteristic of fermions with $V \pm A$ couplings.

The total $W \rightarrow e\nu$ data sample was used in the search for supersymmetric decays of the W, including the following requirements: $E_T^e > 15$ GeV, $\not{E}_T > 15$ GeV, and validation cuts on electron signature and $\not{E}_T$. The event sample was compared to the expectations from known processes using the ISAJET Monte Carlo predictions including full simulation of the detector. The percentages of events expected from background sources are: $W \rightarrow e\nu$ (92%), $W \rightarrow \tau\nu$ (6%), and QCD jet fluctuations (2%). The transverse mass and angular distribution observed are well described by the above-mentioned background contributions. Using this agreement

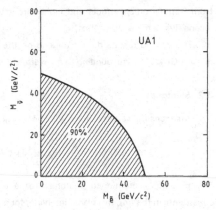

Fig. 3.10 Limits on sleptons and sneutrino masses (90% CL) obtained by UA1 searching for the supersymmetric decay $W \rightarrow \tilde{e}\tilde{\nu}$, with the subsequent decays $\tilde{e} \rightarrow e\tilde{\gamma}$ and $\tilde{\nu} \rightarrow \nu\tilde{\gamma}$ [3.20].

and the expected distributions from a supersymmetric decay of the W, a likelihood fit was performed in the $m_T - q\cos\theta$ plane to extract the excluded domain shown in Fig. 3.10. Of particular interest is the bound[*]

$$m_{\tilde{e}} > 32 \text{ GeV/c}^2 \quad (90\% \text{ CL}) \quad (3.16)$$

---

[*] This limit represents an improvement on the previous result [3.20] and is due to the increased statistics from the 1985 run and to a more detailed Monte Carlo study of the supersymmetric processes.

in the case that $m_{\tilde{e}} \approx m_{\tilde{\nu}}$. Note that this limit applies only to the $\tilde{e}_L$, since the W does not couple to the $\tilde{e}_R$.

The UA1 Collaboration has also searched for the decay of the W $\rightarrow \tilde{\mu}\tilde{\nu}$ and the Z $\rightarrow \tilde{\ell}\tilde{\ell}$ ($\tilde{\ell} = \tilde{e}$ or $\tilde{\mu}$) with $\tilde{\ell} \rightarrow \ell\tilde{\gamma}$. The Z $\rightarrow \tilde{e}\tilde{e}$ mode is less sensitive than is the W $\rightarrow \tilde{e}\tilde{\nu}$ channel described above, resulting in a weaker limit on $m_{\tilde{e}}$. In the present data of the muonic decay channels, the sensitivity is negligible in the Z case and marginal in the W case owing to the inferior momentum resolution (compared with the energy resolution), resulting in a broader transverse or invariant mass distribution.

The UA2 Collaboration has searched for $p\bar{p} \rightarrow Z + X$ followed by $Z \rightarrow \tilde{e}^+\tilde{e}^-$, $\tilde{e}^\pm \rightarrow e^\pm\tilde{\gamma}$ decays [3.15]. In this analysis it was assumed that the $\tilde{\gamma}$ is stable (therefore escaping detection), that $m_{\tilde{e}_L} = m_{\tilde{e}_R}$, and that $m_{\tilde{w}} > m_Z/2$. This supersymmetric process is characterized by final states containing low-mass electron pairs and missing transverse energy $\not{E}_T$. The total data sample from the Z triggers, corresponding to an integrated luminosity of 910 nb^{-1}, has been used in this search. The standard UA2 Z analysis has been modified in such a way that stringent electron validation cuts have been applied to both electron candidates in order to reduce the dominant background from fake electron pairs at low values of the electron-pair mass. Figure 3.11 shows a scatter plot of $m_{e^+e^-}$ versus p_T^{η} for

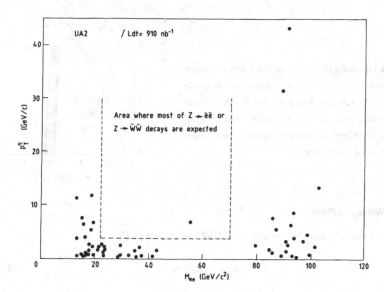

Fig. 3.11 Distribution of the electron-pair mass versus p_T^{η} for the 57 selected events. This sample contains 21 Z $\rightarrow e^+e^-$ candidates and 16 $\pm$ 6 Drell-Yan pairs after background subtraction. Also shown is the dashed area where most of the events coming from Z $\rightarrow \tilde{e}\tilde{e}$ are expected [3.15].

the total of 57 electron-pair candidates selected with $m_{ee} > 10 \, \text{GeV}/c^2$. The variable p_T^η is the projection of $p_T^{e^+e^-}$ on the bisector of the azimuthal angle between the two electron transverse momenta. Note that although $\not{E}_T$ is in principle a signature of this process, it could not be used in the UA2 analysis because of the incomplete angular coverage of the calorimeter. The data with $m_{e^+e^-} < 70 \, \text{GeV}/c^2$ comprised 36 events, of which 20 ± 1 were estimated to be QCD background. The signal from genuine electron pairs is therefore 16 ± 6 events for $m_{e^+e^-} < 70 \, \text{GeV}/c^2$. A Monte Carlo simulation was used to estimate the contribution from $Z \to \bar{e}\bar{e}$ decay to the selected data sample. A likelihood fit to the $(m_{e^+e^-}, p_T^\eta)$ distribution of Fig. 3.11 was performed, taking into account the expected contributions from the Drell–Yan continuum, from QCD background, and from $Z \to \bar{e}\bar{e}$ decays. This analysis resulted in the 90% CL limits in the $(m_{\bar{e}}, m_{\tilde{\gamma}})$ plane shown in Fig. 3.12 together with limits from e^+e^- experiments [3.17]. The shaded area in Fig. 3.12 indicates the variation of the limit if the expected Drell–Yan contribution is increased by 50%.

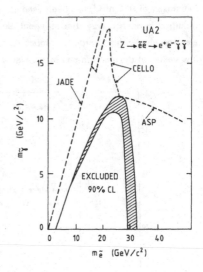

Fig. 3.12 Limits (90% CL) on the $\bar{e}$ and $\tilde{\gamma}$ masses obtained by UA2 [3.15], assuming a stable photino. The dashed area indicates the theoretical uncertainty in Drell–Yan production arising from higher-order terms. Also shown are recent limits from e^+e^- experiments.

3.3 Winos and Zinos

Unsuccessful searches for $e^+e^- \to \widetilde{W}^+\widetilde{W}^-$ tell us that [3.17]

$$m_{\widetilde{W}^\pm} \geq 22 \, \text{GeV}/c^2, \tag{3.17}$$

whilst unsuccessful searches for $e^+e^- \to \tilde{\gamma}\widetilde{Z}$ give us bounds on $m_{\widetilde{Z}}$ which range up to $40 \, \text{GeV}/c^2$ [3.17], but depend on the values assumed for $m_{\tilde{\gamma}}$ and $m_{\bar{e}}$. If $m_{\widetilde{W}^\pm,\widetilde{Z}} \ll m_{\tilde{\gamma},\tilde{q}}$, we expect decay modes and branching ratios similar to those for conventional charged and neutral heavy leptons. If $m_{\tilde{\gamma}} < m_{\widetilde{W}^\pm,\widetilde{Z}} < m_{\tilde{q}}$, we expect the decays $\widetilde{W}^\pm \to \bar{\ell}^\pm \nu$ and $\ell^\pm \tilde{\nu}$ and

$\widetilde{Z} \to \widetilde{\ell}^{\pm} \ell^{\mp}$ and $\widetilde{\nu}\nu$ to dominate. If $m_{\widetilde{\ell},\widetilde{q}} < m_{\widetilde{W}^{\pm},\widetilde{Z}}$, we expect the decays $\widetilde{W}^{\pm} \to \widetilde{\ell}\nu$, $\widetilde{q}\bar{q}$, etc., and $\widetilde{Z} \to \widetilde{\ell}\ell$, $\widetilde{\nu}\nu$, $\widetilde{q}\bar{q}$, etc., to dominate, with branching ratios similar to those of the $W^{\pm}$ and Z into the corresponding sparticles. The relative magnitudes of $m_{\widetilde{\ell}}$, $m_{\widetilde{W}^{\pm},\widetilde{Z}}$, and $m_{\widetilde{q}}$ are uncertain, although in most models $m_{\widetilde{\ell}} < m_{\widetilde{q}}$.

The dominant production mechanisms for the $\widetilde{W}^{\pm}$ and $\widetilde{Z}$ are likely to be:

$W^{\pm} \to \widetilde{W}^{\pm} \widetilde{\gamma}$	giving $\ell^{\pm} + \not{E}_T$, or dijet $+ \not{E}_T$ events;
$W^{\pm} \to \widetilde{W}^{\pm} \widetilde{Z}$	giving $3\ell^{\pm} + \not{E}_T$, or $\ell^{\pm} +$ dijet $+ \not{E}_T$, or
	$\ell^{+}\ell^{-} +$ dijet $+ \not{E}_T$, or quadrijet $+ \not{E}_T$ events;
$Z \to \widetilde{W}^{+}\widetilde{W}^{-}$	giving $\ell^{+}\ell^{-} + \not{E}_T$, or $\ell^{\pm} +$ dijet $+ \not{E}_T$, or quadrijet $+ \not{E}_T$ events;
$Z \to \widetilde{Z}\widetilde{Z}$	giving $4\ell^{\pm} + \not{E}_T$, or $2\ell^{\pm} +$ dijet $+ \not{E}_T$, or quadrijet $+ \not{E}_T$ events.

Clearly, not all these jets and/or leptons would be distinguished by experiment, so we must look for $\widetilde{W}^{\pm}$ and $\widetilde{Z}$ in a combination of $n\ell^{\pm} + m$ jets $+ \not{E}_T$ event classes with $n \leqslant 4$ and $m \leqslant 4$.

The only published limit from this class of searches is by UA2 [3.15], looking for $e^{+}e^{-}$ events with non-zero p_T^{η}, as for the slepton search described in subsection 3.2. A Monte Carlo simulation was performed to estimate the expected contribution from $Z \to \widetilde{W}\widetilde{W} \to ee\widetilde{\nu}\widetilde{\nu}$ decays. It was assumed that the branching ratio for $\widetilde{W} \to e\widetilde{\nu}$ is 33%, that the $\widetilde{\nu}$ is invisible, and that $m_{\widetilde{e}} > m_Z/2$. The dashed area in Fig. 3.11 indicates where most of the events coming from $Z \to \widetilde{W}\widetilde{W}$ are expected. A likelihood fit to the distribution in the $(m_{e^{+}e^{-}}, p_T^{\eta})$ plane analogous to that described in subsection 3.2 led to the excluded domain of the $(m_{\widetilde{W}}, m_{\widetilde{\nu}})$ plane shown in Fig. 3.13 together with limits from $e^{+}e^{-}$ experiments [3.17]. The shaded area in

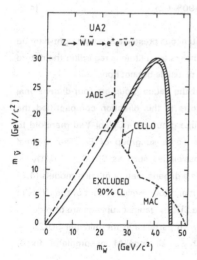

Fig. 3.13 Limits (90% CL) on the $\widetilde{W}$ and $\widetilde{\nu}$ masses obtained by UA2 [3.15]. The dashed area indicates the theoretical uncertainty in Drell–Yan production arising from higher-order terms. Also shown are recent limits from $e^{+}e^{-}$ experiments.

Fig. 3.13 shows the change in the limit if the expected Drell–Yan contribution is increased by 50%, taking into account the larger higher-order QCD corrections. In the $\widetilde{W}$ case, the UA2 results improve significantly the values obtained in e^+e^- experiments so far.

4. ADDITIONAL GAUGE BOSONS

There are many proposals for additional gauge bosons. Here we mention just a few so as to suggest the range of possibilities. The $SU(2)_L \times U(1)_Y$ gauge group of the Standard Model is not particularly handsome, nor has it been derived convincingly from any TOE. Therefore it is natural to consider expansions of the gauge group which might i) make it more elegant, or ii) relate it more directly to some TOE framework. Alternatively, some authors hope to solve the problem of the origins of the $W^\pm$ and Z mass in a composite model, in which case other more massive vector bosons can be expected.

Typical of the first approach are proposals that the gauge group be expanded to $SU(2)_L \times SU(2)_R \times U(1)$ with parity violated spontaneously [4.1]. In this case one would expect three additional gauge bosons — two charged bosons $W_R^\pm{}'$ and one neutral boson Z'. In this framework it is natural to expect that the magnitudes of the $W_R^\pm{}'$ couplings could be similar to those of the known W. Also, in this approach the Z' would have couplings similar to the Standard Model Z. There are important bounds on the mass of the $W_R^\pm{}'$ from μ decay [4.2],

$$W_R^\pm{}' > 432 \, \text{GeV}/c^2 \qquad (90\% \, \text{CL}), \tag{4.1}$$

and from the magnitude of K^0–$\bar{K}^0$ mixing [4.3],

$$W_R^\pm{}' > 1.6 \, \text{TeV}/c^2 \qquad (90\% \, \text{CL}). \tag{4.2}$$

However, the first of these bounds assumes the simultaneous presence of a light right-handed neutrino, which may not be true. The second bound makes assumptions regarding the nature of generalized right-handed Cabibbo mixing which are plausible but not ironclad.

Typical of the TOE approach to augmentation of the gauge group are four-dimensional models inspired by the superstring [4.4]. Early efforts in this direction compactified the ten-dimensional heterotic string with an $E_8 \times E_8'$ gauge group on a Calabi–Yau manifold to obtain a four-dimensional gauge group which was some subgroup of E_6. The minimal possibility was $SU(2)_L \times U(1)_Y \times U(1)_E$; larger possibilities such as $SU(2)_L \times U(1)_Y \times U(1)^{1 \, \text{or} \, 2}$ or $SU(2)_L \times SU(2)_R \times U(1)^{1 \, \text{or} \, 2}$ also exist. It is a generic feature of such models [4.5] that the U(1) gauge couplings are expected to have a magnitude similar to that of the $U(1)_Y$ in the Standard Model, so that $g_1^2/g_2^2 = \tan^2\theta_w$, whilst the $SU(2)_R$ gauge couplings are of order g_2 as usual. The extra U(1) hypercharges Y_E of the known particles in the minimal $SU(2)_L \times U(1)_Y \times U(1)_E$ extension of the Standard Model (which we call Model A) are completely fixed,

Table 4.1

Possible neutral currents in superstring models

	T_{3L}	$\sqrt{(^5/_3)}Y$	$\sqrt{(^5/_3)}Y'$	Y''
$(u, d)_L$	$\pm\,^1/_2$	$^1/_6$	$^1/_3$	0
u_L^c	0	$-\,^2/_3$	$^1/_3$	0
d_L^c	0	$^1/_3$	$-\,^1/_6$	$^1/_2$
$(\nu, \ell)_L$	$\pm\,^1/_2$	$-\,^1/_2$	$-\,^1/_6$	$^1/_2$
e_L^c	0	1	$^1/_3$	0
D_L	0	$-\,^1/_3$	$-\,^2/_3$	0
D_L^c	0	$^1/_3$	$-\,^1/_6$	$-\,^1/_2$
ν_L^c	0	0	$^5/_6$	$-\,^1/_2$
N_L	0	0	$^5/_6$	$^1/_2$
$(H^+, H^0)_L$	$\pm\,^1/_2$	$^1/_2$	$-\,^2/_3$	0
$(\bar{H}^+, \bar{H}^0)_L$	$\pm\,^1/_2$	$-\,^1/_2$	$-\,^1/_6$	$-\,^1/_2$

Model A: Z' couples to Y'
Model B: Z' couples to $[\sqrt{(^3/_8)}Y' - \sqrt{(^5/_8)}Y'']$
Model C: Z' couples to $[\sqrt{(^3/_8)}Y' + \sqrt{(^5/_8)}Y'']$

and are shown in the first column in Table 4.1. In models based on $SU(2)_L \times U(1)_Y \times U(1)^2$ it is often postulated that one of the additional $U(1)$ factors is spontaneously broken by a large field v.e.v. and is hence irrelevant to contemporary phenomenology. The remaining orthogonal $U(1)$ combination is model-dependent, and two possibilities have been favoured in the literature, which we denote as Models B and C [4.6]. The extra $U(1)$ hypercharges in these two cases are listed in the second and third columns of Table 4.1. Bounds on the masses of these various $U(1)$ bosons can be obtained from low-energy ν–q, ν–e, and e–q scattering, from e^+e^- annihilation, and from the observed values of the $W^\pm$ and Z masses. Making a global fit to all these electroweak data but without imposing all the constraints on the vector boson masses which might be expected from the conventional Higgs structure, it has been found that [2.22]

$$m_{Z'} > \left\{ \begin{array}{ll} \text{Model A} & 129\,\text{GeV/c}^2 \\ \text{Model B} & 352\,\text{GeV/c}^2 \\ \text{Model C} & 180\,\text{GeV/c}^2 \end{array} \right\} \quad 90\%\ \text{CL}. \qquad (4.3)$$

We will briefly mention how these limits compare with those from the CERN $p\bar{p}$ Collider. An important ambiguity in obtaining these comes from the ignorance of the Z' decay modes. The Z' may be able to decay into exotic particles and sparticles which are not kinematically accessible to conventional $W^\pm$ and Z decays. This could suppress the observable $Z' \rightarrow e^+e^-$

Table 4.2

Z′ width and branching ratio into e^+e^- pairs for three representative 'superstring-inspired' models: (I) Z′ can decay into only the observed fermions and the top quark; (II) Z′ can decay into three families of **27** fermions and their supersymmetric partners [4.6]

Model	Case	$10^3 \times \Gamma_{Z'}/m_{Z'}$	$BR(e^+e^-)$%
A	I	6.5	3.6
	II	38	0.6
B	I	12	5.9
	II	38	1.8
C	I	6.5	5.4
	II	38	0.9

or $\mu^+\mu^-$ branching ratio by a significant factor, as shown in Table 4.2. Because of the experimental limits on the total $W^\pm$ and Z decay widths reported elsewhere in this volume [2.3], there are constraints on the masses of such exotic particles which could in principle give interesting lower bounds on the Z′ $\to e^+e^-$, $\mu^+\mu^-$ branching ratios. In practice, however, branching ratios almost as low as the minima shown in Table 4.2 are possible. When quoting lower limits on superstring Z′ masses, we will consider both the minimal and maximal cases in Table 4.2.

Both UA1 and UA2 have published negative searches for additional gauge bosons W′ and/or Z′. In the case of UA1 [4.7], no W′ $\to e\bar{\nu}_e$ candidates have been observed with $m_T(p_T^e, \not{E}_T)$ beyond the distribution expected for conventional W $\to e\bar{\nu}_e$ decays, corresponding to

$$\sigma \cdot BR(W' \to e\bar{\nu}_e) < 4.6 \text{ pb} \quad (90\% \text{ CL}). \quad (4.4)$$

If we assume that the W′ has couplings to quarks of the same magnitude as the conventional W and that the W′ $\to e\nu_e$ branching ratio is the same as the W $\to e\nu_e$ branching ratio, and use the quark distributions of Diemoz et al. [4.8], to estimate the W′ production cross-section, then from formula (4.4) we can deduce that [4.7]

$$m_{W'} > 220 \text{ GeV/c}^2. \quad (4.5)$$

This limit should be directly applicable to the W_R of an $SU(2)_L \times SU(2)_R \times U(1)_Y$ model. In the case of Z′ $\to \ell^+\ell^-$, no e^+e^- pair was found beyond the distribution expected for Z $\to e^+e^-$ decays. On the basis of these negative results, it was concluded that [4.7]

$$\sigma \cdot BR(Z' \rightarrow e^+e^-) < 4.7 \, pb \qquad (90\% \, CL). \qquad\qquad (4.6)$$

Using the structure functions of Diemoz et al. [4.8] and assuming that the Z' has quark couplings of similar strength to those of the Z, as well as a similar e^+e^- branching ratio, it was concluded that [4.7]

$$m_{Z'} > 173 \, GeV/c^2 \qquad (90\% \, CL). \qquad\qquad (4.7)$$

This limit could be directly applicable to a neutral gauge boson in an $SU(2)_L \times SU(2)_R \times U(1)_Y$ model, but is not applicable to superstring-inspired models which have smaller couplings and possibly smaller leptonic branching ratios.

In Fig. 4.1 the cross-section times branching ratio ($\sigma \cdot BR$) is plotted as a function of the additional vector boson masses. The solid line (dashed line) corresponds to the limit obtained using structure functions of Diemoz et al. (Duke and Owens set 1), where the curves are

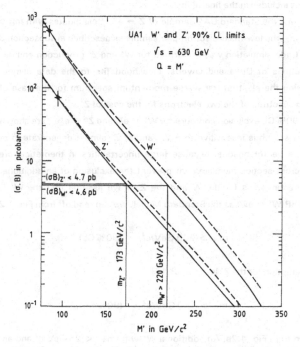

Fig. 4.1 Limits (90% CL) obtained on ($\sigma \cdot BR$) as a function of additional vector-boson masses at $\sqrt{s} = 630$ GeV [4.7] using structure functions [4.8] of Diemoz et al. (solid line) and of Duke and Owens (dashed line). The curves are normalized to the UA1 results at the W and Z mass.

normalized to the UA1 results at the W and Z mass. The muonic decay channels have not been used in the searches for additional vector bosons because of the inferior momentum resolution of the measured muon track; the resulting transverse and invariant mass distributions would give a much weaker limit.

The UA2 Collaboration has performed similar analyses [3.15], but has quoted directly lower bounds on $m_{W'}$ and $m_{Z'}$ as functions of $\lambda_q^2 B_e$, which is defined by

$$\lambda_q^2 \equiv \sum_q g_{q\bar{q}Z'}^2 / \sum_q g_{q\bar{q}Z}^2 \,, \quad B_e \equiv BR(Z' \to e^+e^-)/BR(Z \to e^+e^-). \tag{4.8}$$

Four different data samples have been used to search for additional vector bosons:

i) for $m_{W'} > m_W$, the standard UA2 sample of $W \to e\nu$ candidates with $p_T^e > 20$ GeV/c and $m_T(e\nu) > 50$ GeV/c^2;

ii) for $m_{W'} < m_W$, electron candidates with $p_T^e > 12$ GeV/c and no jet activity opposite in azimuth to the electron; the estimated contribution from the dominant background coming from two-jet events, where one jet fakes the electron signature and the other jet escapes detection, is included in the final limit;

iii) for $m_{Z'} > m_Z$, the standard UA2 sample of $Z \to e^+e^-$ candidates with $m_{ee} > 76$ GeV/c^2;

iv) for $m_{Z'} < m_Z$, the low-mass electron-pair sample as described in subsection 3.2.

A Monte Carlo simulation was performed for W' and Z' production and decay using the structure functions of Duke and Owens. Likelihood fits to the data samples were then performed, using the electron transverse momentum spectrum for the case of W' and the invariant mass spectrum of the two electrons for the case of Z'.

The UA2 90% CL exclusion contours for $W' \to e\bar{\nu}_e$ and $Z' \to e^+e^-$ are shown in Figs. 4.2a and 4.2b. This search is insensitive to a W' and/or Z' almost degenerate in mass with the Standard Model vector bosons, because large uncertainties in theoretical predictions and experimental cross-section measurement prevent the exclusion of an additional contribution above the expected rates for the W and the Z. If we assume that the W' has the same couplings and BR(W' $\to e\bar{\nu}_e$) as the Standard Model, we can read off from Fig. 4.2a that [3.15]

$$m_{W'} > 209 \text{ GeV/c}^2 \quad (90\% \text{ CL}). \tag{4.9}$$

The same assumption for the Z' leads to [3.15]

$$m_{Z'} > 180 \text{ GeV/c}^2 \quad (90\% \text{ CL}), \tag{4.10}$$

as can be seen from Fig. 4.2b. An additional W' with $m_{W'} < 25$ GeV/c^2 and an additional Z' with $m_{Z'} < 50$ GeV/c^2 would have been detected at the ISR and/or PETRA. Therefore additional vector bosons with masses smaller than the standard ones are excluded for

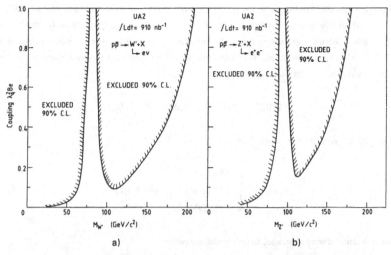

Fig. 4.2 Limits (90% CL) on additional vector bosons W' and Z' depending on $\lambda_q^2 B_e$ (coupling times branching ratio) obtained by UA2 [3.15]: a) for the W' obtained from an analysis of single electron candidates; b) for the Z' obtained from an analysis of electron pairs.

couplings to quarks and leptons similar to the Standard Model ones. For future reference we note that the exclusion contour in Fig. 4.2b and the special case (4.10) correspond, in general, to

$$\sigma \cdot BR(Z' \rightarrow e^+e^-) < 4 \, pb \quad (90\% \, CL). \tag{4.11}$$

As an example of a model with $\lambda_q^2 B_e$ (4.8) less than unity, let us consider what may occur in the minimal rank-5 superstring-inspired model, where the extra U(1) hypercharges Y_E are those specified in column 1 of Table 4.1, and the coupling strength $\alpha_E = 0.016$ is specified by the renormalization group. These predictions imply that $\lambda_q^2 < 1$. As mentioned above, two alternative hypotheses can be formulated about the Z' decays: either it decays only into conventional quarks and leptons — the 'minimal' scenario; or it can decay into three generations of particles in the full 27 representations of E_6, together with their supersymmetric partners — the 'maximal' scenario. In the 'maximal' case, we must nevertheless respect experimental bounds on the masses of sleptons [2.13, 2.21] and other charged particles in e^+e^- annihilation, as well as bounds on the squark masses from the CERN $p\bar{p}$ Collider itself [3.13, 3.15].

These particles could be light enough to appear in Z decays as well as in Z' decays. The most conservative lower bound on the Z' mass is obtained by adjusting the undiscovered particle masses consistently with the experimental bounds, so as to fit the observed $\sigma \cdot BR(Z \rightarrow e^+e^-)$ satisfactorily whilst minimizing $\sigma \cdot BR(Z' \rightarrow e^+e^-)$. This has been done for the minimal rank-5 and other superstring-inspired models. Comparing the resulting minimized $\sigma \cdot BR(Z' \rightarrow e^+e^-)$ with the upper limit obtained by combining preliminary results of UA1 and the UA2 value (4.11), we get [4.9]

$$\sigma \cdot BR(Z' \rightarrow e^+e^-) < 1.8 \, pb \qquad (90\% \, CL). \qquad (4.12)$$

The resulting lower bounds on $m_{Z'}$ are

$$m_{Z'} > \begin{cases} 118 \, GeV/c^2 \\ 140 \, GeV/c^2 \qquad (90\% \, CL) \\ 115 \, GeV/c^2 \end{cases} \qquad (4.13)$$

in the 'maximal' decay scenario, to be compared with

$$m_{Z'} > \begin{cases} 167 \, GeV/c^2 \\ 171 \, GeV//c^2 \qquad (90\% \, CL) \\ 158 \, GeV/c^2 \end{cases} \qquad (4.14)$$

in the 'minimal' decay scenario, where the Z' decays only into conventional quarks and leptons. We see that the more conservative bounds (4.13) are less stringent than those (4.3) already available from the neutral-current data, whilst the bounds (4.14) obtained in the 'minimal' scenario are more stringent. The optimal conservative bound can be obtained by combining the neutral current bounds (4.3) with the 'maximal' decay collider bounds (4.13). This exercise has been carried out for the minimal rank-5 model, with the result [4.9]

$$m_{Z'} > 156 \, GeV \qquad (90\% \, CL), \qquad (4.15)$$

which is considerably stronger than the individual bounds for this model.

Before leaving the superstring-inspired Z', we remind the reader that because of its mixing with the Standard Model Z, the Z' has in general a non-zero coupling to W pairs, which can yield

$$BR(Z' \rightarrow W^+W^-)/BR(Z' \rightarrow \ell^+\ell^-) = O(1). \qquad (4.16)$$

However, as might be guessed from the present lower limits (4.13) and (4.14) on the Z' mass from searches for $Z' \rightarrow \ell^+\ell^-$, the present data samples of UA1 and UA2 are too small to

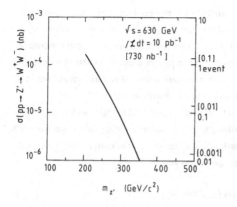

Fig. 4.3 Total cross-section $\sigma(p\bar{p} \rightarrow Z' \rightarrow W^+W^-)$ as a function of $m_{Z'}$ for Model C (see text) at $\sqrt{s} = 630$ GeV. The expected numbers of events are given for total integrated luminosities of 10 pb^{-1} and 730 nb^{-1} [4.6].

expect a signal for $Z' \rightarrow W^+W^-$. However, this is a possible prospect for the larger data sets obtainable with ACOL, as can be seen from the cross-sections shown in Fig. 4.3 for model C [4.6].

5. COMPOSITE MODELS

There are several theoretical motivations for composite models. One of these is the proliferation of quark and lepton flavours, which can give the idea that quarks and leptons may be composite particles [5.1]. Another is the massive nature of the $W^\pm$ and Z bosons, which suggests to some theorists the possibility that they may be composite, like the ϱ vector meson in QCD [5.2]. Alternatively, one response to the problematic quadratic divergences (3.1) in the mass of the elementary Higgs boson has been to suggest that this boson may be composite [5.3]. We are unaware of any experimental probes of this last idea at the CERN $p\bar{p}$ Collider, and we will concentrate on tests of composite models of the quarks and leptons.

There are several possible manifestations of quark and lepton compositeness, including new contact interactions having the new strong interaction scale Λ, quark and lepton form factors, and excited quarks q^* or leptons ℓ^*. A general framework for discussing four-fermion contact interactions has been set up by Eichten, Lane and Peskin [5.4]:

$$\mathcal{L}_{ELP} = g_{eff}^2 \left[(\eta_{LL}/2\Lambda_{LL}^2) (\bar{\psi}_L\gamma^\mu\psi_L) (\bar{\psi}_L\gamma_\mu\psi_L) + \eta_{RR}/2\Lambda_{RR}^2)(\bar{\psi}_R\gamma^\mu\psi_R) (\bar{\psi}_R\gamma_\mu\psi_R) + \right.$$
$$\left. (\eta_{LR}/2\Lambda_{LR}^2) (\bar{\psi}_L\gamma^\mu\psi_L) (\bar{\psi}_R\gamma_\mu\psi_R) + (\eta_{RL}/2\Lambda_{RL}^2) (\bar{\psi}_R\gamma^\mu\psi_R) (\bar{\psi}_L\gamma_\mu\psi_L) \right], \qquad (5.1)$$

where Λ_{ij} is proportional to the binding scale of the composite model and we have allowed for different possible helicity structures of the four-fermion interactions (η_{ij} may be ± 1, and we assume that $g_{eff}^2 = 4\pi$). Experiments at the CERN $p\bar{p}$ Collider are only sensitive to four-quark contact interactions. Their effects may be probed either by looking at the total rate for two-jet events, or by measuring the two-jet angular distributions: the contact terms (5.1) give much more isotropic distributions than does one-gluon exchange. The principal experimental problem in interpreting any measurement of the total two-jet cross-section is the theoretical uncertainty in the initial-state proton distributions and in higher-order QCD effects. Possible limitations on bounding Λ using the angular distributions are a limited angular acceptance for jets and radiative corrections to the QCD tree-level angular distributions.

There is little sensitivity to quark or lepton form factors at the CERN $p\bar{p}$ Collider, and there has not been any experimental interest in probing them, so we will not discuss them further.

To discuss the production of excited quarks q^* and leptons ℓ^* it is convenient to use the following general parametrization of the f–f* gauge-boson vertex,

$$\mathcal{L} = (g/M)\, \bar{f}\sigma_{\mu\nu}f^* F^{\mu\nu}, \qquad (5.2)$$

where again we expect that M is of the order of the compositeness scale. We can use formula (5.2) — where g is the SU(3) gauge coupling, f is a quark, $F^{\mu\nu}$ is a gluon field strength, and f^* is an excited quark — to estimate the total cross-section for $g + q \to q^*$. This cross-section can then be compared with experimental bounds on dijet mass bumps, in order to give a lower bound on M as a function of m_{q^*}. We expect the dominant decay mode of the q^* to be into q + g, but q + γ, $W^\pm$, Z decays are also possible, and could have comparable rates if the corresponding M values are similar. There have also been searches for bumps in γ + jet combinations which could come from $q^* \to q + \gamma$ decay, and for excited leptons in $Z \to \bar{\ell}\ell^*$ and $\bar{\ell}^*\ell$ and $W \to \ell^\pm \nu^*$ and $\ell^{*\pm} \nu$ decays, stimulated by the early observations [5.5] of radiative Z decays, $Z \to e^+e^-\gamma$ and $Z \to \mu^+\mu^-\gamma$. Models for these decays using vertices of the form (5.2) have been proposed. They tend to give angular distributions that are closer to phase space than are the observed angular distributions, which are more similar to those predicted by calculation of QED radiative corrections to $Z \to e^+e^-$ and $Z \to \mu^+\mu^-$. There have also been attempts to interpret the radiative Z decays in the context of composite models of the vector bosons.

The experimental searches at the CERN $p\bar{p}$ Collider for contact interactions of the form (5.1) have been moulded by the characteristics of the detectors. Because of its limited rapidity coverage, the UA2 Collaboration has concentrated on the overall normalization of the large-p_T jet cross-section [5.6]. Finite values of Λ_c would produce an excess of events, compared with ordinary QCD predictions ($\Lambda_c = \infty$) at large p_T. Under the assumption that the main uncertainties (systematic errors, K-factor) are about constant over the full p_T range, deviations in the high-p_T tail can be observed. Taking into account both the theoretical and experimental uncertainties, the following limit was obtained [5.6]:

$$\Lambda_c > 370\,\text{GeV} \qquad (95\%\ \text{CL}). \qquad (5.3)$$

As already commented above, this technique is limited by theoretical uncertainties in the large-p_T jet cross-sections, as well as by systematic errors in the overall energy calibration of the calorimeters. To avoid the latter problem and also to take advantage of the larger rapidity coverage, UA1 preferred to use the two-jet angular distribution to bound Λ_c. Figure 5.1 shows the angular distribution observed for jet pairs with masses between 240 GeV/c^2 and 300 GeV/c^2 [5.7]. The variable χ is defined as

$$\chi \equiv (1 + \cos\theta)/(1 - \cos\theta), \qquad (5.4)$$

where the c.m. scattering angle θ was calculated as the angle between the axis of the jet pair and the beam direction in the jet–jet rest frame [5.8]. The distribution in χ would be almost flat in the absence of perturbative QCD scaling violations. A finite parton size implying substructure will modify this angular distribution such that more events are expected at wide angles relative to the QCD prediction. The leading-order QCD calculation, including scale-breaking effects and using the EHLQ structure functions [5.9] with $\Lambda = 200$ MeV and $Q^2 = p_T^2$, fits the angular distribution very well, as can be seen from the solid line in Fig. 5.1. The

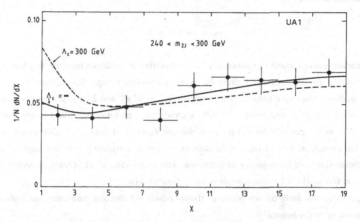

Fig. 5.1 Normalized angular distribution for very high mass jet-pairs as a function of χ. The solid curve is the QCD production ($\Lambda_c = \infty$); the dotted curve corresponds to $\Lambda_c = 300$ GeV, which is clearly excluded by the UA1 data [5.7].

QCD prediction corresponds to $\Lambda_c = \infty$. The dashed line is obtained with $\Lambda_c = 300$ GeV in the effective Lagrangian (5.1). By varying the value of the parameter Λ_c in the fit to the measured angular distribution and taking into account the systematic uncertainty on the jet-energy scale, UA1 obtained the lower limit [5.7]

$$\Lambda_c > 415 \text{ GeV} \qquad (95\% \text{ CL}). \qquad (5.5)$$

This figure is less stringent than the corresponding bounds from e^+e^- annihilation, which however do not probe for compositeness in the $\bar{q}q\bar{q}q$ interaction.

Some excitement was sparked among composite modellers a few years ago by the observation of three radiative Z decays in the 1983 data sample: one $Z \to e^+e^-\gamma$ event from UA2, and one $Z \to e^+e^-\gamma$ and one $Z \to \mu^+\mu^-\gamma$ event from UA1 [5.5]. Key parameters of these three events are gathered in Table 5.1. Figure 5.2 shows Dalitz plot distributions for the observed three events, and also distributions expected from conventional radiative decays calculated using QED, and from some exotic sources [5.10]. The following expressions have been used for the plots in Fig. 5.2:

$$x_L = [\text{lower } m^2(\ell\gamma)]/m^2(\ell^+\ell^-\gamma),$$

$$x_H = [\text{higher } m^2(\ell\gamma)]/m^2(\ell^+\ell^-\gamma), \tag{5.6}$$

$$x = m^2(\ell^+\ell^-)/m^2(\ell^+\ell^-\gamma),$$

which satisfy the relation

$$x_L + x_H + x = 1. \tag{5.7}$$

The observed events looked qualitatively like conventional radiative decays, but the energies and/or angles of the observed photons were surprisingly high for conventional bremsstrahlung. Since the initial excitement, many more $Z \to e^+e^-$ and $Z \to \mu^+\mu^-$ events have been accumulated and only one more $Z \to \mu^+\mu^-\gamma$ candidate has been observed in the UA1 data sample. The event parameters are given in the last column of Table 5.1. Compared with the total event sample, the radiative events seen are no longer surprising from the point of view of conventional QED radiative decay calculations. The probability of observing an event like the one seen in the total UA2 event sample is now about 0.4 [5.11].

Inspired by the early observations of these radiative Z decays, searches were also made for radiative $W \to \ell\nu\gamma$ events.

The UA1 Collaboration searched for massive $e\nu\gamma$ and $\mu\nu\gamma$ final states containing an energetic photon using the 1983 data sample corresponding to 136 nb^{-1} at $\sqrt{s} = 546$ GeV [5.12]. The event selection followed very closely the search for ordinary $W \to e\nu$ and $W \to \mu\nu$ events, with the additional requirement of observing a photon candidate with E_T in excess of 10 GeV. No events that are consistent with the production and decay of a massive $(e^* \nu_e\gamma)$ or $(\mu^* \nu_\mu\gamma)$ state have been found. Using the electron channel, an upper limit on the production of an excited state of charged leptons was derived and expressed in terms of the ratio R:

$$R \equiv BR(W \to e^*\nu_e \to e\gamma\nu_e)/BR(W \to e\nu_e). \tag{5.8}$$

Table 5.1

Properties of the $\ell^+\ell^-\gamma$ events (masses in GeV/c^2)

	$e^+e^-\gamma$		$\mu^+\mu^-\gamma$	
	UA1[a)]	UA2[a)]	UA1[a)]	UA1[b)]
E_γ (GeV)	42.9 ± 1.3	24.4 ± 1.4	28.3 ± 1.1	29.3 ± 1.1
$\Delta\alpha(\ell_1\gamma)$ (°)	11.5	31.4	7.9	31.8
$m(\ell_1\ell_2)$	42.9 ± 3.4	50.3 ± 1.7	66.0 ± 16.3	61.3 ± 52.5
$m(\ell_1\ell_2\gamma)$	104.0 ± 2.2	90.6 ± 2.1	80.5 ± 14.7	94.0 ± 55.7
$m(\ell_1\gamma)$	3.9 ± 0.3	9.0 ± 0.4	5.5 ± 0.4	13.1 ± 2.0
$m(\ell_2\gamma)$	94.7 ± 1.9	74.8 ± 2.3	45.7 ± 10.8	70.0 ± 59.1

a) 1983 data; b) 1985 data.

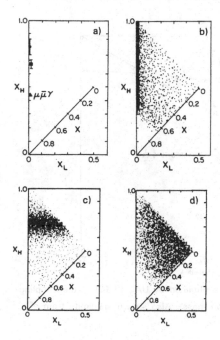

Fig. 5.2 Dalitz plots for $\ell^+\ell^-\gamma$ events [5.10]:
a) for $e^+e^-\gamma$ data (• for UA1 and ■ for UA2);
b) for Standard Model $p\bar{p} \to \ell^+\ell^-\gamma + X$;
c) for $Z \to e^*\bar{e} + e\bar{e}^*$ with $m_{e^*} = 80$ GeV;
d) for $Z \to Z^*\gamma$ or $\gamma^*\gamma$.

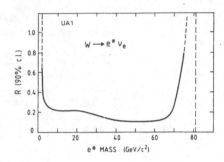

Fig. 5.3 Upper limit (90% CL) for R as a function of m_{e^*} for the process $W \to e^* \nu_e \to e\gamma\nu$ obtained by UA1 [5.12].

Figure 5.3 shows the 90% CL upper limit on R as a function of the e^* mass obtained from this analysis.

The UA2 Collaboration has reported results of a search for excited electrons using the full data sample, corresponding to an integrated luminosity of 910 nb^{-1} [3.15]. Requiring an electron candidate with $p_T^e > 11$ GeV/c, a photon candidate with $p_T^\gamma > 10$ GeV/c and $\not{p}_T > (m_W - M^*)/4$ (where M^* is the electron–photon mass), no event was selected. It is expected that more than 95% of the events coming from $W \to e^*\nu \to e\gamma\nu$ would survive the above cuts for $M^* \geqslant 20$ GeV/c^2. From a Monte Carlo simulation, the ratio R [Eq. (5.8)] as a function of M^* was extracted. The 95% confidence region in the coupling-strength λ–M^* plane, for which the excited electron is excluded, is shown in Fig. 5.4. The UA2 result is compared with the most recent limits from e^+e^- data [5.13].

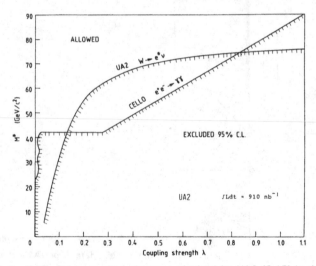

Fig. 5.4 Limits (95% CL) on excited electrons obtained by UA2 [3.15] in the coupling-strength λ–M^* plane (M^* is the excited electron mass). Also shown are limits from e^+e^- experiments.

6. OTHER NEW PARTICLES

There are many other species of new particles which have been proposed and could be searched for at the CERN $p\bar{p}$ Collider. Here we will limit ourselves to three examples where we are aware of specific experimental searches: leptoquarks, monopoles, and free quarks.

6.1 Leptoquarks

These are strongly interacting colour-triplet particles with couplings to lepton–quark combinations. They appear in several different theoretical frameworks, notably superstring-inspired models [6.1] and some scenarios for compositeness of quarks, leptons [6.2], W, and Z detectable at a low-energy scale. In most such models, any light leptoquark bosons would have spin zero. The superstring-inspired models favour charge $|q| = \frac{1}{3}$ [6.3] whilst the composite models favour charge $|q| = \frac{2}{3}$. If the leptoquarks have generation-conserving couplings, as is favoured by upper limits on flavour-changing neutral currents, we might expect the following decays in the superstring case:

$$LQ_1 \rightarrow d\nu_e,\, ue^- ,$$
$$LQ_2 \rightarrow s\nu_\mu,\, c\mu^- , \qquad (6.1)$$
$$LQ_3 \rightarrow b\nu_\tau,\, t\tau^- ,$$

or in the composite model,

$$LQ_1 \rightarrow u\bar{\nu}_e,\, de^+ ,$$
$$LQ_2 \rightarrow c\bar{\nu}_\mu,\, s\mu^+ , \qquad (6.2)$$
$$LQ_3 \rightarrow t\bar{\nu}_\tau,\, b\tau^+ ,$$

In either case, the best signatures are offered by the second-generation leptoquark LQ_2, which can yield $\mu^+\mu^-$ + jet events, μ + jet (+ $\not{E}_T$) events, and jet + $\not{E}_T$ events.

In UA1 a search for pair-produced leptoquarks, i.e. $p\bar{p} \rightarrow LQ_2\overline{LQ}_2$ + X was performed using each of these three signatures [6.4]. The ISAJET Monte Carlo program was used for simulating the production and decay of leptoquark events in the UA1 detector. The predicted production cross-section as a function of the leptoquark mass is shown in Fig. 6.1.

To study a possible signal in the $\mu^+\mu^-$ + jet channel, the UA1 dimuon data sample ($p_T^\mu >$ 3 GeV/c, $m_{\mu\mu} >$ 6 GeV/c^2) has been used [6.5]. In order to obtain the best possible signal-to-background ratio for $\mu^+\mu^-$ + jet events coming from LQ production, the selection cuts were modified in the following way: i) isolated muons with $p_T^\mu >$ 5 GeV/c^2, ii) $\Delta\phi(\mu_1, \mu_2) <$ 170° requiring opposite-sign muons, iii) two jets with $E_T^j >$ 8 GeV, iv) $m_{\mu - jet} >$ 15 GeV/c^2 for each combination, and v) $m_{LQ} = m_{\overline{LQ}}$ to within 40%. One event survived this selection, to be compared with 1.9 ± 0.8 events expected from the $b\bar{b}$, $c\bar{c}$ production, Drell–Yan, and Υ

312

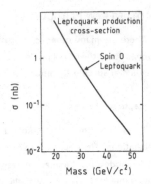

Fig. 6.1 Predicted leptoquark production cross-section at $\sqrt{s}$ = 630 GeV as a function of leptoquark mass obtained with the ISAJET Monte Carlo program [6.4].

decays. Comparison with the rates expected from leptoquarks of different masses (139 events for m_{LQ} = 20 GeV/c^2, 27 for m_{LQ} = 30 GeV/c^2, 4.5 for m_{LQ} = 40 GeV/c^2) gave the 90% CL limits shown in Fig. 6.2, in the plane of m_{LQ} versus the branching ratio for LQ → μ + jet.

The second sample explored was the one containing μ + jet events. In this case we are looking for events $p\bar{p}$ → LQ($\overline{LQ}$) → $\mu\bar{\nu}c\bar{s}$, where a high-p_T muon and two jets are expected in the final state. Accordingly the selection was: i) p_T^μ > 10 GeV/c^2, and ii) two jets with E_T^{j1} > 12 GeV, E_T^{j2} > 8 GeV. Additional technical cuts have been applied, tailored to the expected leptoquark signature. To obtain the best possible limit on the leptoquark mass, a likelihood function was constructed to distinguish leptoquark events from background events. The distributions of p_T^μ, of E_T of the second jet (E_T^{j2}), of $\not{E}_T$, and of $\cos\theta_{j_2}^*$ were used for this likelihood function[*]. Figure 6.3 shows the distribution of the likelihood function for the observed events, compared with that expected for the background (normalized to the data) and for a leptoquark of 30 GeV. The region of L > 0 has been used to obtain the limit from the μ + jet sample which is also shown in Fig. 6.2.

Finally, UA1 used the $\not{E}_T$ event sample discussed in subsection 2.2 to look for double LQ → ν + jet decays. The standard selection, i.e. i) $\not{E}_T$ > 15 GeV/c^2 and N_σ > 4, and ii) E_T^j > 12 GeV, was used, together with iii) E_T^{j} < 40 GeV and iv) a tighter cut on L_τ to eliminate events with τ characteristics. This selection left 7 events, to be compared with 10.5 ± 2.9 ± 0.8 expected from standard physics processes and jet fluctuations, and 13 (14) (2.7) events from a leptoquark of mass 20 (30) (40) GeV/c^2 decaying entirely into ν + jet. The deduced limit on the LQ mass is plotted in Fig. 6.2. As can be seen from the excluded regions obtained from the $\not{E}_T$ + jet sample, leptoquark masses below ~ 25 GeV/c^2 cannot be excluded. At low leptoquark masses, the efficiency for selecting events is very small owing to

---

[*]) $\theta_{j_2}^*$ is the angle between the antiproton beam direction and the axis of jet 2 in the centre of mass of the system consisting of the muon, jet 1, jet 2, and $\not{E}_T$.

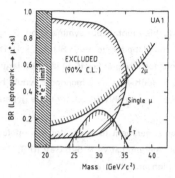
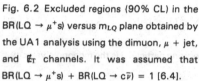

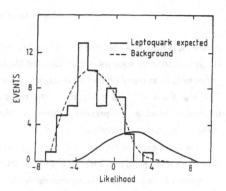

Fig. 6.2 Excluded regions (90% CL) in the BR(LQ $\to \mu^+$s) versus m_{LQ} plane obtained by the UA1 analysis using the dimuon, μ + jet, and $\not{E}_T$ channels. It was assumed that BR(LQ $\to \mu^+$s) + BR(LQ $\to c\bar{\nu}$) = 1 [6.4].

Fig. 6.3 Likelihood function for single μ + jet events calculated for data (histogram), for the expected background normalized to the data (dashed line), and for the expectation from a leptoquark of 30 GeV/c^2 mass [6.4].

the low trigger efficiency. In addition, the 8% systematic error on the jet energy scale prevents the placing of a limit below ~ 25 GeV/c^2. Combining the results of the searches for a second-generation leptoquark in the three different channels discussed, and assuming BR(LQ $\to \mu^+$s) + BR(LQ $\to c\bar{\nu}$) = 1, led to the bound [6.4]

$$m_{LQ} > 33 \text{ GeV/c}^2, \tag{6.3}$$

except for a small window where 21 GeV/c$^2 \leqslant m_{LQ} \leqslant 25$ GeV/c^2 and the branching ratio of LQ $\to \mu^+$s is less than 10%.

6.2 Monopoles

It was pointed out many years ago by Dirac [6.6] that magnetic monopoles could be incorporated into QED if their charges g obeyed a quantization condition:

$$g \cdot e = 2\pi \cdot n \quad \text{(n is an integer)}. \tag{6.4}$$

However, the mass of a Dirac monopole was not predictable. More recently, 't Hooft and Polyakov [6.7] have shown that a magnetic monopole is a generic feature of Yang–Mills gauge theory, appearing when a simple non-Abelian gauge group is broken spontaneously to the U(1) electromagnetic subgroup. In this case the mass of the monopole is

314

$$m_M = O(1) \times m_V/\alpha,\tag{6.5}$$

where m_V is the mass acquired by a gauge boson during this spontaneous symmetry breaking. No such monopole appears in the Standard Model, because the gauge group $SU(2)_L \times U(1)_Y$ is not simple. In many unified theories, the value of m_V appearing in Eq. (6.5) is much larger than m_W. Even with m_W in Eq. (6.5), the mass of a 't Hooft–Polyakov monopole would be beyond the reach of any present accelerator. Therefore one can only search for more general Dirac monopoles at the Collider.

The UA3 experiment has been the only dedicated experiment to search for monopoles. The data-taking period was during the first Collider run operating at $\sqrt{s} = 546$ GeV [6.8]. The experiment aims at identifying the monopoles by their high ionization rate:

$$(dE/dx)/(dE/dx_m) = N^2\beta^2 \, (g_D^2/e^2),\tag{6.6}$$

where dE/dx_m is the ionization rate for a minimum-ionizing particle, β is the monopole velocity, and $g_D = hc/2e$.

Plastic detectors, suitable for searching for monopoles, have been placed inside the vacuum pipe of the Collider around the collision point, around the beam pipe, and around the Central Detector of the UA1 experiment, in order to track the monopoles through the magnetic field. No monopole candidates were found. This experiment established an upper limit on the monopole production cross-section of

$$\sigma \leqslant 2 \times 10^{-31} \, cm^2 \qquad (90\% \, CL),\tag{6.7}$$

for monopoles weighing up to 150 GeV, as seen in Fig. 6.4. Figure 6.4a corresponds to the assumption that fractional charged quarks are elementary $(g = 3g_D)$; Fig. 6.4b corresponds to the electron being elementary $(g = g_D)$.

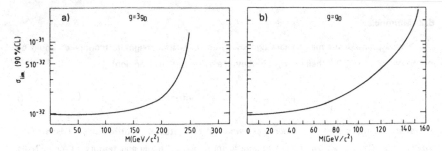

Fig. 6.4 Cross-section limits (90% CL) as a function of the monopole mass obtained by UA3 [6.8]. Curves are given for two choices of the magnetic charge g_D.

6.3 Free Quarks

It is natural to ask whether fractionally charged quarks may become free in nature. According to the standard wisdom in QCD they are confined, but it could be argued that QCD is not quite correct, or that our understanding of QCD is incomplete. If quarks were to become free, some uncertainty would still surround their interaction cross-section with matter, which would affect the strategy for experimental detection. It has often been assumed that quarks would have small interaction cross-sections, but this has been questioned more recently [6.9].

The UA2 Collaboration has searched for free quarks during the first two running periods of the Collider, using a telescope of scintillation counters to detect charged particles with abnormally low ionization densities. Details of the experimental set-up can be found in Ref. [6.10].

Two triggers were used to select events containing low-ionization-density candidates in the quark counters. The first trigger was set up to search for quarks with charge $\pm \frac{1}{3}$, and required a signal in a selected set of counters [imposing an allowed range on the measured amplitude in terms of minimum-ionizing particle (mip) equivalent] in coincidence with a minimum-bias signal. The number of events collected under this trigger condition corresponds to an integrated luminosity of 14.8 nb^{-1}. The second trigger was used to search for quarks with charge $\pm \frac{2}{3}$ requiring a minimum-bias signal together with a signal in all scintillator counters, with an upper limit imposed on their amplitude in terms of mip. This data set corresponds to 3.6 nb^{-1} of integrated luminosity. The most probable ionization I_0 for an event was evaluated from the pulse heights measured in the different counters, using a maximum likelihood method. No candidate events with $I_0 < 0.7$ mip were found for either trigger mode.

The efficiency for detecting quarks with charges equal to $\pm \frac{1}{3}$ and $\pm \frac{2}{3}$ was evaluated by Monte Carlo assuming a momentum distribution of the form $p_T e^{-B m_T}$, where $m_T = (p_T^2 + m^2)^{1/2}$ and $B = 5$ $GeV^{-1} c^2$. The Monte Carlo generated events were subject to the same selection criteria as those used for the experimental data. Figure 6.5 shows the resulting 90% CL upper limits on the ratio R,

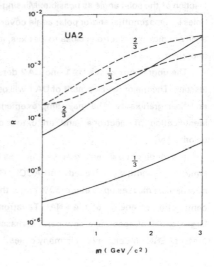

Fig. 6.5 The 90% CL upper limit on R, the number of quarks per single charged particle as a function of the quark mass obtained by UA2. The labels $\frac{1}{3}$ and $\frac{2}{3}$ refer to the absolute values of the quark charges. The dashed curves represent earlier UA2 results, see Ref. [6.10].

the number of quarks per single charged particle, as a function of the quark mass m for both $\pm \frac{1}{3}$ and $\pm \frac{2}{3}$ electric charges. These limits, of order 2.8×10^{-6} and 5.6×10^{-5}, respectively, for very light quarks, increase rapidly with increasing quark mass. The limits are valid under the assumption that free quarks have the same interaction length as that of ordinary hadrons.

7. SUMMARY AND PREVIEW

The CERN $p\bar{p}$ Collider has not yet discovered any unexpected new particles. The cleanliness of large-E_T events, and the ease with which the $W^{\pm}$ and the Z^0 were discovered, disarmed many people who were sceptical of high-energy hadron–hadron colliders. Moreover, it is impressive to see how detectors which were designed with other physics objectives in view have been used to make effective searches for new physics signals which had not been foreseen at the time of construction. Here we have in mind, for instance, physics with a missing-energy signature such as $W \rightarrow \tau\nu$ decays, neutrino counting, and searches for heavy leptons and sparticles. The searches for the latter types of new particles are the most powerful to date, and it is not the fault of the CERN $p\bar{p}$ Collider that these particles do not lie in the mass range accessible to the first generation of Collider experiments.

We extract the following lessons from these searches and from others such as the search for additional heavy quarks. Missing energy will in future be as important a tool for event analysis as are jet energy measurement and lepton identification, and future collider detectors must be able to combine these three features if they seek to explore all the possible new physics sketched in Table 1.1. One corollary is the need for uniform detection over as large a fraction of the solid angle as possible. Missing-energy searches, in particular, are hampered by holes in the azimuthal and/or polar angle coverage. The fine granularity of the detector and the ability to pick out electrons close to jet axes, in addition to muon identification in jets, are also important.

The upgradings of the UA1 and UA2 detectors for running with ACOL reflect the above lessons. The major development of UA1 will be an improved calorimetry with better resolution and finer granularity. The major improvements of UA2 are an improved tracking for better identification of electrons, and an extension in the range of polar angles covered by calorimetry.

The range of new particle masses accessible to these improved detectors depends on the luminosity eventually achieved with ACOL. On the average, one could expect sensitivity to new particle masses up to about 50% larger than those probed so far. The considerably higher centre-of-mass energy of the FNAL Tevatron Collider should eventually pay off with even larger ranges of accessible new particle masses. These should extend up to about twice the present CERN $p\bar{p}$ Collider limit in many cases.

It will be interesting to see whether the e^+e^- colliders which are to operate in the (100 to 200) GeV centre-of-mass energy range (the SLC and LEP) will find any new particles that are inaccessible to the CERN and FNAL $p\bar{p}$ colliders, such as a Higgs boson, or whether the greater cleanliness of the SLC and LEP finds its major application in precision experiments. We can in any case be sure that the CERN $p\bar{p}$ Collider, as well as opening up a new range of energy, has also changed the way physicists perceive different types of accelerators. The first explorer in the next energy range up to 1 TeV will presumably be another hadron–hadron collider.

Acknowledgement

It is a pleasure to thank the Scientific Reports Editing and Text Processing Sections at CERN for their competence and patience in preparing this paper.

REFERENCES

[2.1] Glashow, S.L., Nucl. Phys. **22**, 579 (1961).

Weinberg, S., Phys. Rev. Lett. **19**, 1264 (1967).

Salam, A., Proc. 8th Nobel Symposium, Aspenäsgården, 1968, ed. N. Svartholm (Almqvist and Wiksell, Stockholm, 1968), p. 367.

[2.2] Martinelli, G. and Della Negra, M., 'Heavy flavour production' (this book).

[2.3] Altarelli, G. and Di Lella, L., 'Physics of the Intermediate Vector Bosons' (this book).

[2.4] For a recent review, see Hitlin, D., to appear in Proc. Int. Symp. on Lepton and Photon Interactions at High Energies, Hamburg, 1987.

[2.5] Albajar, C. et al. (UA1 Collaboration), Phys. Lett. **B185**, 233 (1987).

[2.6] Albrecht, H. et al. (ARGUS Collaboration), DESY preprint 87-148 (1987).

[2.7] Sarkar, S. and Cooper, A., Phys. Lett. **148B**, 347 (1984).

[2.8] Yang, J., Turner, M.S., Steigman, G., Schramm, D.N. and Olive, K., Ap. J. **281**, 493 (1984).

Boesgaard, A. and Steigman, G., Ann. Rev. Astr. Astrophys. **23**, 319 (1985).

[2.9] Steigman, G., Olive, K., Schramm, D.N. and Turner, M.S., Phys. Lett. **176B**, 33 (1986).

[2.10] Ellis, J., Enqvist, K., Nanopoulos, D.V. and Sarkar, S., Phys. Lett. **167B**, 457 (1986).

[2.11] Hirata, K. et al., Phys. Rev. Lett. **58**, 1490 (1987).

Bionta, R. et al., Phys. Rev. Lett. **58**, 1494 (1987).

[2.12] Ellis, J. and Olive, K., Phys. Lett. **193B**, 525 (1987).

Schramm, D.N., Proc. 22nd Rencontre de Moriond (in press) (1987).

[2.13] Davier, M., Proc. 23rd Int. Conf. on High-Energy Physics, Berkeley, 1986, ed. S.C. Loken (World Scientific, Singapore, 1987), p. 25.

[2.14] Burke, D. (ASP Collaboration), private communication (1987).

[2.15] Ellis, J. and Peccei, R. (eds.), Physics at LEP (CERN 86-02, Geneva, 1986).

[2.16] Albajar, C. et al. (UA1 Collaboration), Phys. Lett. **B198**, 271 (1987).

[2.17] Altarelli, G. et al., Z. Phys. **C27**, 617 (1985).

[2.18] Paige, F. and Protopopescu, S., ISAJET Program Version SW, BNL 38034 (1986).

[2.19] Albajar, C. et al. (UA1 Collaboration), Phys. Lett. **B185**, 241 (1987).

[2.20] Albajar, C. et al. (UA1 Collaboration), Phys. Lett. **B193**, 389 (1987).

[2.21] Takasaki, F., to appear in Proc. Int. Symp. on Lepton and Photon Interactions at High Energies, Hamburg, 1987.

[2.22] Amaldi, U. et al., Phys. Rev. **D36**, 403 (1987).

Costa, G. et al., CERN preprint TH.4675/87 (1987).

[2.23] Cline, D. and Rubbia, C., Phys. Lett. **127B**, 277 (1983).

Barger, V. et al., Phys. Lett. **133B**, 449 (1983) and Phys. Rev. **D29**, 2020 (1984).

Aurenche, P. and Kinnunen, R., Z. Phys. **C28**, 107 (1985).

[2.24] Mohammadi, M., Ph.D. thesis, University of Wisconsin, Madison (1987).

[2.25] Linde, A.D., Sov. Phys.–JETP Lett. **23**, 64 (1976).

Weinberg, S., Phys. Rev. Lett. **36**, 294 (1976).

[2.26] Lee, B.W., Quigg, C. and Thacker, H., Phys. Rev. **D16**, 1519 (1977).

[2.27] Ellis, J., Gaillard, M.K. and Nanopoulos, D.V., Nucl. Phys. **B106**, 292 (1976).

Barbieri, R. and Ericson, T.D., Phys. Lett. **57B**, 270 (1975).

Vainshtein, A.I. et al., Sov. Phys. Usp. **23**, 429 (1980).

[2.28] Ruskov, R., Phys. Lett. **187B**, 165 (1987).

[2.29] Wilczek, Phys. Rev. Lett. **39**, 1304 (1977).

Vysotsky, M., Phys. Lett. **97B**, 159 (1980).

[2.30] Willey, R.S. and Yu, H.L., Phys. Rev. **D26**, 3086 (1982).

Grzadkowski, B. and Krawczyk, P., Z. Phys. **C18**, 43 (1983).

Pham, T.N. and Sutherland, D.G., Phys. Lett. **151B**, 444 (1985).

See Ref. [2.28].

[2.31] Bloom, E. and Peck, C., in e^+e^- Annihilation, New Quarks and Leptons, ed. R.N. Cahn, (Benjamin–Cummings, Menlo Park, 1985), p. 333.

[2.32] Ellis, J. et al., in Ref. [2.27].

[2.33] Haber, H.E., Schwarz, A.S. and Snyder, A.E., Nucl. Phys. **B294**, 301 (1987).

[2.34] Willey, R.S., Phys. Lett. **B173**, 480 (1986).

[2.35] Towcock, T. (CLEO Collaboration), to appear in Proc. Int. Europhysics Conf. on High-Energy Physics, Uppsala, 1987.

[2.36] Ellis, J., Franzini, P. and Gaillard, M.K., work in progress (1988).

[2.37] Narain, M. et al. (CUSB Collaboration), Cornell preprint (1987).

[2.38] Vysotsky, M., in Ref. [2.29].

[2.39] Glashow, S.L., Nanopoulos, D.V. and Yildiz, A., Phys. Rev. **D18**, 1724 (1978).

[2.40] Georgi, H., Glashow, S.L., Machacek, M. and Nanopoulos, D.V., Phys. Rev. Lett. **40**, 692 (1978).

[3.1] Recent reviews of SUSY include:

Nath, P., Arnowitt, R. and Chamseddine, A., Applied N = 1 Supergravity, ICPT Series in Theor. Physics (World Scientific, Singapore, 1984), Vol. 1.

Nilles, H.P., Phys. Rep. **110**, 1 (1984).

Haber, H.E., and Kane, G.L., Phys. Rep. **117**, 75 (1985).

[3.2] Fayet, P., Unification of the fundamental particle interactions, eds. S. Ferrara, J. Ellis and P. van Nieuwenhuizen (Plenum Press, Inc., New York, 1980), p. 587.

Ellis, J. et al., Nucl. Phys. **B238**, 453 (1984).

[3.3] Nitz, D. et al., Univ. of Michigan preprint UM HE 86–16 (1986).

[3.4] Wolfram, S., Phys. Lett. **82B**, 65 (1979).

Dover, C.B., Gaisser, T.K. and Steigman, G., Phys. Rev. Lett. **42**, 1117 (1979).

[3.5] Ellis, J. et al., in Ref. [3.2].

[3.6] Baer, H., Karatas, D. and Tata, X., Phys. Lett. **183B**, 220 (1987).

[3.7] Naroska, B., Phys. Rep. **148**, 67 (1987).

Behrend, H.J. et al., DESY preprint 87–013 (1987).

[3.8] Ellis, J. and Kowalski, H., Phys. Lett. **142B**, 441 (1984) and Nucl. Phys. **B246**, 189 (1984).

[3.9] Barnett, R.M., Haber, H.E. and Kane, G.L., Phys. Rev. Lett. **54**, 1983 (1985) and Nucl. Phys. **B267**, 625 (1986).

[3.10] Baer, H. and Tata, X., Phys. Lett. **160B**, 159 (1985).

[3.11] Baer, H., Ellis, J., Gelmini, G.B., Nanopoulos, D.V. and Tata, X., Phys. Lett. **161B**, 175 (1985).

[3.12] Cooper-Sarkar, A. et al., Phys. Lett. **160B**, 212 (1985).

Badier, J. et al., Z. Phys. **C31**, 21 (1986).

Bergsma, F. et al., Phys. Lett. **121B**, 429 (1983).

Ball, R.C. et al., Phys. Rev. Lett. **53**, 1314 (1984).

Arnold, R. et al., Phys. Lett. **186B**, (1987) 435.

Alper, B. et al., Phys. Lett. **46B**, 265 (1973).

Cutts, D. et al., Phys. Rev. Lett. **41**, 363 (1978).

Gustafson, H.R. et al., Phys. Rev. Lett. **37**, 474 (1976).

Tuts, P.M. et al., Phys. Lett. **186B**, 233 (1987).

Albrecht, H. et al., Phys. Lett. **167B**, 360 (1986).

[3.13] Albajar, C. et al. (UA1 Collaboration), Phys. Lett. **B198**, 261 (1987).

[3.14] Dawson, S., Eichten, E. and Quigg, C., Phys. Rev. **D31**, 1581 (1985).

[3.15] Ansari, R. et al. (UA2 Collaboration), Phys. Lett. **B195**, 613 (1987).

[3.16] Komatsu, H. and Kubo, J., Phys. Lett. **157B**, 90 (1985).

Haber, H.E. et al., Phys. Lett. **160B**, 297 (1985).

[3.17] See Ref. [2.13].

Whitaker, S., Proc. 23rd Int. Conf. on High-Energy Physics, Berkeley, 1986, ed. S.C. Loken (World Scientific, Singapore, 1987), p. 602.

[3.18] Barnett, R.M., Haber, H.E. and Lackner, K., Phys. Rev. **D29**, 1990 (1984).

Barbieri, R., Cabibbo, N., Maiani, L. and Petrarca, Phys. Lett. **127B**, 458 (1983).

Baer, H., Ellis, J., Nanopoulos, D.V. and Tata, X., Phys. Lett. **153B**, 265 (1985).

[3.19] Barnett, R.M. et al. and Baer, H. et al., in Ref. [3.18].

Cabibbo, N., Maiani, L. and Petrarca, S., Phys. Lett. **132B**, 195 (1983).

[3.20] Arnison, G. et al. (UA1 Collaboration), Europhys. Lett. **1**, 327 (1986).

Albajar, C. et al. (UA1 Collaboration), paper in preparation.

Haywood, S.J., Ph.D. thesis, University of Birmingham, 1987.

[4.1] Langacker, P. et al., Phys. Rev. **D30**, 1470 (1984).

[4.2] Jodidio, A. et al., Phys. Rev. **D34**, 1967 (1986).

[4.3] Bander, M., Beall, G. and Soni, A., Phys. Rev. Lett. **48**, 848 (1982).

[4.4] Green, M.B., Schwarz, J.H. and Witten, E., 'Superstring theory' (Cambridge Univ. Press, Cambridge, 1987).

[4.5] Witten, E., Nucl. Phys. **B258**, 75 (1985).

Dine, M., Kaplunovsky, V., Mangano, M., Nappi, C. and Seiberg, N., Nucl. Phys. **B259**, 519 (1986).

Breit, J.D., Ovrut, B.A. and Segré, G., Phys. Lett. **158B**, 33 (1985).

[4.6] del Aguila, F., Quiros, M. and Zwirner, F., Nucl. Phys. **B284**, 530 (1987) and **B287**, 419 (1987).

[4.7] Albajar, C. et al. (UA1 Collaboration), paper in preparation.

[4.8] Diemoz, M., Ferroni, F., Longo, E. and Martinelli, G., preprint CERN-TH.4751/87 (1987).

Duke, D.W. and Owens, J.F., Phys. Rev. **D30**, 49 (1984).

[4.9] Ellis, J., Franzini, P. and Zwirner, F., CERN preprint TH.4838/87 (1987).

[5.1] Peskin, M.E., Proc. Int. Symp. on Lepton and Photon Interactions at High Energies, Kyoto, 1985, eds. M. Konuma and C. Takahashi (Kyoto Univ., Kyoto, 1985), p. 714.

[5.2] Schrempp, B. and Schrempp, F., Proc. Workshop on Physics at Future Accelerators, La Thuile and CERN, 1987 (CERN 87–07, Geneva, 1987), Vol. 2, p. 339 and references therein.

[5.3] Farhi, E. and Susskind, L., Phys. Rep. **74C**, 277 (1981).

[5.4] Eichten, E., Lane, K.D., and Peskin, M.E., Phys. Rev. Lett. **50**, 811 (1983).

[5.5] Bagnaia, P. et al. (UA2 Collaboration), Phys. Lett. **129B**, 130 (1983).

Arnison G. et al. (UA1 Collaboration), Phys. Lett. **126B**, 398 (1983) and **147B**, 241 (1984).

[5.6] Appel, J.A. et al. (UA2 Collaboration), Phys. Lett. **160B**, 349 (1985).

[5.7] Arnison, G. et al. (UA1 Collaboration), Phys. Lett. **177B**, 244 (1986).

[5.8] Collins, J.C. and Soper, D.E., Phys. Rev. **D16**, 2219 (1977).

[5.9] Eichten, E., Hinchliffe, I., Lane, K. and Quigg, C., Rev. Mod. Phys. **56**, 247 (1984).

[5.10] Barger, V., Baer, H. and Hagiwara, K., Phys. Rev. **D30**, 1513 (1984).

[5.11] Loucatos, S., Proc. 6th Topical Workshop on p$\bar{\text{p}}$ Collider Physics, Aachen, 1986, eds. K. Eggert, H. Faissner and E. Radermacher (World Scientific, Singapore, 1987), p. 3.

[5.12] Arnison, G. et al. (UA1 Collaboration), Phys. Lett. **135B**, 250 (1984).

[5.13] Behrend, H.J. et al. (CELLO Collaboration), Phys. Lett. **168B**, 420 (1986).

[6.1] For reviews, see:
  Ellis, J., preprint CERN–TH.4840/87 (1987).
  Nilles, H.P., preprint CERN–TH.4444/86 (1986).
  Ibáñez, L.E., preprint CERN–TH.4459/86 (1986).
[6.2] Schrempp, B., Proc. 6th Topical Workshop on p$\bar{\text{p}}$ Collider Physics, Aachen, 1986, eds. K. Eggert, H. Faissner and E. Radermacher (World Scientific, Singapore, 1987), p. 642.
[6.3] Angelopoulos, V.D. et al., Nucl. Phys. **B292**, 59 (1987).
[6.4] S. Geer (UA1 Collaboration), preprint CERN–EP/87–163 (1987), to appear in Proc. Int. Europhysics Conf. on High-Energy Physics, Uppsala, 1987.
  Grassmann, H., Ph.D. thesis, University of Aachen (1987).
[6.5] Albajar, C. et al. (UA1 Collaboration), Phys. Lett. **B186,** 237 and 247 (1987).
[6.6] Dirac, P.A.M., Proc. Royal Soc. London **A133,** 60 (1931).
[6.7] 't Hooft, G., Nucl. Phys. **B79,** 276 (1974).
  Polyakov, A.M., JETP Lett. **20,** 194 (1974).
[6.8] Aubert, B. et al. (UA3 Collaboration), Phys. Lett. **120B,** 465 (1983).
[6.9] De Rújula, A., Giles, R.C. and Jaffe, R.L., Phys. Rev. **D17,** 285 (1978).
[6.10] Banner, M. et al. (UA2 Collaboration), Phys. Lett. **121B,** 187 (1983) and **156B,** 129 (1985).

PHYSICS AT THE IMPROVED CERN p̄p COLLIDER

D. Froidevaux
Laboratoire de l'Accélérateur Linéaire,
Orsay, France

P. Jenni
CERN, Geneva, Switzerland

1. INTRODUCTION

After five years of very successful operation during 1981 to 1985 the CERN p̄p Collider has been undergoing an important upgrading programme. Its major items [1] are the addition of a new separate antiproton collector ring (AC) to the existing antiproton accumulator (AA), and the "six-bunch" operation of the SPS in p̄p storage mode. The project has progressed well such that the first Collider run using AC could take place as scheduled in November and December 1987. The peak luminosity is expected to increase by a factor 10, or even more, over the past performance, reaching about 5×10^{30} cm^{-2}s^{-1}. The experiments expect data samples corresponding to $\mathscr{L} \gtrsim 10$ pb^{-1} to result from the future Collider runs at $\sqrt{s} = 630$ GeV during the next few years. The performance reached during the first run at the end of 1987 (peak luminosity $\sim 3 \times 10^{29}$ cm^{-2}s^{-1} and $\mathscr{L} \sim 50$ nb^{-1} only) was however far below design values due to various technical problems both at the antiproton collector and accumulator complex (AAC) and the SPS itself which are expected to be overcome for the forthcoming data taking periods.

In this article, we discuss physics motivations and expectations for experimentation with upgraded detectors at the improved CERN p̄p Collider (referred to as ACOL in what follows). After a description of the upgraded detectors (Section 2), we discuss the expectations for jet physics and QCD studies (Section 3), for the physics related to the Standard Model of electroweak interactions (Section 4), and for speculative physics beyond the Standard Model (Section 5).

2. UPGRADED DETECTORS

The improved CERN $\bar{p}p$ Collider is being exploited in the near future by four experiments. One of them, UA6 [2], is strictly speaking not a collider experiment as it uses either the circulating proton or antiproton beams with an internal hydrogen gas jet target to study inclusive electromagnetic final states at large transverse momentum and Λ production in pp and $\bar{p}p$ interactions at $\sqrt{s}$ = 24.3 GeV. Another experiment, UA8 [3], is studying jet structure in high mass diffractive processes with sets of mini-drift wire chambers installed in so-called Roman-pots on both sides of the interaction region housing UA2. These forward detectors measure the recoil protons (antiprotons) whereas the central energy flow is examined using the information of the UA2 detector.

In the following, we concentrate on the upgrade programmes of the two large-solid-angle general-purpose detectors UA1 and UA2.

2.1 The upgraded UA1 experiment

The main emphasis of the improvement programme for the UA1 experiment [4] is on the replacement of the old central and forward electromagnetic (e.m.) calorimeters (gondolas and bouchons) by a uranium/tetramethyl/pentane (TMP) calorimeter (Fig. 1). The use of uranium allows about 2.6 interaction lengths (λ) to be accommodated in the space occupied by the old e.m. calorimeter alone ($\approx 1\lambda$). The U/TMP calorimeter will therefore not only act as the e.m. calorimeter but, on the average, also absorb a large fraction of the hadronic energy. The calorimeter consists of 4250 cells, each subtending an angle $\Delta\theta \times \Delta\phi = 5° \times 6°$. The electromagnetic section of this calorimeter consists of 2 mm U plates, and is subdivided longitudinally into four sections of 3, 6, 9, and 6 radiation lengths, whilst the hadronic part has 5 mm thick U plates and two samplings. Ten supergondolas will replace the previous 48 lead/scintillator gondolas (see Fig. 1).

The use of TMP readout - a radiation-hard, room-temperature-liquid ionization-chamber technique - will provide high granularity without dead

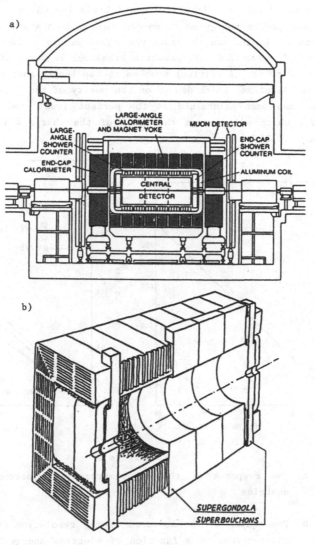

Figure 1

a. The UA1 detector in its original form showing the central and forward (end-cap) e.m. calorimeters (shower counters) that are to be replaced by U/TMP calorimeters.

b. General layout of the new central (supergondolas) and forward (superbouchons) U/TMP calorimeters.

spaces due to cryogenics. It will potentially allow UA1 to reach accurate, uniform, and stable energy measurements (minimize systematic errors on W and Z mass determinations). Prototype measurements with an e.m. U/TMP calorimeter [5] have shown the expected behaviour in terms of linearity and resolution (Fig. 2). A critical problem is the length and the stability of the electron lifetime, which depend on the purity of the liquid. A value of about 3 μs has been maintained at the percent level over a three month period [5], which will ensure that most of the charge will be collected (drift time ≈ 0.25 μs).

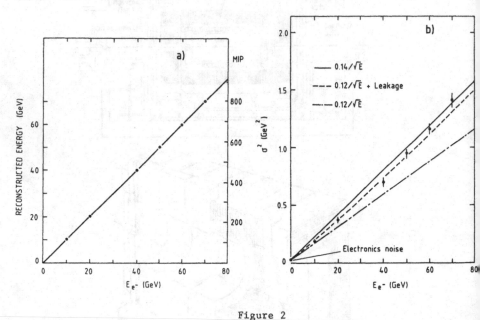

Figure 2

a. The response of the calorimeters to electrons of various energies.

b. The square of the measured energy resolution of the combined calorimeters as a function of electron energy. The solid and dot-dashed lines are for $\Delta E/E = 0.12/\sqrt{E}$ and $0.14/\sqrt{E}$ respectively where E is in GeV. The dashed line is the expected result if the resolution of a calorimeter made entirely of 2 mm plates were $0.12/\sqrt{E}$.
(Both from Ref. [5]).

Other components of the UA1 detector will be upgraded as well, in particular :

a. The muon detection is improved by increasing the hadron absorber thickness both in the central and in the forward regions with additional iron instrumented with planes of limited-streamer tubes and with additional forward drift tubes. Complete muon detection is then provided in the angular range from 15° to 165° with respect to the beam axis.

b. A new data acquisition system with a multilevel trigger structure using information both from the calorimetry and the muon detection system allows UA1 to cope with the increased collision rate expected for the future runs.

The upgraded UA1 detector is expected to be fully operational in 1989.

2.2 The upgraded UA2 experiment

The upgrading of the UA2 experiment, described in detail elsewhere [6], aimed at mainly two aspects: i) full calorimeter coverage, and ii) better electron identification at low transverse momenta (p_T). Full e.m. and hadron calorimeter coverage is achieved with the addition of new end-cap modules covering the angular regions $6° < \theta < 40°$ with respect to the beam directions. These end-cap calorimeters consist of lead/scintillator samplings for the e.m. part and iron/scintillator samplings for the hadronic part, both read out by the wavelength shifter (WLS) technique. They are assembled in wedge-shaped modules subtending 30° in azimuth (Fig. 3). The construction, performance, and granularity ($\Delta\phi \times \Delta\eta \simeq 15° \times 0.2$; $\Delta\eta$ is the pseudorapidity) are chosen to match well the old central calorimeter [7]. This matching is particularly important for the new multilevel triggering scheme, which provides selective triggers for e.m. clusters, hadronic jets, and missing transverse momentum (p_T) over the relevant η range ($-2 < \eta < 2$) and all azimuthal angles ϕ with minimal cracks. In order to use the scintillator/WLS technique, much effort had to be expended on calibrating all the cells in a test beam.

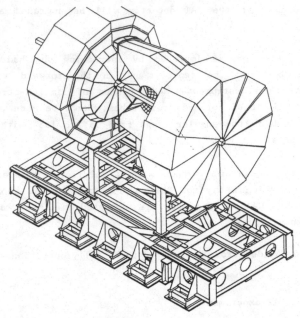

Figure 3

3. General layout of the UA2 calorimeter. The movable support structures which allow to recess the end-caps from the beam pipe are not shown.

This is also valid for the central calorimeter, for which all the scintillator plates in the hadronic compartments were replaced. The UA2 Collaboration is aiming to achieve a systematic energy scale error of $\leq$ 1% for the e.m. calorimeter and $\leq$ 2% for the hadronic one.

Electron identification in UA2 is improved by the use of a completely new central detector assembly [6] consisting of the following cylindrical layers around the beryllium vacuum pipe (Fig. 4): the particles are first tracked through a drift chamber vertex detector (jet chamber); a silicon-pad detector then measures the particle dE/dx to reject e^+e^- pairs from photon conversions and Dalitz decays; two layers of transition radiation detectors follow, which provide an independent electron identification in addition to the calorimeter, suppressing backgrounds from accidental overlaps of a γ with a charged pion; finally, a novel

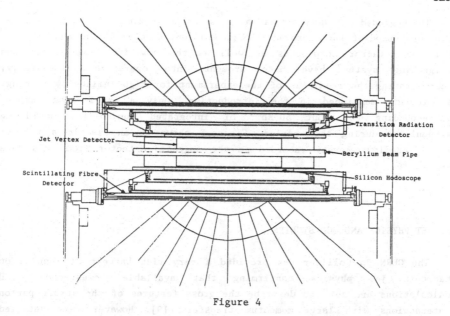

Jet Vertex Detector

Scintillating Fibre Detector

Transition Radiation Detector

Beryllium Beam Pipe

Silicon Hodoscope

Figure 4

Longitudinal view of the new UA2 central detector.

scintillating-fibre detector gives, in a very limited radial space of 6 cm, a second track segment followed by preshower detection after 1.5 X_0 of lead absorber. It is planned to install a further layer of silicon-pad detectors [8] just around the beryllium beam pipe during 1988 in order to enhance the tracking capability of the detector in events with multiple interactions, which are expected to occur frequently for the highest Collider luminosity. The track finding and the electron identification are complemented in the forward regions by proportional tube chambers, including a 2 X_0 lead radiator for preshower detection, mounted on the calorimeter end-caps. All these detector components have given very satisfactory test-beam results. A background rejection of at least a factor of 20 is expected for the inclusive electron measurements as compared with the previous UA2 performance. This will be very important at low p_T (typically 10-15 GeV), where most of the signal electrons from a possible t-quark coming from $W \rightarrow t\bar{b}$ decays are expected (see 4.6).

Finally time-of-flight counter arrays located on both sides of the interaction region provide a fast vertex localization and discriminate against beam-gas backgrounds.

The upgraded UA2 detector has been assembled in time to meet the initial running period of the improved CERN $\bar{p}p$ Collider complex at the end of 1987. All the detector components as well as the new data acquisition and triggering system have been brought into operation successfully, demonstrating the readiness of the experiment for high luminosity running. As already mentioned in the introduction, only a small data sample, corresponding to about 50 nb^{-1} of integrated luminosity, could be accumulated during this period. Figure 5 shows as an example a typical $W \rightarrow e\nu$ candidate event recorded in the upgraded detector from this data sample.

3. JET PHYSICS AND QCD STUDIES

The CERN $\bar{p}p$ Collider has provided a very rich harvest of results on hadronic jet physics confirming that available lowest-order QCD calculations are able to describe the gross features of the strong parton interactions with large momentum transfers [9]. However very detailed quantitative tests of QCD with Collider data suffer from many systematic limitations. From the theoretical side there is a lack in understanding of higher order effects (in the strong coupling constant α_s) and of all "soft" effects due to low momentum transfers which cannot be treated perturbatively. The momentum transfer scale Q^2 is not unambiguously defined, and the fragmentation of scattered partons into hadronic final states is not rigorously described. From the experimental point of view uncertainties arise at some level from the ambiguous assignment of low momentum final state particles to a given jet which may partially overlap with other jets and with spectator hadrons not directly related to the sub-process of interest. Further systematic uncertainties are due to instrumental effects, for example in the detailed knowledge of the calorimeter response to hadronic jets.

The largely increased statistical precision from the jet data samples expected from the improved CERN $\bar{p}p$ Collider will therefore not directly reflect into much higher precision QCD tests in most cases where the systematic uncertainties are already important. Nevertheless significant improvements can be expected over the present data. A few typical jet and

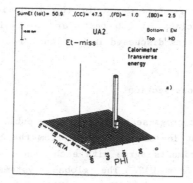

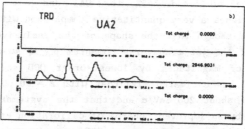

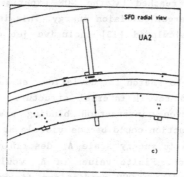

Figure 5

5. Example of a W → ev candidate event recorded in the upgraded UA2 detector.

     a. transverse energy distribution in the calorimeters

     b. transition radiation signals along the electron track

     c. tracking and preshower signals in the scintillating fibre detector for the electron candidate.

QCD physics topics are considered in the following to illustrate these facts. A different class of QCD tests are possible by the study of IVB production properties as discussed in Section 4.1.

3.1 Inclusive jet cross-section

The inclusive jet cross-sections have been published by UA1 [10] and UA2 [11] already, without including the data from the 1985 running period. The data from both experiments agree with each other, and Fig. 6a shows as an example the data from UA2. The global systematic uncertainties are evaluated to be ±70% and ±45% for UA1 and UA2 respectively. This uncertainty precludes a very quantitative comparison with QCD calculations. It is remarkable though that the shape of the inclusive jet cross-section as a function of the jet transverse momentum p_T is so well described over several orders of magnitudes by lowest-order QCD calculations. For the future runs, UA2 has estimated [12] that with $\mathscr{L} = 10$ pb^{-1} they will extend the p_T range to about 200 GeV/c and that the systematic errors could be reduced to ±35% by an improved control of calibration effects. However much higher p_T's will be reached by the CDF experiment at the TEVATRON $\bar{p}p$ Collider due to the larger collision energy. This is illustrated in Fig. 6b which compares the calculated [13] inclusive jet cross-sections at $\sqrt{s}$ of 0.63 and 2 TeV.

It is interesting to search for deviations at large jet p_T from the QCD behaviour which could reveal internal structure of the quarks due to a hypothetical super-strong interaction binding preons inside the quarks [14]. A contact interaction could become visible at momentum transfers well below the characteristic energy scale Λ_c describing the strength of this conjectured interaction. Finite values of Λ_c would produce an excess of events compared to ordinary QCD predictions (corresponding to $\Lambda_c = \infty$) at large p_T. From the inclusive jet cross-sections the present 95% CL lower limit is $\Lambda_c > 400$ GeV [10,11]. This limit is expected [12] to improve to about 650 GeV with the future runs at the CERN $\bar{p}p$ Collider, but it cannot compete with the limit $\Lambda_c \geq 1.6$ TeV, which can be obtained with even a modest data sample ($\mathscr{L} = 1$ pb^{-1}) at $\sqrt{s} = 2$ TeV.

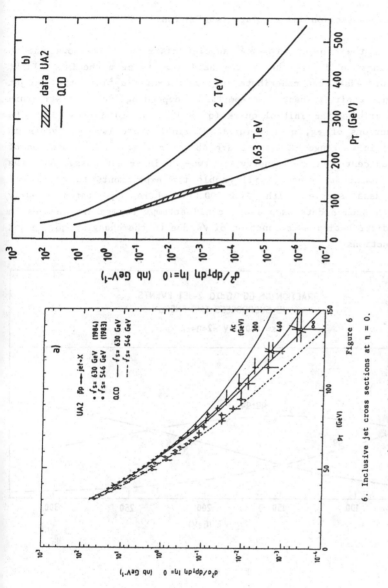

6. Inclusive jet cross sections at $\eta = 0$.

Figure 6

a. As an example the UA2 data [11] at $\sqrt{s} = 546$ GeV and 630 GeV. The effect of a hypothetical new contact interaction is also shown (see text).

b. A comparison of the expected inclusive cross sections at $\sqrt{s} = 630$ GeV and 2 TeV.

334

3.2 Two-jet cross-section and angular distribution

A study of the two-jet mass and angular distributions allows one to investigate in more detail the parton hard scattering mechanism. Eight basic $2 \rightarrow 2$ processes can contribute in lowest order (α_s^2) to the two-jet cross-section, their respective importance depending on the structure functions describing the initial quark (q) and gluon (g) fluxes [15]. The expected fractions of gg, $q\bar{q}$ and (qg + $\bar{q}$g) final state two-jet events in the pseudo-rapidity range -2 < η < 2 are shown in Fig. 7 as a function of the subprocess centre of mass energy $\sqrt{\hat{s}}$ (two-jet invariant mass). Only the future high luminosity running will enable the experiments to accumulate significant data samples with $\sqrt{\hat{s}} \geq 300$ GeV where $q\bar{q}$ states clearly dominate. With such a data sample one could attempt to measure changes in the angular distribution as a function of $\sqrt{\hat{s}}$ due to the changing $q\bar{q}$, gg and (qg + $\bar{q}$g) fractions.

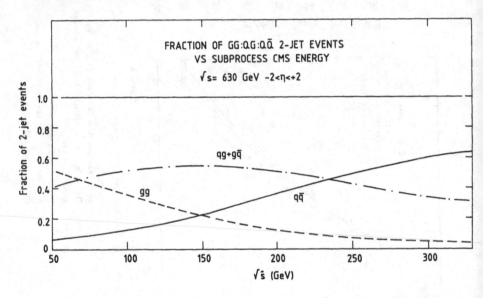

Figure 7

7. The fractions of qq, gg and qg + $\bar{q}$g final state two-jet events in the range -2 < η < 2 as a function of the subprocess centre of mass energy $\sqrt{\hat{s}}$.

UA1 [16] and UA2 [17] have used their two-jet cross sections and angular distributions to extract an "effective" structure function F(x) for the colliding nucleons, assuming that the contributions from the two nucleons factorize and that the angular distribution is described by a weighted average over all elementary processes [18]. F(x) is a sum of all quark and gluon densities with x being the fraction of the longitudinal momentum carried by the interacting parton. The results, shown in Fig. 8, are in reasonable agreement with measurements from deep inelastic lepton-hadron scattering evolved to the Q^2 of the Collider jets. The low x region (x ≤ 0.2) is dominated by the gluon distribution. The high x region (≥ 0.5), dominated by the quark distributions, can be probed only with significantly larger data samples as will become available at the improved Collider. However one has to keep in mind that the systematic errors on the data are about ±30%, and that the experimental results $\sqrt{K} \cdot F(x)$ depend on the assumptions made for the higher order (α_s^3) corrections (K-factors), Q^2 and Λ_{QCD}.

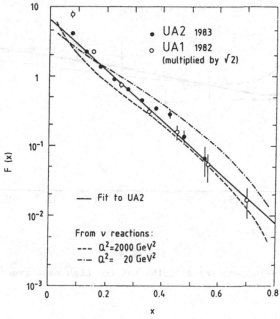

Figure 8

8. Effective structure function. The data are from UA1 [16] and UA2 [17]. See Ref. [17] for further details.

336

The two-jet angular distribution is also sensitive to hypothetical contact interactions [14] already mentioned in the previous sub-section, which would cause an enhancement at large scattering angles θ over the pure QCD distribution. Figure 9 shows the UA1 data [19] in the variable $\chi = (1 + \cos\theta)/(1 - \cos\theta)$ and illustrates the effect of $\Lambda_c = 300$ GeV. Their present 95% CL lower limit on Λ_c from the χ distribution is 415 GeV, slightly better than the one from the inclusive jet cross-section alone (see 3.1). Similarly, somewhat improved limits over the ones quoted in the previous section can be expected from the angular distributions from the future large data samples.

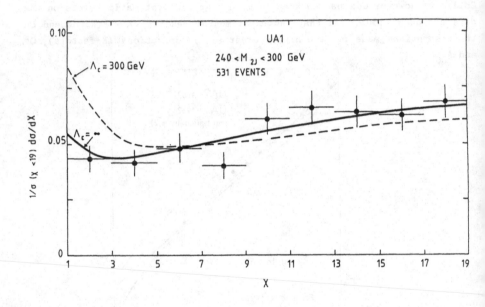

Figure 9

The two-jet angular distribution for high mass events from UA1 [19].

3.3 Three-jet events and α_s

The study of three-jet events allows us to extend the QCD tests to order α_s^3 processes, dominated by gluon bremsstrahlung. Various angular and energy distributions of three-jet events have been successfully compared to QCD expectations by UA1 [20] and UA2 [21]. Further detailed studies will be possible with the increased statistics and the larger acceptances of the upgraded detectors (in particular of UA2). As an example it will become possible to pursue a study of the $\cos\xi = (p_2^* - p_3^*)/ p_1^*$ distribution [21], which describes the asymmetry of the energies found in the softer two jets (p_1^*, p_2^*, p_3^* are the three jet momenta in their common centre of mass system in decreasing order). The distribution of $\cos\xi$ is characteristic of gluon bremsstrahlung off a $q\bar{q}$ pair in e^+e^- annihilations [22], and provides a test of the spin of the gluon. Though in the $\bar{p}p$ case many different subprocesses contribute to the three-jet production, the $\cos\xi$ distribution remains a sensitive probe of the QCD dynamics.

The most topical interest of the three-jet studies however lies in the possibility to extract α_s. The basic idea follows from the fact that the two - and three-jet cross-sections are approximatively proportional to α_s^2 and α_s^3 respectively, and that therefore the ratio, for which many systematic uncertainties in the cross-section measurements are expected to cancel, should yield a determination of α_s. The calculated cross-section ratio can be adjusted to agree with the experimental value by varying α_s. The calculations are however affected by higher-order α_s corrections (expressed as K_2 - and K_3 - factors for the two - and three-jet cross-sections respectively) implying that the three - to - two jet ratio is proportional to α_s (K_3/K_2). Another theoretical uncertainty arises from the ambiguous choice of the Q^2 scale for the two - and three-jet events. Initial $2 \to 2$ and $2 \to 3$ parton scatterings do not contribute alone to the experimentally observed two - and three-jet cross-sections due to incomplete detector acceptance and overlapping jets. This means that further systematic uncertainties in the data analysis arise from model dependent corrections like fragmentation effects, four-(or more) jet contributions, underlying (spectator) event fluctuations, structure functions and so on.

The present results on $\alpha_s(K_3/K_2)$ using the same Q^2 scale for two - and three-jet events are $0.23 \pm 0.02 \pm 0.04$ from UA1 [22] and $0.23 \pm 0.01 \pm 0.04$ from UA2 [21], where the first error is statistical and the second one describes the systematic uncertainties. The Q^2 used is the maximum jet transverse momentum squared, giving average values of 4000 GeV^2 and 1700 GeV^2 for the two experiments respectively. Changing this Q^2 definition for example to the average jet transverse momentum squared, thereby reducing the mean three-jet Q^2 with respect to the mean two-jet Q^2, would reduce the UA2 result [21] for $\alpha_s(K_3/K_2)$ by 25%.

The value of $\alpha_s(K_3/K_2)$ as a function of $\sqrt{\hat{s}}$ has been studied by UA1 and UA2. As an example Fig. 10 shows the most recent data of UA2 [23]. The expected variation of α_s is shown as well (obtained by computing the variation of α_s as a function of Q^2 at the value of the mean Q^2 for each $\sqrt{\hat{s}}$ bin and assuming that K_3/K_2 remains constant). The present accuracy of the data is obviously not sufficient to demonstrate the "running" of α_s. The expected improvement of the statistical error at high values of $\sqrt{\hat{s}}$ in the future data sample is illustrated for the upgraded UA2 experiment for two 10 GeV wide $\sqrt{\hat{s}}$ bins in Fig. 10. Provided that the data span a sufficiently large $\sqrt{\hat{s}}$ range, the Q^2 dependence should become observable whereas the absolute value will likely remain dominated by the systematic uncertainty. It is interesting to note that these data allow one in turn to extract the ratio K_3/K_2 as a function of $\sqrt{\hat{s}}$, if the QCD scale parameter is known from other measurements.

3.4 Multi-jet production

Four-and more parton final state events are of particular interest because there are many "new physics" processes for which such events are an important background [24]. Multi-jet production occurs not only due to higher order (α_s^4) corrections to the two-jet processes but there is also an interesting new class of interactions, namely the multi-parton scattering, where two pairs of incoming partons from the same nucleons interact [25].

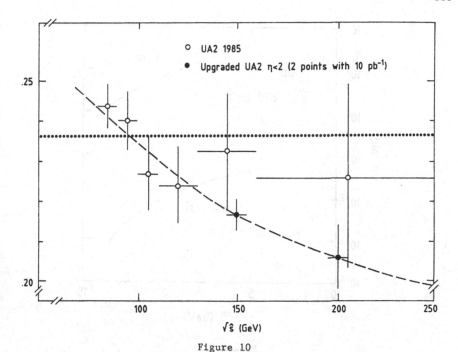

Figure 10

10. The values of $\alpha_s(K_3/K_2)$ as a function of $\sqrt{\hat{s}}$ from UA2 [23]. The
expected improvement on the statistical precision with future runs
($\mathscr{L} = 10$ pb^{-1}) is indicated for two $\sqrt{\hat{s}}$ bins of 10 GeV width.

As an illustrative example the process $\bar{p}p \to W \to t\bar{b} \to 4$ jets is
considered at $\sqrt{s} = 630$ GeV with $m_t = 40$ GeV. The results of a detailed
study by Ref. [24] are shown in Fig. 11. Even after restricting the search
to the central region only ($|\eta| < 1$) and requiring that the four-jet
invariant mass lies within ± 10 GeV/c^2 of the W mass, the QCD $2 \to 4$
scattering processes dominate the signal by more than two orders of
magnitudes, independent of the jet p_T threshold, p_T^{min}. It is worth noting
that the flavour combination of the signal differs in general from the one
of the QCD background, but there is no way at present to select jets
efficiently according to their parton origin. Other examples have been
considered in Ref. [24], and there remains the general conclusion that QCD
multi-jet events dominate searches for new physics when considering purely
hadronic final states.

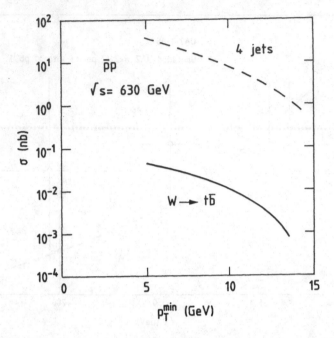

Figure 11

11. Dependence on the jet transverse momentum threshold p_T^{min} for the four jet signal (solid curve) and background (dashed curve) for $W \rightarrow t\bar{b}$ production in $\bar{p}p$ collisions at $\sqrt{s} = 630$ GeV, reproduced from Ref. [24].

No multi-parton signal has been observed yet at the CERN $\bar{p}p$ Collider, the present four-jet data [26] is well described by a $2 \rightarrow 4$ scattering QCD model alone. The new multi-level triggering system of the upgraded UA2 experiment aims at reaching lower jet transverse momenta where the multi-parton signal is expected to be enhanced with respect to normal QCD. It is thereby hoped that a signal can be revealed in the future high statistics data sample.

3.5 Direct photon production

The direct photon production $\bar{p}p \rightarrow \gamma + X$ is of particular interest for QCD tests because of two main advantages. First, it is experimentally possible to measure directly the photon transverse momentum in the electromagnetic calorimeters with much better accuracy than for jets, and second fragmentation effects are absent. A complete QCD calculation in next-to-leading order (α_s^2) is available [27]. However the rates expected are about three orders of magnitude smaller than for inclusive jet production. The present data [28,29] on direct photon production are indeed limited by their statistical precision and are confined to transverse momenta well below 100 GeV/c. The systematic uncertainties on the cross-section are about ± 20%. The data are expected to improve significantly with the increased integrated luminosity in the near future.

4. STANDARD MODEL PHYSICS

In this section, we discuss the physics which can be studied at ACOL, in terms of the Standard Model of electroweak interactions. The extension of the physics results obtained from the study of W, Z decays to the higher luminosities expected at ACOL is discussed in subsections 4.1 to 4.3. Pair production of IVBs is reviewed in 4.4, heavy flavour physics (charm, beauty) in 4.5 and the search for new flavours in 4.6 and 4.7. Finally, for completeness, the search for a neutral Higgs boson is discussed in 4.8. Wherever relevant, the physics potentials of ACOL versus TEVATRON will be compared.

For all comparisons with the past measurements of UA1 and UA2, we refer the reader to the corresponding chapters in this book.

4.1 IVB production properties.

4.1.1 IVB production cross-sections.

The main features of IVB production at hadron colliders are described elsewhere [30] and summarized in Fig. 12. In spite of the clear rise in cross-section for W, Z production (σ_w, σ_z) when $\sqrt{s}$ increases, we expect that precision measurements in the W, Z sector will become increasingly difficult at higher $\sqrt{s}$. Comparing, for example, the TEVATRON ($\sqrt{s}$ = 1.8 TeV) with ACOL ($\sqrt{s}$ = 630 GeV), we expect a factor of 3 increase in σ_w and σ_z, but also a factor of 10 increase in QCD two-jet production, σ_{jet}, with $p_T^{jet} \approx 40$ GeV/c and $|y_{jet}| < 2$ (where p_T^{jet}, y_{jet} are the jet transverse momentum and rapidity). This results in a worse signal to background ratio at TEVATRON compared to ACOL, which might affect precision measurements using W $\rightarrow$ eυ or W $\rightarrow$ $\mu\upsilon$ decays (the background from fake electrons or muons under the Z $\rightarrow$ ee or $\mu\mu$ peak is negligible). In addition, as shown by Figs. 12b and 12c, the gain in rate at higher $\sqrt{s}$ is ruined by a broader p_T^W distribution and by less central production of the W's. In particular, the mean p_T^W at TEVATRON is expected to be almost double that at ACOL, thus broadening significantly the Jacobian peak observed in W $\rightarrow$ eυ decay, and affecting the precision measurements of m_w (see section 4.3).

The accuracy on the measurement of σ_w, σ_z at ACOL will not improve substantially the present measurements which are already dominated by systematic uncertainties on the luminosity for σ_w. More importantly, the QCD predictions of Ref. 30 are affected by large uncertainties (choice of scale for α_s, structure functions, value of the top quark mass), most of which are not likely to improve in the near future. It will be interesting, however, to compare the measurement of σ_w, σ_z at TEVATRON with the values obtained at lower $\sqrt{s}$.

4.1.2 Lepton universality.

The measurement of W $\rightarrow$ eυ, W $\rightarrow$ $\mu\upsilon$ and W $\rightarrow$ $\tau\upsilon$ decays by UA1 has allowed a quantitative test of lepton universality at large momentum transfer, $Q^2 \approx m_W^2$. In the future, UA1 will probably remain the best equipped experiment to improve the accuracy of this test, without, however, being able to reach the precision of experiments at lower Q^2 (PETRA, PEP ...).

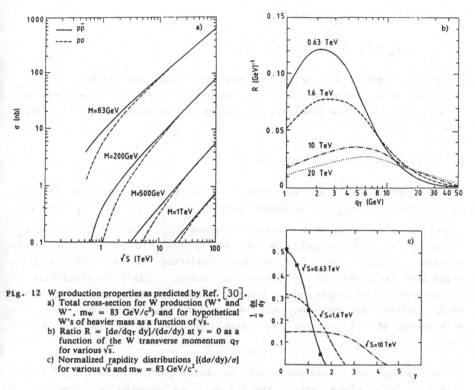

Fig. 12 W production properties as predicted by Ref. [30].
a) Total cross-section for W production (W⁺ and W⁻, $m_W = 83$ GeV/c²) and for hypothetical W's of heavier mass as a function of √s.
b) Ratio $R = [d\sigma/dq_T \, dy]/(d\sigma/dy)$ at $y = 0$ as a function of the W transverse momentum q_T for various √s.
c) Normalized rapidity distributions $[(d\sigma/dy)/\sigma]$ for various √s and $m_W = 83$ GeV/c².

4.1.3 Measurement of the W width.

Most of the experimental and theoretical uncertainties discussed in 4.1.1 disappear when measuring cross-section ratios, and it is well known by now that the determination of the σ_Z/σ_W ratio is the best method of extracting the Z width [31] through the relation

$$\Gamma_Z/\Gamma_W = R_{exp}(\sigma_Z/\sigma_W)[\Gamma(Z \rightarrow ee)/\Gamma(W \rightarrow e\nu)],$$

where $R_{exp} = (\sigma B)_{W \rightarrow e\nu}/(\sigma B)_{Z \rightarrow ee}$ is measured experimentally, and the other quantities are predicted from theory. The main uncertainty on the theoretical predictions arises from the uncertainty on the structure function parametrization used to predict σ_Z/σ_W. This uncertainty is not expected to improve in the near future, and we rewrite the above relation as [32] :

$$\Gamma_z/\Gamma_w = R_{exp}(0.114 \pm 0.008)$$

or $$\Gamma_w = \Gamma_z (8.77 \pm 0.62)/R_{exp}$$

where the errors take into account the theoretical uncertainty on σ_z/σ_w. From this we can expect to extract a measurement of Γ_w to better than 10%, once Γ_z will have been precisely determined at the forthcoming Z factories (SLC and LEP).

4.1.4 IVB production in association with jets.

The prediction of the exact shape of the W transverse-momentum spectrum, p_T^w, is the result of a somewhat unique QCD calculation [30] involving all-order effects from soft gluons for values of p_T^w much smaller than m_w. In principle, this calculation can be compared to the experimentally measured distributions. In the data collected before ACOL, such a comparison was unfortunately impossible because of lack of statistics in the case of $Z \rightarrow ee$ decays, where however p_T^z is precisely measured from the decay electron and positron, and because of large uncertainties on the measurement of the neutrino transverse momentum, in the case of $W \rightarrow e\nu$ decays.

In practice the comparison should be restricted to values of p_T^w larger than ~ 15 GeV/c, where most of the W transverse momentum is balanced by a hadron jet (or jets) which can be more reliably measured by the calorimetry.

In this subsection, we shall first extrapolate a recent measurement of the strong coupling constant, α_s, using a UA2 sample of 230 $W \rightarrow e\nu$ decays, of which 28 are produced in association with one hard jet ($p_T^{jet} > 10$ GeV/c) [33]. This measurement uses approximate expressions of the K-factors [34] necessary to extract α_s from the data. We assume for the following that, in the coming years, these approximate expressions will become exact, thus reducing a main source of uncertainty in these measurements.

Table 1 shows the cross-sections for W production in association with 1 or 2 hard jets ($p_T^{jet} > 15$ GeV/c), with subsequent W → eν decay, at ACOL and TEVATRON, including the relative contributions from the relevant graphs. As expected from Fig. 12b, the fraction of W's produced in association with hard jets is much higher at TEVATRON, and the contribution from gluon graphs has substantially increased.

Table 1

Production of W's in association with hard jets at ACOL and TEVATRON. Electron selection efficiencies and jet reconstruction effects are included.

$p_T^{jet} > 15$ GeV/c	(W → eν) + 1 jet		(W → eν) + 2 jets	
	Cross-section	Contributions	Cross-section	Contributions
ACOL √s = 630 GeV	12.5 pb (3.3% of all W → eν)	$q\bar{q}$: 79% qg : 21%	1.0 pb (0.3% of all W → eν)	$q\bar{q}$: 76% qg : 23% gg : 1%
TEVATRON √s = 1800 GeV	69.0 pb (6.1% of all W → eν)	$q\bar{q}$: 52% qg : 48%	15.0 pb (1.3% of all W → eν)	$q\bar{q}$: 43% qg : 55% gg : 2%

With an integrated luminosity of 10 pb^{-1}, we can expect 125 reconstructed W → eν decays with one jet of p_T larger than 15 GeV/c (this number would increase by almost a factor 2 if the threshold were 10 GeV/c). The experimental error on α_s will arise mainly from the effect of the underlying event and the jet energy measurement. One can hope to extract α_s using this method at ACOL, with similar experimental and theoretical (structure functions, higher orders) uncertainties of approximately 10%.

Another approach would be to use the precise measurement of the p_T^z distribution to extract a measurement of Λ_{QCD}, using the sensitivity of the p_T^z-spectrum to Λ_{QCD} at low values of p_T^z, as shown in Fig. 13. Unfortunately, even though the experimental measurement is accurate in the case of Z → ee decays, the theoretical uncertainties on the calculation of Ref. 30 remain large for $p_T^z < 3$-4 GeV/c. They are mainly due to the definition of Q^2 used for $\alpha_s(Q^2)$, and to the exact value used as a cutoff when Q^2 (or p_T^W) becomes very small. From a simple simulation, we find that we would need about 300 reconstructed $Z \to ee$ decays to extract Λ_{QCD} to within $\pm$ 200 MeV using this method. As we shall see in the next section, such a sample will be available at ACOL or even at TEVATRON in the next few years. Furthermore, as discussed in 4.3, a precise knowledge of the p_T^Z spectrum, extrapolated to the p_T^W spectrum, will greatly reduce systematic errors on the W-mass measurement.

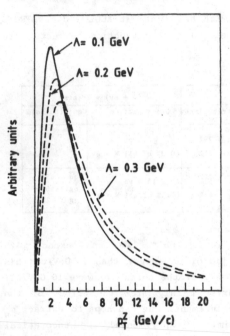

Fig. 13

Sensitivity of the Z transverse-momentum spectrum to Λ_{QCD}.

4.2 IVB decay modes.

4.2.1 Leptonic decay modes.

We first recall that, for approximately 1 pb^{-1} of integrated luminosity at the Sp$\bar{p}$S collider, a total of 600 W $\to$ (e,μ,τ)ν decays and 100 Z $\to$ ee or $\mu\mu$ decays have been observed in UA1 and UA2 combined. The bulk of the statistics is dominated by $W \to e\nu$ and $Z \to ee$. The $W \to \mu\nu$, $W \to \tau\nu$, and $Z \to \mu\mu$ decay modes have been observed in UA1 with much less statistics, because of smaller acceptance and lower cut efficiencies. This situation is likely to remain in the future, both at ACOL and TEVATRON, because of the difficulty in extracting muon and tau signals from the background. In this subsection, we shall therefore concentrate on the $W \to e\nu$ and $Z \to ee$ decay modes, in order to evaluate realistically the expected samples for mass measurements and tests of the standard model (see 4.3).

Table 2 shows the expected W and Z rates in UA2 at ACOL and in CDF at TEVATRON, assuming somewhat optimistically integrated luminosities of 10 pb^{-1} at ACOL and 5 pb^{-1} at TEVATRON by the end of 1989. Taking into account the detector acceptances and electron selection efficiencies (assumed to be 70% for $W \to e\nu$ decays, and 90% for $Z \to ee$ decays), we expect a total of 3000 (5000) reconstructed $W \to e\nu$ decays and 350 (800) reconstructed $Z \to ee$ decays in UA2 (CDF).

Table 2
W and Z rates at ACOL and the TEVATRON

	UA2 (ACOL)	CDF (Tevatron)	Comments
$\sqrt{s}$ (GeV)	630	1800	
$\int L\,dt$ by end of 1989 (pb^{-1})	10	5	Optimistic estimates
$(\sigma B)_{W \to e\nu}$ (nb)	0.6	1.8	Rate better at the Tevatron
e/jet ratio at $p_T = 40$ GeV/c	10^{-3}	3×10^{-4}	Signal/background better at ACOL
(p_T^W) (GeV/c)	9	14	
No. of reconstructed $W \to e\nu$	~ 3000	~ 5000	After acceptance and selection
No. of reconstructed $Z \to ee$	~ 350	~ 800	
No. of $W \to e\nu$ for mass fit	2000-2500	3500-4000	Depends upon fiducial cuts (calorimeter cracks, etc.)
No. of $Z \to ee$ for mass fit	150-250	400-600	
Stat. error on m_W (MeV/c^2)	~ 200	~ 200	
Stat. error on m_Z (MeV/c^2)	~ 200	~ 130	

These numbers are substantially reduced if one retains only events within a fiducial volume, which excludes calorimeter cracks, where the uncertainties on the energy measurements are large.

4.2.2 Hadronic decay modes.

The W and Z bosons are expected to decay into $q\bar{q}$ pairs with sizeable rates, and be observed as two-jet final states. Examples of these decays are $W^+ \rightarrow u\bar{d}$ and $c\bar{s}$, and $Z \rightarrow u\bar{u}$, $d\bar{d}$, $c\bar{c}$, $s\bar{s}$, $b\bar{b}$. As noted in Ref. [24], where the specific process $p\bar{p} \rightarrow W^+ \rightarrow u\bar{d}$ is considered, the background from QCD processes, $p\bar{p} \rightarrow u\bar{d}$, is not overwhelming in the two-jet mass range where the W signal is expected. In fact, the two processes have comparable rates at central rapidities, at least at $\sqrt{s}$ = 630 GeV. Unfortunately, there is at present no known method for distinguishing jets of a specific flavour on an event-by-event basis. Therefore, the QCD backgrounds to a $W,Z \rightarrow q\bar{q}$ signal are summed over all flavours of quarks, and also over gluons, which are the dominant background process in the W,Z mass range.

As discussed elsewhere in this book, UA2 has observed a signal for the $W,Z \rightarrow q\bar{q}$ decays over the dominant QCD background, but this signal is far from providing overwhelming statistical evidence for this very important IVB decay mode. It turns out that jet-jet spectroscopy is one of the crucial experimental features at future hadron colliders, and the present $p\bar{p}$ colliders are an ideal testing ground for the feasability of high precision and high granularity calorimeters, within the high luminosity environment of future hadronic machines.

Therefore it will be very interesting to see whether UA1, UA2, CDF and D0, with their different calorimeters, will be able to clearly establish the presence of the $W,Z \rightarrow q\bar{q}$ decays above the QCD background, with the much larger integrated luminosities expected in the near future.

4.3 IVB mass measurements and Standard Model tests.

Future precision mass measurements at hadron colliders will clearly concentrate more and more on m_W, because experiments at the e^+e^- colliders operating at the Z pole (SLC and LEP) will determine m_Z to an accuracy of better than 50 MeV/c² [35], which will never be approached by any experiment at ACOL or TEVATRON. This arises from the fact that present mass measurement errors are already dominated by the systematic uncertainties on the energy scale, which are directly linked to the absolute calibration of the electromagnetic calorimeters. The goal of the existing calorimeters (UA2, CDF) and of the forthcoming more ambitious ones (UA1,D0) is to reach the level of an uncertainty of ±1% on the absolute energy scale, which would result in errors of the order of ± 1 GeV/c² on the W, Z mass measurements. In the following, we shall assume that the Z mass and width will have been precisely measured at SLC and LEP by 1990.

At that time, the experiments at ACOL and TEVATRON will have hopefully accumulated 2000 to 4000 well-measured reconstructed $W \to e\nu$ decays, as discussed in 4.2.1 and as shown in table 2. Table 2 also shows that the estimated statistical errors on m_W and m_Z are similar, about 200 MeV/c², owing to the fact that the W mass cannot be completely reconstructed.

This leads to a discussion of the systematic uncertainties inherent in the method used to extract the W mass. Known theoretical and experimental uncertainties concern the knowledge of the W width (discussed in 4.1.3), of the W transverse momentum p_T^W (discussed in 4.1.4), as well as the measurement of the neutrino transverse momentum. UA2 estimates, after a careful study, that the systematic error on the W mass, due to the method used to extract its value, will not be smaller than ± 200 MeV/c².

We therefore expect to obtain δm_W = (±0.20 ± 0.20 ± 0.80) GeV/c² (errors due to statistics, method, and calibration, respectively), assuming an uncertainty of ± 1% on the absolute energy scale. This uncertainty cancels in the measurement of $R = m_W/m_Z$, which will be measured with an error of δR = ± 0.003(stat.) ± 0.002 (syst.). This leads to $\delta m_W \approx ± 0.35$ GeV/c², for a known value of m_Z. This accuracy on m_W is unlikely to be improved significantly until LEP operates above the threshold for W pair production.

350

The direct measurement of the weak mixing angle, $\sin^2\theta_W = 1 - (m_W/m_Z)^2 = 1 - R^2$, will have a statistical error of ± 0.006 and a systematic error of ± 0.004. The effect of the radiative corrections to the IVB masses will be determined as illustrated in Fig. 14. The improvement with respect to the present situation will be substantial, allowing such a measurement to provide a stringent test of the Standard Model, in particular with respect to the mass of the top quark, if it has not been measured by then.

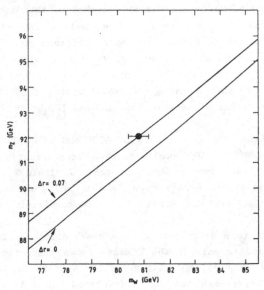

Fig. 14 Expected precision in the knowledge of m_W and m_Z after the completion of the next round of Collider experiments with ACOL, and following a precision measurement of m_Z at e^+e^- machines. The horizontal error bar assumes that the UA1 and UA2 results are statistically combined. The exact position of the measured point in the m_W, m_Z plane is arbitrary.

4.4 IVB pair production.

The physics potential of studies of electroweak gauge-boson pair production was recognized a long time ago [36]. In the Standard Model, there are important cancellations in the amplitudes for WW and WZ production, which depend upon the gauge structure of the WWZ trilinear coupling. The rate for Wγ production is sensitive to the W magnetic moment. Unfortunately, as shown in table 3, the rates for boson pair production, at ACOL and even at the TEVATRON, are too low to allow detailed studies of these potentially very interesting events.

As described in Ref. 24, multijet backgrounds from QCD will swamp any purely hadronic signal for boson pair production. Therefore, in table 3, only events where at least one gauge boson decays into $e\nu$ or ee are considered. Even in this case, it has been pointed out [37] that the background from Standard Model W,Z + 2 jets production, where the two-jet invariant mass is required to be within $\pm$ 10 GeV/c^2 of the W,Z mass, will be almost two orders of magnitude above the boson pair signal, both at ACOL and TEVATRON. Since the requirement of a double leptonic decay ($p\bar{p} \rightarrow WW \rightarrow e\nu e\nu$, $p\bar{p} \rightarrow WZ \rightarrow e\nu ee$ or $p\bar{p} \rightarrow ZZ \rightarrow eeee$) will greatly reduce the already small rates shown in table 3, we conclude that the observation of heavy gauge-boson pair production at ACOL or TEVATRON is excluded.

Table 3

Boson pair production rates at ACOL and TEVATRON

At least one boson into $e\nu$ or ee	$W^{\pm}\gamma$	WW	WZ	ZZ
Events per 10 pb^{-1} at ACOL ($\sqrt{s}$ = 630 GeV)	11	0.5	0.02	0.03
Events per 5 pb^{-1} at the Tevatron ($\sqrt{s}$ = 1.6 TeV)	36	2.1	0.1	0.2

These arguments do not apply to Wγ production, provided one is confident that standard W + jet production, where the jet fragments into single or multiple π^0's, can be sufficiently reduced. These events are of particular interest, since they would allow a measurement of the W anomalous magnetic moment κ, which is predicted to be $\kappa = 1 + O(\alpha_{em})$ in the Standard Model. We recall that κ is derived from

352

$$\mu_w = e(1 + \kappa)/2\, m_w,$$

where μ_w is the W magnetic moment.

Fig. 15 shows the expected angular distribution of the photon with respect to the incoming d-quark, in the (Wγ) centre of mass, for events with a transverse mass of the (evγ) system larger than m_w and with a photon transverse momentum larger than 10 GeV/c.

We first note that, because of the low rate and of the absence of charge measurement in UA2, the measured distributions of this angle, $\cos\theta^*_{d\gamma}$, averaged over W^+ and W^-, will not provide any information on the value of κ. However, as shown by the strong dip expected in the distribution for W^- alone, at $\cos\theta^*_{d\gamma} = 1/3$, perhaps CDF or D0 at the TEVATRON will collect enough data to measure κ to $\pm$ 50% before LEP II becomes operational.

Figure 15

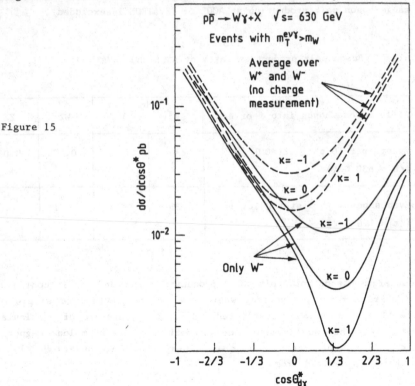

Measurement of the W anomalous magnetic moment, κ, through the angular distribution of photons produced in association with W bosons.

4.5 Heavy flavour physics.

The UA1 results from the analysis of single and dimuon data [38] have shown that, contrary to electrons, muons are identifiable inside or near to hadron jets, thus opening up an exciting field of heavy-flavour physics, leading in particular to the first measurement of B_s-$\overline{B}_s$ mixing. This field should clearly be pursued, in particular with the upgraded UA1 muon detector, which will be by far the most performant muon detector at $p\overline{p}$ colliders in the future.

4.6 Search for the top quark.

The search for the top quark is one of the highest priorities at ACOL and TEVATRON, especially since the most recent limits on the top quark mass, m_{top}, seem to exclude the possibility of observing $Z \rightarrow t\overline{t}$ decays at SLC or LEP [39]. Fig. 16 shows the cross-sections for top-quark production through $p\overline{p} \rightarrow W + X \rightarrow t\overline{b} + X$ and $p\overline{p} \rightarrow t\overline{t} + X$ as a function of m_{top}, for $\sqrt{s}$ = 630 GeV (ACOL) and $\sqrt{s}$ = 1800 GeV (TEVATRON). The curves shown are lowest-order cross-sections and we refer the reader to a recent review of higher-order corrections and theoretical uncertainties in heavy flavour production at high-energy hadronic machines [40] for a more detailed discussion of this subject. For the purpose of the present discussion, we just point out that $t\overline{t}$ production through gluon fusion dominates at TEVATRON, independent of m_{top}, whereas, at ACOL, $t\overline{b}$ production through W-decay dominates for 50 GeV $< m_{top} <$ 70 GeV, and $t\overline{t}$ production through quark fusion dominates for $m_{top} >$ 70 GeV. The rates at TEVATRON are between one and two orders of magnitude higher than at ACOL, depending upon m_{top}.

From Fig. 16, it is clear that hadronic decays of the top quark would be abundant at ACOL and TEVATRON for the integrated luminosities which are hoped to be achieved by the end of 1989. Unfortunately, as shown in Ref. 24, a possible multijet signal from hadronic top-quark decay ($p\overline{p} \rightarrow W + X \rightarrow t\overline{b} + X \rightarrow$ 4 jets + X or $p\overline{p} \rightarrow t\overline{t} + X \rightarrow$ 6 jets + X) will probably be impossible to extract from the very large QCD background. We therefore restrict ourselves from now on to describing the search for semileptonic decays of the top quark; more specifically, we shall discuss a

354

quantitative estimate of the signal to background ratios expected in UA2 at ACOL, in a search for t → beν decays.

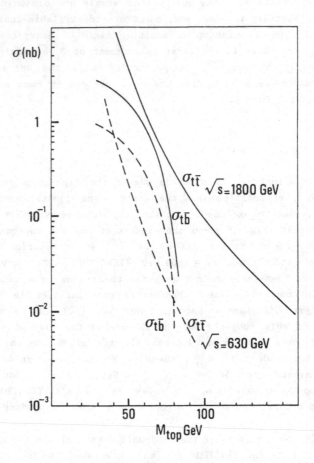

Figure 16

Top-quark production cross-section as a function of the top-quark mass. The lowest-order cross-sections for $\bar{p}p \rightarrow W + X \rightarrow t\bar{b} + X$ and $\bar{p}p \rightarrow t\bar{t} + X$ are shown for ACOL ($\sqrt{s}$ = 630 GeV) and TEVATRON ($\sqrt{s}$ = 1800 TeV).

Fig. 17 shows the expected rates for events with one reconstructed electron (p_T^e > 12 GeV/c) and at least two reconstructed jets (p_T^j > 10 GeV/c) in UA2, for an integrated luminosity of 10 pb^{-1}, as a function of m_{top} [41]. Detector acceptance, jet energy measurements, and electron selection efficiency (ε_e = 23%) have been taken into account. In particular, the electron selection efficiency is low, because of the tight cuts required to reduce the background from fake electrons to an acceptable level (about 10 events), as shown in Fig. 17. Also shown are the backgrounds expected from real electron sources, which are dominated by $p\bar{p} \rightarrow$ W + 2 jets + X, with W $\rightarrow$ eν and mainly by $p\bar{p} \rightarrow b\bar{b}g$, with b $\rightarrow$ ceν. The dominant background from $b\bar{b}$ semileptonic decays has a very steeply falling p_T^e spectrum, but suffers from large systematic uncertainties, since it is greatly reduced by the isolation cuts implicit in any electron selection and the simulation of these cuts requires a detailed knowledge of the fragmentation process and of the calorimeter response to low-energy particles.

Fig. 17 shows that the signal to background is larger than one, for m_{top} < 70 GeV/c^2, which is close to the kinematic limit for W $\rightarrow$ t$\bar{b}$ decay. It is hoped that topological cuts and invariant mass reconstruction will allow us to improve this signal-to-background ratio and extract a clean top quark signal for m_{top} < 70 GeV/c^2. As already stressed previously, good muon identification, as achieved in UA1, is an important asset in the understanding of heavy flavour production, and, of course, would increase the signal event rate by a factor 2.

We also note that, at TEVATRON, the expected rates are 5 times larger than at ACOL, for m_{top} < m_w, but the backgrounds are even larger and it has yet to be demonstrated that leptons of low p_T are observable above background at these higher energies.

If the top-quark mass becomes close to the W-mass, the electrons from top decay will look more and more like electrons from W decay, and the extra jet (or jets) tend to have smaller and smaller p_T. Therefore it will be difficult to distinguish topologically W $\rightarrow$ t$\bar{b}$, with t $\rightarrow$ beν decays from W $\rightarrow$ eν decays, and t$\bar{t}$, with t $\rightarrow$ beν decays from W $\rightarrow$ eν decays, associated with jets.

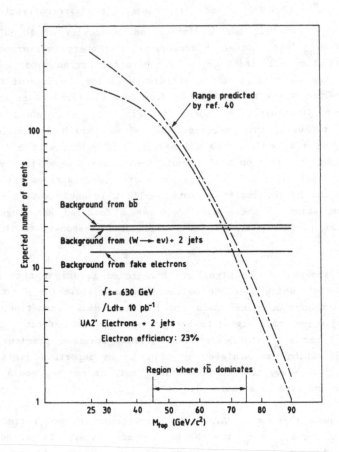

Fig. 17

Expected rates for reconstructed semi-leptonic top decays in UA2 for an integrated luminosity of 10 pb^{-1} as a function of the top-quark mass. Also shown are the expected background levels from fake electrons and from real electron sources (W decay, semi-leptonic b-decay). Acceptances and selection efficiencies are taken into account.

4.7 Search for heavy quarks and leptons.

4.7.1 Search for heavy quarks.

In this subsection, we turn to the search for quarks, with mass, m_Q, above the W mass, whether it be the top quark, which would decay predominantly into a real W boson and a b quark, or a member of a new family, which we assume, in the following, to decay into a real W and a light quark. Such objects will only be observable at ACOL and mainly TEVATRON in the foreseeable future and they can be detected in distinctive final-state topologies, containing one $W \rightarrow e\nu$ decay and four (or more) jets.

Fig. 16 shows that we can only hope to detect such events with observable rates at ACOL, for $m_Q < 100$ GeV/c^2, whereas, at TEVATRON, masses of up to 180 GeV/c^2 should be within reach, since the rates are almost two orders of magnitude larger than at ACOL, for such large masses.

The main background to such a signal, $p\bar{p} \rightarrow Q\bar{Q} \rightarrow WWq\bar{q}$, with at least one $W \rightarrow e\nu$ decay, arises from W production in association with jets. Since the signal to background ratio is of order 1 for the range in p_T^e where the signal is expected, it seems reasonable to assume that such heavy quarks could be easily isolated from the background after topological cuts.

4.7.2 Search for heavy leptons.

In this subsection, we consider the case of a fourth-generation charged lepton L, associated with a new massless neutrino ν_L. Production of L^+L^- or $L\nu_L$ through the Drell-Yan mechanism suffers from a very small cross-section [42], so the most promising channel for observing L production, if $m_L < m_W$, is through the decay $W \rightarrow L\nu_L$, followed by $L \rightarrow q\bar{q}\nu_L$ or $L \rightarrow e(\mu)\nu\nu_L$.

The hadronic decay, $L \rightarrow q\bar{q}\nu_L$, leads to final states with one or two jets and missing transverse momentum, where the jets are not compatible with the main physics background from $W \rightarrow \tau\nu_\tau$ decays. As shown by the UA1 analysis [43], the present limit is $m_L > 41$ GeV/c^2 at the 90% confidence level. It is straightforward to extrapolate from this to ACOL and TEVATRON

358

luminosities, where the experiments should reach the kinematic limit of $m_L \sim 70$ GeV/c², allowed by phase space in W decay.

The leptonic decays, $L \to e(\mu)\nu\nu_L$, are heavily contaminated by background from the dominant $W \to e\nu$ and $W \to \tau\nu_\tau \to e\nu\nu_\tau\bar{\nu}_\tau$ decay modes. The shapes of the electron transverse momentum, p_T^e, and angular distributions can be used to separate the signal from the backgrounds. For 3500 reconstructed $W \to e\nu$ decays with $p_T^e > 15$ GeV/c, we expect 240 electrons with $p_T^e > 10$ GeV/c from $W \to \tau \to e$ decays, 127 from $W \to L \to e$ for $m_L = 40$ GeV/c², and 76 for $m_L = 60$ GeV/c². These numbers are illustrated in Fig. 18, where it can clearly be seen that the electron signal from $W \to L \to e$ will be difficult to separate from the background, especially since uncertainties on the knowledge of Γ_w and p_T^w result in uncertainties on the exact shape of the p_T^e spectrum for the dominant $W \to e\nu$ background.

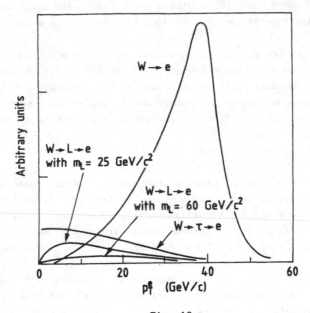

Fig. 18

Electron transverse momentum spectrum from $W \to L\nu_L$, with $L \to e\nu\nu_L$ decays, normalized to the $W \to e\nu$ spectrum, for different values of m_L.

4.8 Search for the Higgs boson.

The search for the neutral Higgs (H) boson is one of the main topics of physics studies at future accelerators [42]. The various H production mechanisms and experimental signatures are the subject of intensive studies in view of detector design for multi-TeV machines.

At ACOL and TEVATRON, the dominant production mechanism for H production is through gluon-gluon fusion via a heavy quark loop, with subsequent decay of the Higgs boson into the heaviest fermion pair, i.e. into $b\bar{b}$, if the Higgs mass is larger than 10 GeV/c^2. Fig. 19 shows the H production cross-section at ACOL as a function of the Higgs mass. The rates are observable up to Higgs masses of about 50 GeV/c^2. At TEVATRON, the range increases to masses of around 100 GeV/c^2. However, the enormous background from hadron jets (a factor of 10^5 above any possible H $\rightarrow$ b$\bar{b}$ signal) precludes the observation of H production through gluon-gluon fusion, even with an experiment optimized for b-quark tagging.

Figure 19

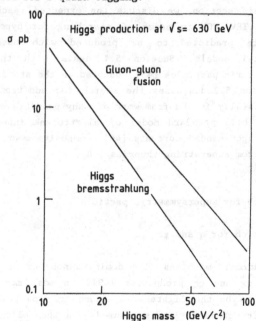

Higgs production at ACOL as a function of the Higgs mass.

A more promising production process is that in association with a W (or Z) boson, the so-called Higgs bremsstrahlung mechanism [44], which is the dominant source of H production at SLC and LEP. At ACOL and TEVATRON, this process is smaller in cross-section than the gluon-gluon fusion mechanism, as shown in Fig. 19, and yields observable rates only up to Higgs masses of 30 (resp. 50) GeV/c^2 at ACOL (resp. TEVATRON). However, in order to get rid of the background from hadron jets, one would be led to require associated production of a Higgs boson and a real Z, with the latter decaying into a lepton pair, thus reducing the observable rate by an order of magnitude.

Therefore, it seems very unlikely that the Higgs boson could be discovered at existing hadron colliders.

5. PHYSICS BEYOND THE STANDARD MODEL

In this section, we discuss the expected reach of the experiments at ACOL and TEVATRON in terms of searching for hypothetical new particles, which are predicted to be produced with observable rates by many theoretical models. Section 5.1 deals with the search for possible supersymmetric particles [45], denoted in the standard way as $\tilde{e}$, $\tilde{\nu}$, $\tilde{\gamma}$, $\tilde{q}$, $\tilde{g}$ ... Section 5.2 discusses the search for additional vector bosons, which arise naturally in the framework of many possible extensions of the minimal $SU(2)_L$ x $U(1)$ Standard Model of electroweak interactions, whether it be through right-handed currents [46], composite models [47], or various models derived from superstring theories [48].

5.1 Search for supersymmetric particles.

5.1.1 Search for $\tilde{q}$ and $\tilde{g}$.

In hadronic machines, the dominant sources of supersymmetric particles are $\tilde{q}\tilde{q}$, $\tilde{q}\tilde{g}$ and $\tilde{g}\tilde{g}$ production [49]. In most models, the photino, $\tilde{\gamma}$, is expected to be the lightest supersymmetric particle, and therefore stable if R-parity conservation is assumed. In the following, we therefore assume that the $\tilde{g}$ decays into $q\bar{q}\tilde{\gamma}$, and the $\tilde{q}$ dominantly into $q\tilde{g}$ (if $m_{\tilde{q}} > m_{\tilde{g}}$), or into $q\tilde{\gamma}$ (if $m_{\tilde{q}} < m_{\tilde{g}}$).

The hadronic production of $\tilde{q}$ and $\tilde{g}$ results, in this case, in final states containing two to six jets and missing transverse momentum. The main backgrounds arise from QCD multijet production, where one or more jets are not properly reconstructed in the detector, owing to fragmentation, calorimeter response, or holes in the apparatus, and from W,Z production with jets, where the W,Z decays leptonically and is not seen in the detector ($Z \to \nu\bar{\nu}$, $W \to ev$ with the electron inside a jet etc...). Whereas the selection cuts in the past analysis had efficiencies of the order of 1% in the relevant $\tilde{q}$ and $\tilde{g}$ mass range, we hope to achieve, with the upgraded detectors, efficiencies of up to 10%, while keeping the backgrounds at a manageable level.

From this we conclude that $\tilde{q}$, $\tilde{g}$ production will be observable with integrated luminosities of 10 pb^{-1}, provided the production cross-section are larger than 10 pb, given the 10% selection efficiency. Fig. 20 shows the summed production cross-section for $\tilde{q}$, $\tilde{g}$ at ACOL and TEVATRON as a function of the $\tilde{q}$ and $\tilde{g}$ masses. The upgraded detectors at ACOL should be sensitive to $\tilde{q}$ and $\tilde{g}$ masses below 120 GeV/c^2, whereas the detectors at TEVATRON should reach $\tilde{q}$ and $\tilde{g}$ masses of about 200 GeV/c^2.

Figure 20

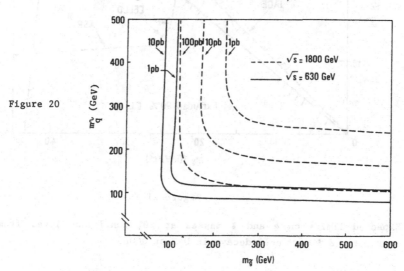

Cross-sections for production of $\tilde{q}$ and $\tilde{g}$ at ACOL ($\sqrt{s}$ = 630 GeV) and TEVATRON ($\sqrt{s}$ = 1800 GeV), summed over $\tilde{q}\tilde{q}$, $\tilde{q}\tilde{g}$ and $\tilde{g}\tilde{g}$ production, as a function of the $\tilde{q}$ and $\tilde{g}$ masses.

362

5.1.2 Search for scalar leptons and gauginos.

Scalar leptons $(\tilde{e}, \tilde{\mu}, \tilde{\nu})$ and gauginos $(\tilde{W})$ can be produced with observable rates in W, Z decays. UA1 and UA2 have published limits on $W \to \tilde{e}\tilde{\nu}$ [50], $Z \to \tilde{e}\tilde{e}$ and $Z \to \tilde{W}\tilde{W}$ [51] decays. The final states are characterized by one or two leptons and missing transverse momentum. The background from $W \to e\nu$, $W \to \tau \to e\nu\nu_\tau$ and continuum Drell-Yan production can be overcome with kinematic cuts. Therefore we expect these searches to be pushed to their kinematic limits in the near future. Fig. 21 shows as an example the limits which will be obtained by UA2 at ACOL from the decay $Z \to \tilde{e}\tilde{e} \to ee\tilde{\gamma}\tilde{\gamma}$ as a function of the scalar electron and photino masses. Also shown in Fig. 21 are the present existing limits from experiments at PETRA and PEP.

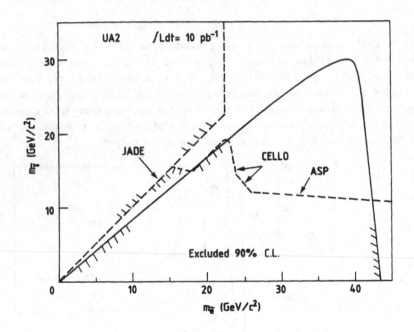

Figure 21

Expected limits on $\tilde{e}$ and $\tilde{\gamma}$ masses at 90% confidence level from a search for $Z \to \tilde{e}\tilde{e} \to ee\tilde{\gamma}\tilde{\gamma}$ decays in UA2 at ACOL.

5.2 Search for additional vector bosons.

Fig. 22 shows the production cross-sections for additional charged, W', or neutral, Z', vector bosons at √s = 630 GeV and √s = 1.8 TeV, assuming standard couplings to leptons and quarks, and for W' → eν or Z' → ee decays. Fig. 22 shows the obvious benefit of higher √s. Whereas the mass limits at ACOL will be 300 GeV/c² for a W' and 250 GeV/c² for a Z', they increase to 650 GeV/c² for a W' and 550 GeV/c² for Z' at TEVATRON. Since there are no known sources of background to the very high transverse momentum electrons and very high mass electron pairs expected from W' and Z' decays, this is clearly a field where high-energy hadron colliders play a unique role, by making a large mass range accessible to experiments.

Figure 22

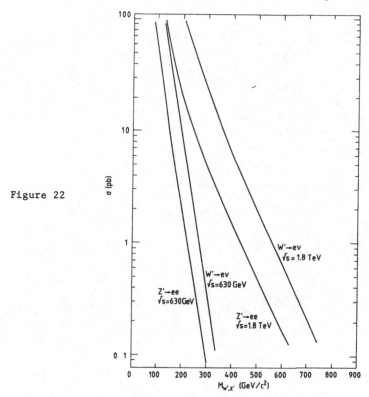

Production of additional charged (W') and neutral (Z') vector bosons at ACOL and TEVATRON as a function of their mass, assuming standard couplings to leptons and quarks.

6. CONCLUSION

The upgraded experiments at ACOL will eagerly take data in the next few years, in order to test more precisely the prediction of the Standard Model, and hopefully to find its missing elements (top quark and Higgs boson), but mainly with the hope of discovering new physics, which will in most cases also be accessible to TEVATRON, SLC and LEP.

REFERENCES

1. Evans, L., et al., "The CERN Proton-Antiproton Collider", this volume.

2. UA6 Collaboration, "Experiments at CERN in 1987", Geneva, p. 97, (1987).

3. UA8 Collaboration, "Experiments at CERN in 1987", Geneva, p. 101, (1987).

4. UA1 Collaboration, Dowell, J.D., "Proc. 6th Topical Workshop on Proton-Antiproton Collider Physics", Aachen, 1986, eds. K. Eggert et al., (World Scientific, Singapore, 1987), p. 419, and references therein.

5. Albrow, M., et al., Nucl. Instrum. Methods A265, 303 (1988).

6. UA2 Collaboration, Booth, C., same as Ref. [4], p. 381, and references therein.

7. Beer, A., et al., Nucl. Instrum. Methods 224, 360 (1984).

8. UA2 Collaboration, "Proposal for the Installation of a Second Silicon Array in the UA2 Detector", CERN/SPSC 87-14, SPSC/P93 Add. 4 (1987).

9. Ellis, R.K. and Scott, W.G., "The Physics of Hadronic Jets", this volume.

10. UA1 Collaboration, Arnison, G., et al., Phys. Lett. 172B, 461 (1986).

11. UA2 Collaboration, Appel, J.A., et al., Phys. Lett. 160B, 349 (1985).

12. Jenni, P., "Proc. Workshop on Physics in the 90's at the SPS Collider, Zinal", 1985, ed. R. Budde (CERN, Geneva, 1985), p. 64.

13. Structure functions from Eichten, E., et al., Rev. Mod. Phys. 56, 579 (1984) have been used.

14. Eichten, E., et al., Phys. Rev. Lett. 50, 811 (1983).

15. Combridge, B.L., Kripfganz, J. and Ranft, J., Phys. Lett. 70B, 234 (1977);
    see also : Cutler, R. and Sivers, D., Phys. Rev. D17, 196 (1978).

16. UA1 Collaboration, Arnison, G., et al., Phys. Lett. 136B, 294 (1984).

17. UA2 Collaboration, Bagnaia, P., et al., Phys. Lett. 144B, 283 (1984).

18. Cohen-Tannoudji, G., et al., Phys. Rev. D28, 1628 (1983);
    Combridge, B.L. and Maxwell, C.J., RL-83-095 (1983) ;
    Halzen, F. and Hoyer, P., Phys. Lett. 130B, 326 (1983).

19. UA1 Collaboration, Arnison, G., et al., Phys. Lett. 177B, 244 (1986).

20. UA1 Collaboration, Arnison, G., et al., Phys. Lett. 158B, 494 (1985).

21. UA2 Collaboration, Appel, J.A., et al., Z. Phys. C30, 341 (1986).

22. UA1 Collaboration, Arnison, G., et al., Phys. Lett. 158B, 494 (1985).

23. UA2 Collaboration, Meier, K., to appear in "Proc. Advanced Research Workshop on QCD Hard Hadron Processes", St. Croix, Virgin Islands, Oct. 8-13, 1987.

24. Kunszt, Z. and Stirling, W.J., "Multijet cross-sections in hadronic collisions", Proc. Workshop on Physics at Future Accelerators, La Thuile 1987, CERN 87-07, Vol. II, p. 548 (1987).

25. Paver, N. and Treleani, D., Nuovo Cimento 70A, 215 (1982);
    Humpert, B., Phys. Lett. 131B, 461 (1983).

26. UA2 Collaboration, Einsweiler, K., "Proc. XXI Rencontre de Moriond", Les Arcs, 16-22 March 1986, Editions Frontières, p. 9 (1986).

27. Aurenche, P., et al., Phys. Lett. 140B, 87 (1984);
    Aurenche, P., et al., Nucl. Phys. B297, 661 (1988).

28. UA2 Collaboration, Appel, J.A., et al., Phys. Lett. 176B, 239 (1986).

29. UA1 Collaboration, Albajar, C., et al., CERN-EP/88-45 (1988), submitted to Phys. Lett. B.

30. Altarelli, G., et al., Nucl. Phys. B246, 12 (1984);
    Altarelli, G., et al., Z. Phys. C27, 617 (1985).

31. Cabibbo, N., Proc. Third Topical Workshop on Proton-Antiproton Collider Physics, Rome, 1983, CERN 83-04, Geneva. p. 567 (983);
    Halzen, F., and Mursula, K., Phys. Rev. Lett. 51, 857 (1983) ;
    Hikasa, K., Phys. Rev. D29, 1939 (1984).

32. DiLella, L., Proton-Antiproton Collider Physics : Experimental Aspects, CERN-EP/88-02.

33. Ruhlmann, V., Thèse de Doctorat, Université de Paris VI (1988);
    Jakobs, K., Ph. D. Thesis, University of Heidelberg (1988).

34. Stirling, W.J., and Bawa, A.C., DPT/87/42 (1987).

35. Altarelli, G., et al., in Physics at LEP, eds. J. Ellis and R. Peccei, CERN 86-02, Geneva, p. 3 (1986).

36. Brown, R.W., and Mikaelian, K.O., Phys. Rev. D19, 922 (1979);
    Eichten, E., et al., Rev. Mod. Phys. 56,579 (1984).

37. Stirling, W.J., et al., Phys. Lett. 163B, 261 (1985).

38. UA1 Collaboration, Albajar, C., et al., Phys. Lett. 186B, 237 (1987); and Phys. Lett. 186B, 247 (1987).

39. UA1 Collaboration, Albajar, C., et al., Z. Phys. C37, 505 (1988);
    Ellis, J., et al., Phys. Lett. 192B, 201 (1987).

40. Altarelli, G., et al., CERN preprint, CERN-TH/4978/88.

41. Moniez, M., Thèse d'Etat, LAL preprint, LAL 88-17 (1988).

42. Eichten, E., et al., Rev. Mod. Phys. 56, 579 (1984).

43. UA1 Collaboration, Albajar, C., et al., Phys. Lett. 185B, 241 (1987).

44. Glashow, S.L., et al., Phys. Rev. C18, 1724 (1978).

45. For a review see : Fayet, P. and Ferrara, S., Phys. Rep. 32, 249
    (1977).

46. Langacker, P., et al., Phys. Rev. D30, 1470 (1984).

47. Baur, U., et al., preprint MPI-PAE/PTH 29/85 (1985).

48. Cohen, E., et al., Phys. Lett. B165, 419 (1987);
    del Aguila, F., et al., Nucl. Phys. B287, 419 (1987) ;
    London, D. and Rosner, J., Phys. Rev. D34, 1530 (1986).

49. Dawson, S., et al., Phys. Rev. D31, 1581 (1985).

50. UA1 Collaboration, Arnison, G., et al., Europhys. Lett. 1, 327
    (1986).

51. UA2 Collaboration, Ansari, R., et al., Phys. Lett. B195, 613 (1987).

Physics for Future Supercolliders*

GORDON L. KANE

Randall Laboratory of Physics
University of Michigan, Ann Arbor, Michigan 48109

1. Introduction

There has been remarkable progress in understanding the fundamental laws of nature in the past two decades, so that now a real theory of all observed particle interactions exists, called the Standard Model. Finally, after hundreds of years of the development of physics starting from careful descriptions of velocity and acceleration, it is possible to ask broad general questions about such topics as the origins of mass, or why some particles or symmetries exist and not others. Today no experiment in particle physics disagrees with expectations, and there are no puzzles of the traditional sort. Rather, the next stage of experimentation will, it is hoped, help provide answers to very general, fundamental, questions.

Throughout the history of particle physics, people have basically built experimental devices when the technology allowed it. Today, the U.S. physics community has proposed that a high energy ($\sqrt{s} = 40$ TeV), high luminosity ($L = 10^{33} cm^{-2} sec^{-1}$) pp collider (the SSC) be constructed. The European and Japanese communities, and perhaps others are considering proposals similar to the SSC, or perhaps equivalent $e^+ e^-$ colliders. No previous accelerator has had its justification developed and studied nearly to the extent that has been the case for the SSC. The first major study ocurred at Snowmass, Colorado, for three weeks in 1982 (Snowmass 82).[1] Then three further efforts occurred: (1) the PSSC studies[2] over a period of months, (2) a workshop[3], and (3) the calculation[4] of a set of updated cross sections using new structure functions. Next came the Snowmass 84 study[5], the second large, international meeting which included extensive analysis of the physics possibilities. Both for their own interest, and

* Research supported in part by the U. S. Department of Energy.

in preparation for Snowmass 86, a series of workshops were arranged on several physics topics, beginning in summer of 1985 (at Oregon[6]), FNAL[7]), UCLA[8]), and Madison[9])). All of these studies provide both a panorama of the kinds of new physics that might be found at the SSC, and some documentation of the feasibility of doing the physics. Most recently, the Snowmass 86[10]) study concentrated on the utilization of the SSC, and the Berkeley Workshop[11]) on the requirements for SSC detectors.

There are several reasons why the physics of the SSC has been so carefully examined.

(1) It was necessary to confirm in detail that the energy and luminosity chosen for the SSC were consistent with accomplishing the physics goals that led to the choice of the SSC as the basic project for U.S. particle physics in the next decade.

(2) It was necessary to confirm that the requirements on detectors in order to do the kinds of physics envisioned were achievable with technology that would allow construction of detectors on the same time scale as the accelerator, at a reasonable cost.

A large number of people felt that because of the large cost and relative importance of the SSC in the U.S. particle physics program, these matters were too significant to leave to be decided by anyone's intuition. Similar intensive studies have occurred[12),13)] in Europe in 1986-87, to help distinguish among the physics possibilities from e^+e^-, ep, and pp colliders, and to help set $\sqrt{s}$ and the luminosity.

Does it make sense to believe we can talk about the physics goals of future colliders? How do we know what will happen? Nobody knows, of course, what new discoveries will be made, if any, at future colliders. But careful analysis shows that we can indeed talk about it, and deal with points (1) and (2) above. Let me review the situation.

We can split the possible kinds of physics into four categories, (i) Higgs physics, (ii) hypothetical gauge theory extensions, (iii) possible strongly interacting structure of quarks and leptons, and (iv) further tests of the Standard Model. Here I will briefly describe each of these. In separate sections below I will give some details about the status of the analysis of each category. The basic approach is to understand what would occur at a particular future facility in each case, to see what range of physics parameters can be covered, and, more importantly, to be sure the new physics would not be missed if it were there.

(i) Higgs physics is necessarily a part of the Standard Model. Without some form of Higgs physics the Standard Model is inconsistent and necessarily incomplete. Some scalar interaction must exist or the electroweak interaction of vector bosons would not be described[14] by a perturbative calculation. Further, either the scalar interaction will turn up as a particle, or it will appear as a new and detectable interaction in the W pair system. Its effects cannot be pushed to arbitrarily high mass. Exactly what range in M_{ww} has to be studied, at what luminosity, in order to be sure an effect is detected is not completely settled for any particular collider; it is approximately $M_{ww} \lesssim 1.5$ TeV. At the end of this chapter some comparative analysis is given.

The Higgs field could appear as a fundamental field, to go into the basic Lagrangian of the theory. If so, it could occur as in the minimal Standard Model with a single SU(2) doublet of complex scalar fields, in which case a single, electrically neutral, particle could exist. Or the Higgs field could occur as required in supersymmetric theories, with two doublets, leading to three neutral bosons and a charged pair $H^{\pm}$. In the minimal Standard Model the Higgs width grows as its mass cubed, $2\Gamma_H(TeV) = [M_h(TeV)]^3$, so Γ_H gets very large. In conventional supersymmetric theories the widths of the Higgs bosons stay small compared to their masses.

Or the underlying scalar field responsible for mass might occur not as a fundamental field but as a composite of some other new fundamental objects, as happens with superconductivity. The particle physics version of this approach that has been studied most extensively is technicolor.[15] It leads to a variety of experimental signals that are detectable[1),4),5)] at the SSC and colliders that are starting to do physics in 1987-88. More recently the composite Higgs approach has been pursued[16] by H. Georgi and collaborators.

While no one knows what form the Higgs physics will take, the essential point is that all of the above alternatives can be analyzed. Each alternative can be examined in turn, for all allowed masses of the Higgs boson(s). Although it is a subtle problem to get the theory right, and there are many Standard Model backgrounds that mimic the signal, it appears likely that for any M_h above about 40 GeV, Higgs physics can be eventually studied at the SSC. Everything can be calculated because the beams available are known and their couplings to Higgs bosons are a part of the Standard Model. To finally determine how far above one TeV the SSC can go will require careful simulation of the events, with backgrounds added. This analysis is underway.

Returning to our initial concern, we see that regardless of our ignorance of the final outcome for the physics of the Higgs sector, we can decide what is needed to obtain the relevant experimental information. For the fundamental Higgs we can consider all approaches and all masses and calculate the rest. It is a lot of work, but it is worthwhile

372

to help attack the central problem of particle physics. If there is not a fundamental Higgs boson, it is probably easy to see experimental signals in the TeV region. The various approaches are summarized in Figure 1.

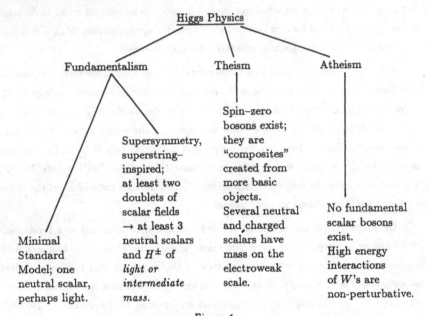

Figure 1
Approaches to Higgs physics. The essential point for our purposes
is that <u>all alternatives are distinguishable experimentally</u> .

(ii) The second kind of new physics that could occur is extensions of the Standard Model. Here the situation can also be analyzed but for quite different reasons than in the Higgs case. The possible extensions of the Standard Model that could occur, such as new flavors of quarks and/or leptons, supersymmetry, further unifications, and so on, are all hypothetical. Several of them might occur, or none of them. If they occur, they could occur on any mass scale. No scale is preferred by any compelling argument; 100 GeV is as likely as 1 TeV.

Because they are extensions in the gauge theory framework, the couplings of all hypothetical new particles to the known gauge bosons and fermions are determined. Once we choose a particular hypothetical model we can calculate all production cross sections and all decay rates. We can explore all signatures. We can work out the range of masses that could be studied at any accelerator. All of this analysis is under good control. Many of the cross sections were studied earlier; in 1986-87 some have been recalculated with all of the factors of two (normally suppressions) due to mixing effects, etc. Past analyses often studied a particularly clean decay branching ratio as a signature, assuming it dominated, in order to establish that some modes could be handled. Now realistic branching ratios have been included in some cases. Analyses have been done for fermions and gauge bosons from larger gauge groups such as E_6, left-right symmetric theories, and horizontal symmetries.

The most difficult problems to deal with are the Standard Model backgrounds that mimic the signal. For example, the cross sections to make heavy leptons up to quite large masses are significant.[17] A lepton L^+ heavier than a W could decay via $L^+ \to W^+ L^\circ$ where L° is a neutral lepton that escapes. Then a pair $L^+ L^-$ gives $W^+ W^-$ plus missing energy. The total missing energy is not useful at a hadron collider, so this kind of event has to be distinguished from the normal $W^+ W^-$ production by using information from missing E_T, plus acolinearity and acoplanarity distributions. It is necessary to distinguish the signal from a variety of sources of real W pairs plus the fake pairs when there is one real W plus a pair of jets that look like a W. This work is also in progress. So far, no one has published a complete analysis showing what range of masses of $L^\pm$ can be studied at future colliders, and present claims[18] are not in agreement.

It is not possible to calculate or even guess all possible extensions of the Standard Model, of course. The strategy here is to make several judicious choices and study them carefully. Then examine several more and see if any new features enter, particularly the kind of feature that may affect detector design. It appears that concentrating on supersymmetry and on new quarks and leptons covers most of the ground. A few new features do appear, such as fermions that are electroweak singlets and thus have quite different signatures. At this level of analysis it appears reasonable to hope that a responsible job has been done in finding out what the SSC and other colliders can do and in specifying the detector requirements needed to do it.

It is probably worth emphasizing that there is little uncertainty in this approach. Supersymmetry is as predictive as the Standard Model. The masses are not known, but otherwise the relevant Lagrangian is known. Just as there is no knowledge of the t quark mass or the Higgs mass in the Standard Model, there is no knowledge of the superpartner masses or of the masses of the Higgs states in the supersymmetric theory. But if the superpartners are to be found on the weak scale there are not significant corrections to the couplings, and so all the interactions of the superpartners are known.

Similar remarks hold for almost all gauge theory extensions. Once we postulate a form for the theory, everything needed for calculations is determined except the masses.

(iii) The third category is usually referred to as "Maybe there is some new strong interaction that starts up, so suddenly cross sections get large. Maybe it involves constituents of quarks and leptons or something." But life cannot be so simple.

Our category (i), Higgs physics, *must* contain something. Category (ii) has reasonable extensions of the gauge theory approach that has worked so well for the Standard Model. An approach looking for constituent effects, or new strong interaction, cannot be viewed as an extension of the historical trend from molecules to atoms to nuclei to nucleons to quarks and leptons, because always before the new scale appeared when the kinetic energy of a bound object was at most of the order of that object's mass. But now electrons and quarks have been probed at momentum scales over 10^4 times their masses, and no sign of structure has appeared. Further, if new physics of the category (iii) type should appear, by the time of the SSC we will have probed such possibilities to mass scales approaching a TeV, so the constituent cross section associated with the appearance of a new interaction will be at most of order $\sigma \sim 1/M_{new}^2 \sim 10^{-35} cm^2$. The cross section for colliding protons is typically two orders of magnitude smaller. If a new effect exists, separating its consequences from ordinary Standard Model rates will not be hard, as almost all distributions for jets and leptons will be different, so long as the event rate is large enough.

(iv) Whatever new physics will help make sense of the Standard Model, and its parameters, may show up first as small deviations from Standard Model predictions for scattering or production or decays of quarks, gluons, leptons, or gauge bosons. In all cases, new higher luminosity or higher energy colliders will extend the range of tests of the Standard Model, both to larger values of variables, and in precision. Since the new tests are in a perturbative regime of the theory, it is possible to do meaningful calculations to compare experiment and theory. One example of a major Standard Model prediction is discussed in the next section.

The implications of our analyses can be discussed from two points of view. First, they can be expressed in terms of what range of masses can be explored at any future collider. Second, they lead to a set of requirements on detectors in order to achieve the physics goals. These are obviously related aspects of the problem, and some negotiation is involved. By requiring more of the detectors it is usually possible to study a larger range of masses, and conversely.

This introduction, it is hoped, has convinced the reader that (a) it is necessary to plan ahead and discuss the physics for future colliders, and (b) we know how to discuss the physics for the SSC and other colliders in the ways that are needed.

2. The Non-Albelian W^+W^-Z Vertex

An important test of the Standard Model includes the non-Albelian ZW^+W^- vertex. Traditionally, the test is expected to take place in the process $e^+e^- \rightarrow W^+W^-$, with contributing diagrams as shown in figure 2, at LEP II, starting in 1994. The vertex of interest is circled. Note that it occurs with a neutral current vertex, so it is suppressed relative to the ν exchange that gives the dominant numerical contribution. Also, the cross section peaks at about 200 GeV, the top energy of LEP II, so the decrease of the cross section cannot be studied there.

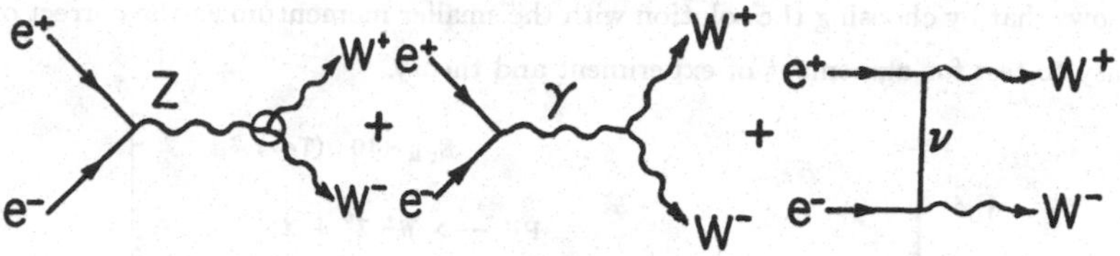

Figure 2
Diagrams for $e^+e^- \rightarrow W^+W^-$.

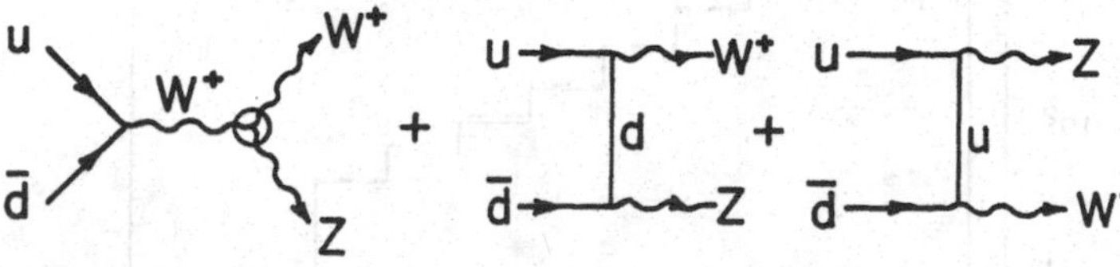

Figure 3
Diagrams for $u\bar{d} \rightarrow W^+Z$.

At a hadron collider the process $q\bar{q} \rightarrow W^+W^-$ occurs and tests the same physics. Unfortunately, the QCD process $q\bar{q} \rightarrow W^{\pm}jj$, with the jj effective mass restricted to the region M_W, has a larger cross section by an order of magnitude;[42] probably there is no way to pick out $q\bar{q} \rightarrow W^+W^-$.

Fortunately, there is an alternative.[19],[57] The same physics is tested by $u\bar{d} \rightarrow W^+Z$. Here we can allow $W^+ \rightarrow \ell^+\nu, Z \rightarrow \ell^+\ell^-$ since there is only one ν, and fully determine the event. The diagrams are now shown in figure 3 and note that the neutral current vertex occurs with the less interesting quark exchange; this process is more sensitive to what we want to test. Since the final state is $\nu\ell^+\ell^-\ell^+$, with no hadrons, it is essentially background free. The rates are not bad.[19] A cross section calculated from the above diagrams has to be multiplied by about

$$2 \times \frac{1}{6} \times 0.06 \times \frac{1}{4} = 5 \times 10^{-3}$$

where the factors are respectively from adding $W^+ + W^-$ cross sections, the branching ratio for $W \to l\nu$, the branching ratio for $Z \to \ell^+\ell^-$, and an allowance for acceptance. At $\int \mathcal{L} \, dt = 10^{40} cm^{-2}$ a 40 GeV bin near the peak has $\sim$500 events and a 200 GeV bin for 800-1000 GeV has $\sim$25 events. Then the full shape of the curve can be measured, giving a large lever arm to test the theory to about a TeV. The cross section for $pp \to W^\pm Z^\circ +$ anything is shown in figure 4. Because $W^\pm \to \ell^\pm\nu$, there is an ambiguity in reconstruction of $\sigma(q\bar{q} \to WZ)$ from the unknown ν longitudinal momentum. Figure 4 also shows that by choosing the solution with the smaller momentum as the correct one, it is easy to test for agreement of experiment and theory.

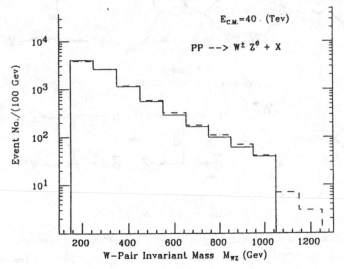

Figure 4
This shows that the cross section for W^+Z can be reconstructed even though there is a missing neutrino. The solid line is the actual cross section, and the dotted line is what is obtained using the solution where the ν has the smallest rapidity. From reference 19.

Studies[19] of the backgrounds for this are in progress. Since the transverse neutrino direction is well defined, the decay planes of W, Z are well determined and a number of polarization tests can also be carried out for $M_{WZ} \sim 500$ GeV. These tests will probably not be completed at the SSC until 1998 or later. That will be over 25 years after the importance of studying them was understood.

3. Higgs Physics

Since we do not know what form the Higgs physics will take, we have to consider a range of rather complete possibilities. As described in the introduction, all alternatives can be distinguished experimentally. Complete information on the spectrum of scalar bosons should be obtained. In different regions of Higgs mass, different techniques have to be used. In this section we first examine what can be done at existing machines or those under construction. Then we consider e^+e^- supercolliders, and finally hadron supercolliders.

A. e^+e^- Colliders

a. $M_h \lesssim 6$ GeV The branching ratio for $\Upsilon \rightarrow \gamma h^\circ$ is about $2.2 \times 10^{-4}(1 - M_h^2/M_\Upsilon^2)$ at tree level[20], but the radiative corrections are large[21], and reduce this by about a factor of 2. Thus at the level of a little over $10^5 \Upsilon$ decays, which could be achieved at CESR or DORIS in the next couple years, a search could be done until $(1 - M_h^2/M_\Upsilon^2)$ gets too small. This region will be covered by the search at Z° factories as well. The CUSB group has just published[22] the first limits on Higgs; they would have seen a signal unless $M_h \gtrsim 3.9$ GeV.

b. $M_h \lesssim 35$ GeV (light higgs) As far as is known, the best way to search for h° at SLC and LEP is the process of figure 5, where the $\ell^+\ell^-$ pair provides the signature. There is a peak in $M_{\ell^+\ell^-}$ when the pair is produced opposite a single particle. With $10^5 Z^\circ$ an h° up to about 15 GeV could be detected, and with $10^6 Z^\circ$'s the search can extend up to about 30 GeV. SLC hopes to have $10^5 Z$'s by the end of 1988 and 10^6 by the end of 1989; LEP hopes to achieve 10^6 by the end of 1990.

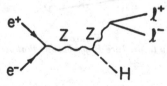

Figure 5
Standard process for $e^+e^- \rightarrow ZH \rightarrow \ell^+\ell^-H$.

c. Toponium If the $\bar{t}t$ $^3S_1$, quarkonium state (called θ) happens to occur at a mass where it can be studied at SLC or LEP, its branching ratio to γh° is expected to be about $2.5 \times (m_\theta/m_\Upsilon)^2$ times greater than the Υ branching ratio. This factor could be 250, giving a branch ratio of about 2.5%. Then it would be easy to search for M_h up to nearly M_θ.

Remarkably, these values for M_h [$M_h \lesssim 35$ GeV unless toponium is accessible] are about as far as we can go at machines under construction. To do better, some new collider is needed.

d. 35 GeV $\lesssim M_h \lesssim$ 85 GeV (intermediate mass Higgs) If the second stage of LEP is built, with $\sqrt{s} \simeq 200$ GeV, it will be possible to search for h° with mass up to about 85 GeV. To do so will require the full design luminosity.

Kinematically a larger value of M_h is accessible, but the backgrounds are too large. The main source of h° will again be

$$e^+e^- \rightarrow Z \rightarrow Zh^\circ, Z \rightarrow \ell^+\ell^-$$

and the serious background is due to $e^+e^- \rightarrow ZZ$ with either Z decaying to $\ell^+\ell^-$. The current schedule is to finish LEP II in 1994, though that would presumably be delayed if a hadron collider is constructed in the LEP tunnel as has recently been proposed.

If FNAL is upgraded to rather high luminosity, $\mathcal{L} \sim 10^{32} cm^{-2} sec^{-1}$, it can produce h° in interesting quantities. Whether any can be detected is a subtle question, depending on the capabilities of the detectors.

e. e^+e^- super colliders At higher energies not only $e^+e^- \rightarrow Zh^\circ$ but also the process[23] of figure 6,

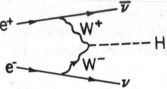

Figure 6
Additional way to produce Higgs bosons that becomes important at high energies.

produces Higgs bosons. This process, $e^+e^- \rightarrow \nu\bar{\nu}h^\circ$, becomes dominant if $\sqrt{s} \gtrsim 300$ GeV. At sufficiently high luminosities, the search for h° at a given e^+e^- collider can be carried out up to $M_h/\sqrt{s} \lesssim \frac{2}{3}$. Figure 7 shows a scatter plot of the *minimum* luminosity (in units of $10^{33} cm^{-2} sec^{-1}$)[24] needed to search at a given M_h and $\sqrt{s}$; once a signal is found a higher luminosity is needed to study it.

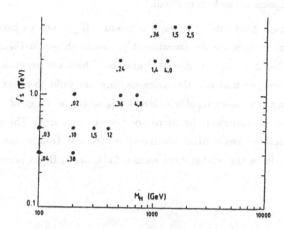

Figure 7
This shows the __minimum__ luminosity required to detect a given mass Higgs boson at a given $\sqrt{s}$.

Several studies of the possibility of finding scalar bosons, or obtaining data on WW scattering, at e^+e^- colliders have been done[25] in the past year or so are underway. It is not so straightforward as might be expected to work at high energy e^+e^- colliders. Backgrounds from WW production and $\gamma\gamma$ collisions become large, and the interesting cross sections mostly decrease. The bottom line that appears to be emerging from the studies (which are mostly not completed at the time of writing) is that with sufficient luminosity, the Higgs physics can be done. As figure 7 shows, the needed $\mathcal{L}$ is rather large, typically over $10^{32} cm^{-2} sec^{-1}$ for discovery of a signal, and presumably an order of magnitude larger to actually study what is present once something is detected.

Unfortunately, it is not known at present whether it is possible to build such a high energy, high luminosity, e^+e^- collider, or how much it would cost to do so, or when such a collider could be completed. In the following, when we need to describe this machine we will call it NLC for "next linear collider".

B. SSC

At the SSC it seems to be possible, at full energy and luminosity, to search directly for Higgs bosons with masses from about $M_Z/2$ to perhaps 1200 GeV, and indirectly beyond that. In different regions, the techniques are very different. Since the SSC is the only supercollider that is presently approved (though not yet funded), and since the Higgs physics must be the main goal of any future collider, we will go through the analysis of Higgs physics at the SSC in detail.

a. **Production Cross Sections** The ways to produce Higgs bosons have been well understood. The main ones are the constituent processes shown in Figure 8. Each process, of course, has a variety of associated graphs. The main uncertainties are in the structure functions, so that resulting cross sections are valid to about a factor of two; the cross sections are more reliable at larger M_h since the range of $x = M_h/\sqrt{s}$ is in the region where the structure functions are directly measured. The current best estimates for production cross sections are shown in figure 9, from reference 26. For later use, figure 10 shows the related cross section for charged Higgs production, also from reference 26.

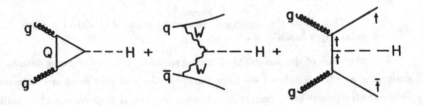

Figure 8
Ways to produce Higgs bosons at a hadron collider.

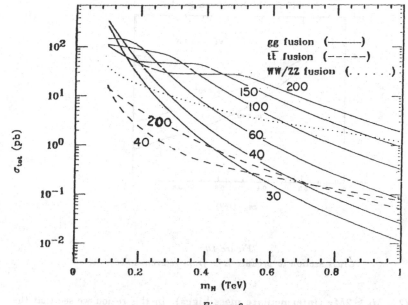

Figure 9
Cross section to produce H^0, from reference (26). Lines are labeled with M_t values.

When M_h gets large, and $H \to WW$ is the dominant decay, a large set of diagrams must be included[27),28)] to have a correct, gauge invariant set for $WW \to WW$. At first these processes were computed in the "real-W" approximation [29)] convoluted with the known W structure functions for W's emitted from quarks or leptons. Now the calculational technology has reached the stage[30)] where the perturbative contributions to $q\bar{q} \to q\bar{q}WW$ can be fully evaluated. The real-W approximation is still essential[31)] to consider any non-perturbative contributions that might arise if WW interactions become strong, in which case no perturbative calculation is possible. With the real-W approximation predictions for strong WW scattering could be translated into experimental predictions, or deviations from the perturbative predictions could be extracted from experimental data. The real-W approximation has been studied in some detail and is now rather well understood[32)], although further analysis is needed.

Using these production cross sections, let us go through the various regions of mass for possible studies of scalar bosons at the SSC.

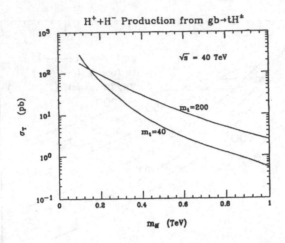

Figure 10
Cross section to produce $H^{\pm}$, from reference (26).

b. $M_Z/2 \lesssim M_h \lesssim 2M_Z$ (intermediate mass higgs) In this region we see that the production cross sections are large. The main decay of h° is to the heaviest fermions, of course; if $M_h > 2M_t$ then $h \to t\bar{t}$ while if $M_h < 2M_t$, then $h \to b\bar{b}$. Unfortunately, the normal Standard Model production of $t\bar{t}$ or $b\bar{b}$ is overwhelmingly large, so an h° signal probably could not be picked out from the backgrounds.

Fortunately, however, the production cross section is sufficiently large that rare decay modes can be used.[33),34)] The branching ratios for the standard Model for one value of M_t are shown in Figure 11. Probably the best mode is $h \to \gamma\gamma$. It occurs[35)] at one loop, and the actual rate is quite interesting since it depends on important couplings. The branching ratio to $\gamma\gamma$ will be just under 10^{-4} if $M_h > 2M_t$, and over 10^{-3} if $M_h < 2M_t$. In the latter case, there is a signal of over 10^3 events, so detecting h° will be easy.

In general, in this region the results depend on M_t through the production cross section, some of the rare decay rates, and the branching ratios. Results have to be discussed as a function of M_t until it is measured.

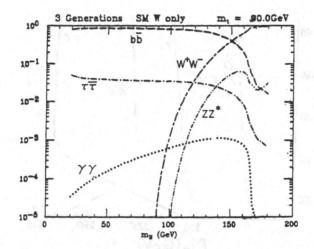

Figure 11
Branching ratios for a standard model h° into several channels,
for $M_t = 90$ GeV.

There are also backgrounds for the $\gamma\gamma$ mode. Some are irreducible Standard Model backgrounds, from $q\bar{q} \to \gamma\gamma$. The cross section for this is much larger than the rate for $qq \to h + X \to \gamma\gamma + X$, of course. If the $\gamma\gamma$ mass resolution is no bigger than a few percent, the background can probably be dealt with[34] when $M_h < 2M_t$, but the signal to noise ratio is considerably less than one if $M_h > 2M_t$. In the latter case, to find a signal will require a luminosity somewhat above $\mathcal{L} = 10^{33} cm^{-2} sec^{-1}$. In addition, there are backgrounds that are detector dependent, from a jet occasionally acting like a single γ. To have no problem with such backgrounds, it should be sufficient to discriminate γ/jet to a few parts in 10^5; several arguments suggest that this is possible, but some further Monte Carlo analysis is needed.

The mode $H \to ZZ^*$ followed by both Z's decaying to lepton pairs can be used[34] for $M_h < 2M_Z$. Background can be effectively eliminated by cutting out low mass charged lepton pairs. Figure 12 shows the regions of the M_t–M_h plane that can be covered by the $\gamma\gamma$ and ZZ^* modes. The $\tau\bar{\tau}$ mode of H may be useable in some regions of the M_t–M_h plane. The main problem is to get good mass resolution for $\tau\bar{\tau}$, which can be accomplished either by producing H opposite[33],[36] a gluon or Z, or by using[37] 3–prong τ decays with less momentum carried off by neutrinos. Studies are in progress; at the present time the recoil method does not appear to provide sufficient resolution.[38]

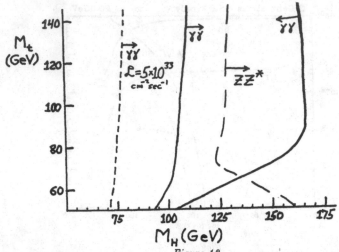

Figure 12
This shows the regions in the M_t-M_h plane where a Standard Model H^0 can be detected at the SSC (from reference 34).

Other rare modes of value are $h \rightarrow Z\gamma$ and $h \rightarrow \theta\gamma$ where θ is the $\bar{t}t$ bound quarkarium $^3S_1$ state. These may not be better to find a signal, but if one is found, they can probably be detected and the branching ratios contain a great deal of useful information about Higgs coupling. For $\theta\gamma$ the rate is[34]

$$\Gamma(h \rightarrow \theta\gamma) \,/\, \Gamma(h \rightarrow \gamma\gamma) \simeq 200/M_h(GeV)$$

so the detectability depends on what fraction of θ decays can be used, and on the backgrounds.

For $M_h < 2M_t$ the $b\bar{b}$ mode should also be detectable if h is produced in conjunction with a W. Two studies have been done,[58] using a semileptonic decay of b or $\bar{b}$, with a P_T cut of the lepton relative to the jet axis, plus a jet cut $\Delta R \lesssim 0.7$, plus a back-to-back cut on W and h. It was concluded that a reasonable signal and signal to noise could be obtained for $M_Z/2 \lesssim M_h \lesssim 125 GeV$.

The conclusion, partly shown in figure 12, is that for a neutral h° of intermediate mass, all [or all but a little] of the M_t, M_h plane can be covered at the standard SSC. With a few times the standard luminosity, the whole region can be certainly be covered. If $M_t \gtrsim 70$ GeV, the whole region can be covered by a combination of modes for $\mathcal{L} = 10^{33} cm^{-2} sec^{-1}$.

c. $2M_Z < M_h \lesssim 600$ GeV (heavy Higgs) In this region, it is easy[39] to deal with a Standard Model $H°$. The rates from

$$H \to ZZ$$
$$\quad\ \ \downarrow\ \rightarrow \ell^+\ell^-$$
$$\quad\ \ \ \downarrow \ell^+\ell^-$$

are large enough to be detected, above background. The main detector requirements will be geometrical. For a supersymmetric theory or some other non-Standard Model, different approaches have to be used; we will mention some briefly below.

d. 600 GeV $\lesssim M_h \lesssim 1$ TeV (heavy Higgs) Here it is necessary to use some more complicated modes. One possibility is

$$H \to ZZ$$
$$\quad\ \ \downarrow\ \rightarrow \ell^+\ell^-$$
$$\quad\ \ \ \downarrow \nu\bar{\nu}$$

which has three times the rate of the decay where both Z give charged leptons. It has been studied analytically in ref. 40. Monte Carlo calculations will be reported in reference 11, with background studies. It may also be possible to use one $Z \to \tau^+\tau^-$.

In ref. 41, it has been argued that an effect can be detected up to $M_h \simeq 1$ TeV using the polarization analysis of ref. 28 on $H \to ZZ$ with both Z's decaying to charged leptons.

If one turns to the mode $H \to W^+W^-$ with one $W \to \ell\nu$ and the other $W \to jj$, the rate is large enough to go well beyond a TeV. Unfortunately, the Standard Model background from $gg \to Wjj$ with $M_{jj} \simeq M_W$ is large compared[42] to the signal, so extracting a signal is difficult. Various cuts help distinguish[43] the signal from the background, but no Monte Carlo calculations have been fully completed to determine how well it can be done. Triggering procedures have been worked out to get a trigger rate that can be handled without loss of a large fraction of the signal. References 44,45 should be consulted.

e. $M_h \gtrsim 1$ TeV (obese Higgs) In this region the width of the Standard Model Higgs is so large that one is studying WW scattering, and there is essentially no help from a

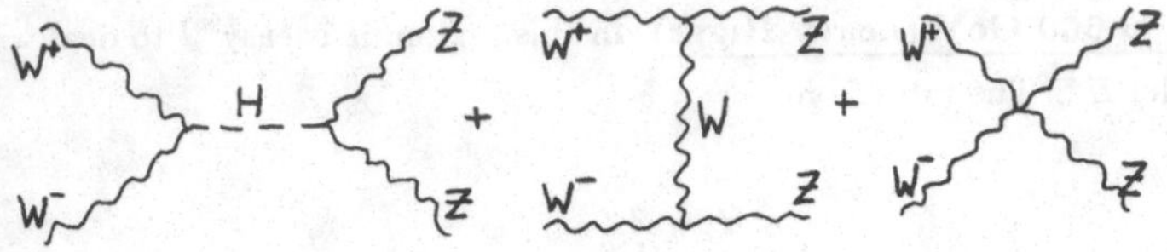

Figure 13
Diagrams for $W_L^+ W_L^- \to Z_L Z_L$.

resonance bump. As M_{WW} increases the ratio of signal to QCD background improves, so it will get easier to extract a signal. The cross section, of course, decreases, but does not become negligible until $M_h > 1.5$ TeV.

Further, because the Higgs mechanism operates to convert Goldstone bosons into the longitudinal modes of $W^\pm Z$, the longitudinally polarized W's are where the new physics[28] will be. By taking advantage of this, and studying the dependence of the fraction of longitudinally polarized W's on M_{WW}, it may be possible to detect Higgs physics as deviations from the tree-level perturbative theory, even for very large M_h.

It can be argued[14),27),28),46)] that even if $M_h \to \infty$ it is possible to find out experimentally. Basically, the argument is as follows. Recall that the polarization vector of a longitudinal vector boson becomes proportional to the momentum at large momentum,

$$\epsilon_\mu^{(o)}(p) \xrightarrow[P_o \gg M_W]{} P_\mu / M_W$$

Consider, for example, the process $W_L^+ W_L^- \to Z_L Z_L$. The contributing diagrams are shown in figure 13.

The HWW coupling is $g_2 M_W g_{\mu\nu}$, so the matrix element for this process is of the form

$$M = \frac{g_2^2 M_W^2 (\epsilon_W \cdot \epsilon_W')(\epsilon_Z \cdot \epsilon_Z')}{M_H^2 - \hat{s}} + f(M_W^2, \hat{s})$$

where f is some function of $\hat{s}$ and M_W^2 arising from the second and third diagrams; f does not depend on M_h.

Since $\epsilon_\mu \sim P_\mu$, if we let the four-momenta be $k_\mu, k_\mu', p_\mu, p_\mu'$ then $\epsilon_W \cdot \epsilon_W' = k \cdot k' / M_W^2$ and $\epsilon_Z \cdot \epsilon_Z' = p \cdot p' / M_W^2$, so with $2p \cdot p' \simeq 2k \cdot k' \simeq \hat{s}$,

$$M = \frac{g_2^2}{4 M_W^2} \frac{\hat{s}^2}{M_W^2 - \hat{s}} + f(M_W^2, \hat{s})$$

Now examine two situations. First, suppose $\hat{s} \to \infty$. Then

$$M \sim \frac{g_2^2}{4M_W^2}\hat{s} + f(M_W^2, \hat{s})$$

But the H contribution to M is purely in the s-wave and the s-wave contribution to M is of the form $e^{i\delta}\sin\delta$, bounded by a constant (this is called a "unitarity bound"). Therefore, the growth with $\hat{s}$ of the first term must be cancelled by $f(M_W^2, \hat{s})$, so we have learned that f grows like $\hat{s}$ if $\hat{s} \gg M_W^2$. Thus once $\sqrt{\hat{s}} \gtrsim \frac{1}{2}$ TeV, we can use

$$f \simeq \frac{g_2^2}{4}\frac{\hat{s}}{M_W^2}$$

Next, suppose $\sqrt{\hat{s}} \lesssim \frac{1}{2}$TeV, and $M_h \gg \hat{s}$. Then the first term of M is suppressed by the factor of $1/M_H^2$, so M must behave like f,

$$M \simeq \frac{g_2^2}{4}\frac{\hat{s}}{M_W^2}$$

Thus M grows like a power of $\hat{s}$, violating the unitarity bound mentioned above. That is not allowed, of course, so additional contributions to $WW \to ZZ$, beyond the tree level, must become important. This is referred[14] to as a breakdown of perturbation theory, or as WW scattering becoming strong. In the regime where M is given by $g_2^2\hat{s}/4M_W^4$, $\hat{\sigma}$ is of order $10^{-34}cm^2$, a large cross section; it should be possible to observe the effect.

This argument shows that it is not possible to push M_h to an unobservably large value without indirect effects occurring in WW interactions, and that the indirect effect should be large enough to observe. Detailed calculations and more general arguments confirm this simple approach.

Calculations have been completed[30] giving the full perturbative amplitude for $ff \to f\bar{f}WW$. They can be used in place of the real-W approximation. Such a procedure is valid if a Higgs particle exists with $M_h \ll 1$TeV. If there is no light Higgs boson, perturbative calculations are not useful, and the real-W approximation is the only way known[31] to extract information on WW scattering from experiment or to make testable predictions from theories about WW scattering and heavy Higgs bosons.

Considerable background analysis has been done for the W^+W^- and heavy Higgs system, some of which was described in Section 3.B.d. Since the fake "W" does not

388

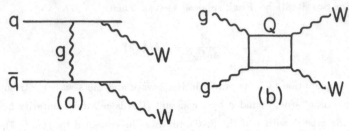

Figure 14
Some irreducible backgrounds to W pair production.

behave like a true spin-1 particle, there are other cuts that can be made to improve
the situation. One important cut, which has not been thoroughly studied because of
the absence of a theory of fragmentation, but has been considered by several people[47],
takes advantage of the expected large multiplicity difference between the decay of a real
W and that of two real jets of several hundred GeV.

Two more backgrounds to a W pair signal that were not calculated before have
now been evaluated. Figure 14a was calculated[48] by Gunion and Kalinowsky; it does
not cause trouble, though it is not negligible. Figure 14b has been of concern for some
time, since it could easily double the background of real W pairs in some regions. It
was finally[10] calculated approximately by Pumplin, Repko, and Kane. It is actually
connected by gauge invariance to the more familiar contribution $gg \rightarrow H \rightarrow WW$
where the first step proceeds by quark loop, since both occur in the process $gg \rightarrow WW$.
They interfere in both the real and imaginary parts of the amplitude. Their contribution
grows like the square of the heaviest quark mass, so they cause little problem if there are
three families with $M_t \leq 40$ GeV, but if $M_t > 60$ GeV or if there is a fourth generation,
then these backgrounds can give a large contribution. Because they interfere with
$gg \rightarrow H \rightarrow WW$, if circumstances are just right they can provide interesting variations
with M_{WW}, and phase variation.

C. Fundamental Higgs Bosons; Supersymmetric Theories

Some description of the spectrum and properties of the scalar bosons in a supersymmetric theory was given above. Here we provide[49] additional detail about the
differences from the Standard Model scalar.

a. Neutral scalars Let h be the light scalar of a supersymmetric theory, φ the Standard Model scalar, H the heavier scalar, and A the neutral pseudoscalar. Then in a

minimal supersymmetric theory[49] at tree level

$$\frac{g(ZZh)}{g(ZZ\phi)} = \frac{M_1^2(M_1^2 - M_z^2)}{(M_1^2 - M_2^2)(M_1^2 + M_2^2 - M_z^2)}$$

where M_1 and M_2 are mass parameters of the supersymmetric theory, with $M_2 \gtrsim M_z, M_1 \lesssim M_z$. Consequently, the rate for the usual way to look, $e^+e^- \rightarrow Z \rightarrow Zh$, can be very different. Similar relations hold for $\theta \rightarrow h\gamma$.

For the heavier scalar H, there is a major change — the coupling to WW is greatly reduced so Γ_H does not grow as M_H^3, and in fact Γ_H stays very small for any M_h. The reason is that the HWW coupling picks up a factor $cos(\beta - \alpha)$, where β measures the ratio of the two vacuum expectation values, and α is determined by some parameters of the Higgs potential. When expressed in terms of masses,

$$cos(\beta - \alpha) = \pm \left[\frac{M_h^2(M_z^2 - M_h^2)}{(M_H^2 - M_h^2)(M_H^2 + M_h^2 - M_z^2)} \right]^{\frac{1}{2}}$$

Then the factor M_H^3 in Γ_H is multiplied by $cos^2(\beta - \alpha)$, which would vary as $1/M_H^4$ at fixed M_h, so that $\Gamma_H < 10$ GeV for any M_h. In general,[50] supersymmetric theories without Higgs singlets have Higgs couplings proportional to gauge couplings, so widths do not grow; in theories with singlets, the widths can grow, but if they are kept at the level expected for perturbative unification the widths never get large as in the Standard Model.

Further, since H has largely decoupled from WW, it decays mainly into supersymmetric channels if they are open. One channel that has been studied [59] is $H \rightarrow \tilde{Z}\tilde{h}, \tilde{Z} \rightarrow Zh$. If the lightest supersymmetric partner (LSP) is $\tilde{\gamma}$, then $\tilde{h} \rightarrow \gamma\tilde{\gamma}$, so this mode gives isolated hard photons plus $\ell^+\ell^-$ from Z, so its signature is good and backgrounds are small. If $\tilde{h}$ is the LSP, backgrounds are worse since WW with each $W \rightarrow \ell\nu$ gives a similar signature.

Production cross sections are different[26] in general from the Standard Model cross section for h, H, but not very much. They can be a few times larger or smaller.

b. Charged scalars Since charged scalars occur in all approaches to spontaneous symmetry breaking beyond the minimal Standard Model, it is very important to be able to detect them. Because the charged Higgs occur in theories with two or more doublets, additional parameters appear, and the masses and couplings are too model

390

dependent to allow firm conclusions. The charged scalars can decay to $t\bar{b}$, $c\bar{b}$, $c\bar{s}$, $\tau\nu$,...
depending on their mass. Presumably the coupling will be proportional to masses so
the heavier states will dominate.

The production cross sections are large, and were shown in figure 10, from reference
26.

No analysis of the $t\bar{b}$ channel, including background effects, has been completed.
The QCD background is not so large, since no simple two body diagrams from large
structure functions produce this final state. With some ability to enhance selection of
b-quarks, it may be well be possible to search for $H^{\pm}$ in the $t\bar{b}$ channel. Some work is
in progress on this problem. [51]

The $\tau\nu$ mode is cleaner, but is appears to have serious difficulties with background.
At the present time it is not known how to use the $\tau\nu$ mode to find a signal.[52]

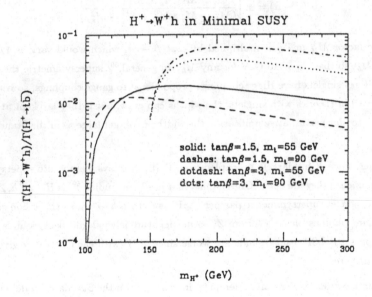

Figure 15.
The branching ratio for a charged Higgs scalar from a minimal
supersymmetric theory to decay to W plus the lightest Higgs neutral
scalar is shown, from reference 34.

Rare modes[34] such as $H^\pm \to W^\pm \gamma$, $W^\pm Z$, $W^\pm h^\circ$, have model dependent rates. In some approaches they can be the best way to find $H^\pm$. For example, in minimal supersymmetric models the tree level decay $H^\pm \to W^\pm h^\circ$ has a factor $\cos^2(\beta - \alpha)$ that can be of order 10^{-2}, while in general two doublet models there is no such factor. Figure 15 shows[34] the branching ratio including a suppression from $\cos^2(\beta - \alpha)$ in the minimal model of reference 49. At the present time there is too much model dependence, and too little is known about backgrounds, to draw firm conclusions about detecting $H^\pm$. While it is not easy to do so, in many theories the parameters are suitable to make detection possible. Among the detector properties that are very helpful are good b tagging and good resolution on missing momentum and on jet momentum.

4. Windows for Extended Gauge Theories

At future colliders the region of parameter space searched for new phenomena will be large. People have imagined many hypothetical new possibilities. Unfortunately, there are no compelling arguments as to what mass the new phenomena should have.

Any particle with electric charge or weak isospin or color charge will be produced at the new machines (and only such particles). New quarks or leptons, new gauge bosons, supersymmetric partners, partners of existing particles in larger representations of some group, are all possibilities. To see how well some new facility could do at producing them, one begins by calculating the production cross section. Then decay signatures are examined, and for each one the Standard Model background must be compared. As the mass of the new object increases, its cross section generally decreases, until at some mass a signal will not appear above backgrounds. Usually that mass is well below the kinematic limit for the production.

One of the most important reasons for analyzing many cases is that the ways to separate signal from background are very detector dependent. By identifying a useful signal and studying what features help to identify it, the detector design can be optimized in that direction. Some examples that have appeared in the Snowmass studies are the importance of very good mass resolution for a pair of jets, and the need to detect single, isolated photons, distinguish them from jets that are mainly electromagnetic, and obtain very good resolution on the effective mass of two photons or an $\ell^+ \ell^- \gamma$ set. Here I summarize some of the progress from workshops, etc.

Monte Carlo programs have been evaluated and improved in a variety of useful ways. A critical view is provided in the Snowmass 86 proceedings in the report of

Collins, Gottschalk, and Webber. From our point of view, Monte Carlo programs such as ISAJET, PYTHIA, and others provide two important kinds of contributions. First, they will eventually contain all Standard Model processes, so that they can be used to study how the Standard Model can be tested and provide a standard of comparison to look for deviations. They can tell us what Standard Model processes will become backgrounds for either other Standard Model processes (e.g., W_{jj} events are a background for WW production) or for possible new physics. There has been good progress here, with many processes now included that were not two years ago. Production of $W + n$ jets is included. The gauge invariant set of diagrams for production of heavy Higgs or W pairs was not included at the time of Snowmass 86, but is now in ISAJET. Whenever possible, the Monte Carlos are tested against data from the CERN collider. Two other checks are applied: (a) two different Monte Carlo programs should give the same results, and (b) the Monte Carlo result should agree with an independent analytic result where they overlap. Even with all the checks, caution is required when extrapolating Monte Carlo results to regions of variables where they have not directly been compared to data. Nevertheless, because the Standard Model theory should apply in the regions of interest, so long as only perturbative results are considered the Monte Carlo results should be reliable.

The second contribution of the Monte Carlos will be to allow the modeling of various alternatives for the way Higgs physics might appear, and for various hypothetical possibilities for new physics. For Standard Model Higgs physics the situation is already or will quite soon be very good, with essentially all alternatives included and studied or in the process of being studied. For hypothetical new physics the situation with new sequential quarks and leptons is good in that the necessary material exists in the Monte Carlos to carry out studies, but they have not been completed.[53] A significant number of supersymmetric processes have been included. The production mechanisms for gluinos and squarks are included, though when they are heavier than a W, some additional decay branching ratios involving W's and neutralinos have to be added. Many of the neutralino production mechanisms are not yet included. The Higgs sector of supersymmetric theories is not implemented in the main Monte Carlo at all. The required theory exists but it has not been put into the simulations. Altogether, while the main Monte Carlo writers have done a great deal in these areas, they could use a lot of help from the rest of us in implementing hypothetical new physics.

The structure functions for heavy quarks have been recalculated and changed;[54] they are 2-3 times larger. This can affect several analyses, particularly for Higgs physics, since the Htt coupling is relatively strong. The t can be a useful trigger for events that could be enriched in H°'s.

The major remaining question may be the effect of Standard Model backgrounds that can mimic the hoped-for signals. It is extremely important to realize that this is a solveable problem. The SSC is designed primarily to study collisions with products that have some jets or leptons or missing momenta larger than about 100 GeV. All that we observe indicates that beyond P_T's of a few GeV the Standard Model describes collisions rather accurately. At CERN over 99% of thousands of events with some $P_T \geq 10$ GeV are well described by the Standard Model. That is what is expected. Short distance collisions should be described perturbatively in both the electroweak theory and in QCD. Once hadron structure effects have become small, as P_T increases beyond a few GeV, or perhaps beyond 10 GeV or so when interferences are involved, the theory should work. Any deviations are the new physics we are looking for.

Indeed, because the background is calculable Standard Model physics rather than "junk", the best approach[55] to finding a signal for something new is to calculate the full structure of the expected results, as a function of a variety of variables, and look at whether there is agreement with what is measured. This approach utilizes the largest possible event rate, as opposed to making cuts to eliminate the background to a hypo-thetical signal, in which case often the signal is reduced below a viable level even though the background is reduced more. Such an approach can clearly be used in looking for new heavy leptons, for example, where many sources of background involve Standard Model W's or jets faking W's. When looking for an effect in WW production, the expected sources of W's, which are known, can be summed to find the expected rates as functions of M_{WW} and angle, the P_T of the W pair, the correlation between the W decay planes, etc. Gunion and Soper[55] have done an analysis to examine the statistical gain from such an approach.

In past studies a number of Standard Model backgrounds have been identified, but not many have been studied. In general, the approach to Standard Model backgrounds, either to other Standard Model processes such as Higgs physics or to possible new physics, has become rather systematic. While one can never be certain that one has thought of everything, it is now likely that all of the major kinds of backgrounds have been identified. On the other hand, considerably more work will be needed before the impact of all of the backgrounds is understood, and before cuts or strategies to avoid limitations caused by backgrounds can be formulated.

We have seen that a high luminosity hadron collider allows one to probe the TeV region in some channels. There is another potential advantage[56] that large luminosity can provide. Quantities of the known particles $W^{\pm}, Z^{\circ}, b, t, \tau$ are produced that greatly exceed the rates available at any other source. If it turns out to be necessary to study the decays of heavy flavors in order to gain insight into the flavor problem, they are available at SSC and perhaps only there.

It may be that to understand the origin of flavors we will have to study the (rare) transitions among flavors that are forbidden by the Standard Model. Possibly CP violation will be related to such transitions as well. To do so will require either extremely intense dedicated e^+e^- colliders with $\sqrt{s} \simeq 2M_\tau - 2M_c$ or $\sqrt{s} \simeq 2M_b$, or higher energy hadron colliders with luminosity as high as is useable ($> 10^{32} \ cm^{-2} \ sec^{-1}$ at least). Probably 10^8 useful b's and 10^{10} useful c's or τ's are the sort of numbers one should aim for, to compare with 10^{12} K's or μ's now being studied at BNL, SIN, TRIUMF, and Los Alamos. These numbers are available to a high energy hadron collider if ways can be found to study them. No detectors so far have been able to extract information about charmed particles in large quantities, but there is hope that with improved vertex detectors and modern techniques some experiments on rare decays of heavy quarks and leptons will be possible.

For example, $B \to \mu e$ is as important to study as $K \to \mu e$, and conceivably could have a much larger rate because a third generation particle is involved. Since the mixing angle for $b \to c$ is much smaller than the Cabibbo angle, one automatically gains a large factor in sensitivity. Experimentally it may not be so difficult to look for a μe pair produced isolated at large P_T, having an effective mass of M_b, and perhaps observed to arise from the decay of a long-lived particle. Other interesting decays, and the possibility of studying CP violation in the B system, have been considered at workshops. While it is not obvious that it can be done, because of problems with triggering, with using vertex detectors in a high rate environment, and with readout, the situation appears promising enough to deserve considerable further effort.

An important physics point is that in the Standard Model CP violation in the B system is supposed to be quite small, so at least 10^7 B's would be needed to see the predicted effect. However, there is so far no reason to assume that CP violation (or $B^{\circ} - \bar{B}^{\circ}$ mixing) is correctly described by the Standard Model (as opposed to most observables, where strong constraints from other measurements imply the theory should be quite good), so it could well happen that some CP violating effects are quite large in the B system. As a consequence, even limits are quite important here.

This possibility has been studied (mainly for b quarks) at the previous Workshops, with limited optimism that some rare decays and CP violation could be studied. At Snowmass 86 considerably more enthusiasm emerged, as reported by the Heavy Flavor working group. They have tried to design a spectrometer that triggers on the sequence $B \rightarrow \psi X, \psi \rightarrow \mu^+\mu^-$. Since over 10^{12} b quarks are produced in 10^7 sec., they can capture of order 10^9 tagged B decays for further study of rare decays and CP violation. The spectrometer, called TASTER, takes advantage of the larger rates near the forward direction by occupying the space from a few degrees to about 20°. Considerably more advanced ideas were studied by a large group of people at the 1987 Detector Workshop, and are described in reference 11.

A. Specific Examples of New Particles

Here I will summarize the present (September 1987) state of understanding of the mass ranges that can be covered for various kinds of new physics at future machines. The situation is actually not very clear, since reliable studies have not been completed for any process where separation from backgrounds is essential. Most studies up to the present time have made unrealistic simplifying assumptions about cross sections or branching ratios or backgrounds or all of these. A number of analyses are underway, for pp and e^+e^- colliders, so perhaps in a year or so better results will be available. It should be noted that the problems are similar at hadron and at e^+e^- colliders in the new energy ranges. At hadron colliders, some processes such as production of a new Z' followed by its decay to $\mu^+\mu^-$ will be background–free, and discovery is only limited by the decreasing production cross section as mass increases. Other processes have to be extracted from Standard Model backgrounds with similar signals, and the discovery limits are determined by the size and characteristics of the background. At e^+e^- colliders at high energies, the background from $\gamma\gamma$ and WW collisions become very large and cannot be ignored. In addition, radiation from the beam affects resolution and knowledge of energies in ways that hurt background separation. In practice a number of discoveries may be background limited at e^+e^- colliders as well.

a. Heavy quarks (Q) and leptons (L) These are basically similar channels, since once M_Q or M_L exceeds M_W the decays will be to W's:

$$Q \to qW, \tag{1}$$

$$L^{\pm} \to L^0 W^{\pm}. \tag{2}$$

The cross sections are known and are large. The signatures are W's plus either jets or missing momentum. Analyses taking backgrounds into account have only been done recently, and only at hadron machines. Reported results[53] are somewhat contradictory.

b. Supersymmetry Even if nature is supersymmetric, there is no precise argument as to what masses the super partners should have. We would have been very lucky if any super partner with a detectable production cross section had been light enough to discover so far; all present limits are well below M_W. Although it is not possible to set upper limits on what masses super partners can have, above some mass scale of order 1 TeV the possibility of explaining the weak scale through the added symmetry of supersymmetry would disappear, and much of the motivation for discovering supersymmetry below the Planck scale would be gone. For practical purposes supersymmetry would become irrelevant to understanding the world we see.

At colliders under discussion, the production cross sections are large enough to allow study and detection of super partners up to about a TeV. In practice the situation is not so simple, for two reasons. First, the super partners that have the largest production cross sections are the colored ones, squarks and gluinos. Once they are heavier than $W^{\pm}$ and Z, decay modes involving these become available and often dominant. In addition, decay modes involving partners of $W^{\pm}$ and Z become comparable; often chains of decays are involved so that signatures become less good, and the product of cross section times several branching ratios becomes rather small. Unfortunately, most studies reported so far in both Europe and the U. S. have made unrealistic assumptions about $\sigma \times$ BR and about signatures, so they are of little value for deducing meaningful absolute numbers. Further, as $\sigma \times$ BR decreases, more backgrounds from other Standard Model processes become very important, and they have not yet been included in the analyses.

The indications from studies[11] are that some or most super partners can be detected, but the absolute ranges in masses that can be studied at any specific machine are not yet available. In all cases the detection limits occur at masses well below those that are produced, as the useful event rate gets too small from effects of branching ratios, triggering, cuts, *etc.* Effectively, luminosity sets the detectable limits. Several studies are in progress that include realistic assumptions and background considerations, so reliable numbers should be available in the next year or two.

c. New Z-bosons Many approaches to understanding and extending the Standard Model lead to symmetry groups on the TeV scale that have additional $U(1)$ groups, and thus additional Z's. A number of studies have been done, using realistic coupling to initial quarks, and to final leptons or W pairs. In almost all cases there are no backgrounds that dominate the signal, so the studies give a good indication of the final result for what mass ranges can be studied. Figure 17 in the next section shows, for one model, the ranges that can be studied as a function of accelerator $\sqrt{s}$ and luminosity.

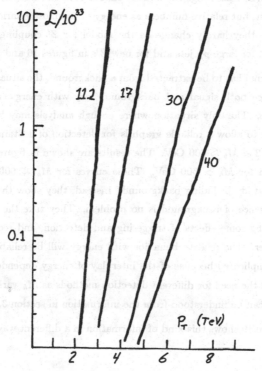

Figure 16
This shows the luminosity (in units of $10^{33}\,cm^{-2}\,sec^{-1}$) needed at a hadron collider to study QCD jets at a given P_T, at the values of $\sqrt{s}$ labeling each line (in units of TeV).

5. Some Comparisons

It is interesting to compare the physics that can be done with various alternatives for energy and luminosity. Two very different situations occur: signals that have to be extracted from a background and those that do not. For the latter, there are two cases. For an example such as the number of jets at some very large p_T, the expected rate is calculable from QCD and the only assumption needed is to decide how well the rate can be measured, in order to detect a deviation. On the other hand, for a hypothetical new case, such as a new Z-boson, there is model dependence in deciding what are the couplings to quarks and leptons; these determine the cross section and useful branching ratios and thus what mass range can be studied, For such an example, absolute numbers are of less interest, but relative numbers as energy or luminosity varies are still of quite general validity—they hardly change as the model for Z' couplings changes. Some results are shown for large p_T jets and for new Z's in figures 16 and 17.

When the signal has to be extracted from a background, the situation is much more complicated, since both signal and background vary with energy, and generally not at the same rate. The only situation where enough analysis may have been done at the present time to allow a reliable graph is for detection of a Standard Model Higgs boson with $M_Z/2 \lesssim M_h \lesssim 600$ GeV. The results are shown in figure 18. Some curves have been added for $M_h > 600$ GeV. These curves for $M_h \gtrsim 600$ GeV are not yet the results of a study including background. Instead, they show the *best* that can be done, if the presence of background is no problem. They take the number of events expected including some effects of triggering and detection, and require a reasonable minimum number. The relative behavior with energy will be reliable. The curves in figure 18 are complicated because of the interplay of energy dependence of signal and background, and the need for different detection methods as M_h varies. All the details of the variation can be understood from the information in section 3.

Finally, figure 19 shows this kind of information in a different way, for several kinds of physics.

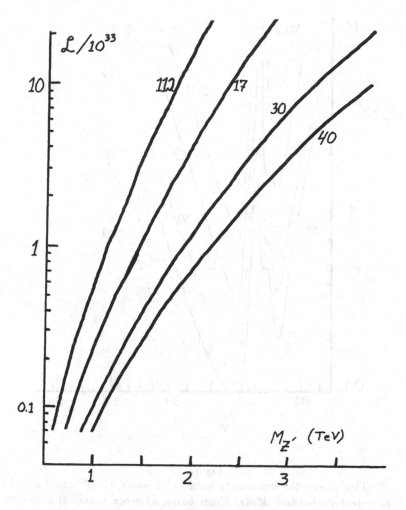

Figure 17

This shows the luminosity (in units of 10^{33} cm^{-2} sec^{-1}) needed at a hadron collider to detect a new Z of a given mass, at the value of $\sqrt{s}$ (in TeV) labeling each curve. This example uses a Z' with E_6 couplings, from reference 60. If stronger couplings, such as Standard Model ones, are used, larger masses can be studied at a given $\sqrt{s}$, but the comparison of different luminosities and energies is hardly affected.

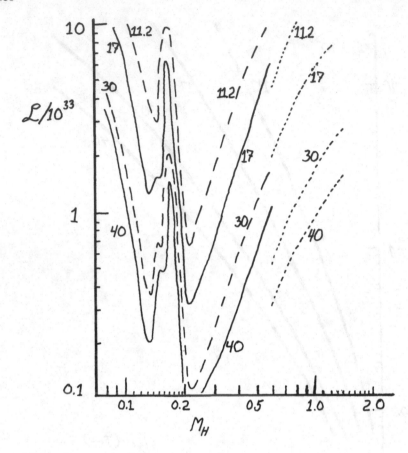

Figure 18
This shows the luminosity needed (in units of $10^{33} \ cm^{-2} \ sec^{-1}$)
to detect a Standard Model Higgs boson of given mass, at a hadron
collider with the value of $\sqrt{s}$(in TeV) labeling the curve. The curves
are complicated because different signatures must be used for different
values of M_h, and different backgrounds must be separated for different
values of M_h. For $M_h > 600 GeV$, no complete analysis including
backgrounds has been finished, so the curves show the best that could
be done if background separation were not a problem. Reading up at a
given M_H allows one to see what $\mathcal{L}$ is needed at a given $\sqrt{s}$ to study
that M_H; reading across at a given $\mathcal{L}$ allows one to see what ranges
of M_H can be studied.

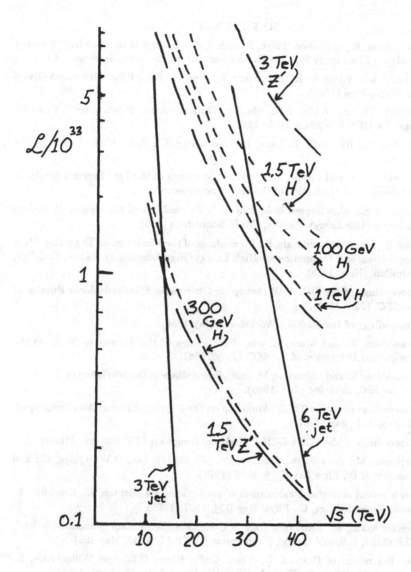

Figure 19
Above a given curve (i.e., at values of luminosity and energy greater than those along the curve) the physics labeling the curve could be studied at a hadron collider.

REFERENCES

1. Donaldson, R., Gustafson, Paige, F., (eds.), Proceedings of the 1982 DPF Summer Study on Elementary Particle Physics and Future Facilities, Snowmass, Co.

2. Hale, P. and Winstein, B., eds., Summary Report of the PSSC Discussion Group Meetings, June (1984).

3. Pilcher, J.E. and White, A.R., eds., "$p\bar{p}$ Options for the Supercollider", Proceedings of a DPF Workshop, (Feb. 1984).

4. Eichten, E., Hinchliffe, I., Lane, K., and Quigg, C., Rev. Mod. Phys. <u>56</u> 579 (1984).

5. Donaldson, R. and Morfin, J.G., eds., Proceedings of the 1984 Summer Study on the Design and Utilization of the SSC, Snowmass, Co.

6. Soper, D.E., ed., "Supercollider Physics", Proceedings of the Oregon Workshop on Super High Energy Physics, World Scientific, (1986).

7. Cox, B., Fenner, R. and Hale, P., Proceedings of the Workshop on Triggering, Data Acquisition, and Computing for High Energy/High Luminosity Hadron Colliders, Fermilab, (Nov. 1985).

8. Proceedings of the UCLA Workshop on Observable Standard Model Physics at the SSC, (Jan. 1986).

9. Proceedings of the Madison Workshop, (May 1986).

10. Donaldson, R. and Marx, J., eds., Proceedings of the Snowmass Study on the Design and Utilization of the SSC, (July 1986).

11. Donaldson, R. and Gilchriese, M., eds., Proceedings of the Workshop on Detectors for the SSC, Berkeley, (July 1986).

12. Proceedings of the La Thuile Workshop on the Physics of Future Accelerators, ed. Mulvey, J.H., Jan. 1987.

13. Proceedings of the ECFA-CERN Working Groups on LEP 200, ed., Mulvey, J..

14. Veltman, M., Acta Phys. Polonica <u>B8</u>, 475 (1977); Lee, B.W., Quigg, C., and Thacker, H.B., Phys Rev. <u>D16</u>, 1519 (1979).

15. For a recent review and calculation of predictions, see Eichten, E., Hinchliffe, I., Lane, K., and Quigg, C., Phys. Rev <u>D34</u>, 1537 (1986).

16. See for example, "Phenomenology of Composite Technicolor Standard Models", Chivukula, R.S. and Georgi, H., preprint HUTP-87/A036, May 1987.

17. See the report of Dawson, S., Yuan, C.-P., Kane, G.L., and Willenbrock, S., reference 10; Dicus, D., Willenbrock, S.W., Phys. Lett. <u>156B</u>, 429 (1985).

18. See Froidevaux, D., ref. 12, and Barger, V., Han, T., and Ohnemus, J., preprint MAD/PH/331, July, 1987.

19. Yuan, C.-P., Raja, R., and Kane, G.L., work in progress.

20. Wilczek, F., Phys. Rev. Let. <u>39</u>, 1304 (1977).

21. Vysotsky, M.I., Phys. Lett. <u>97B</u>, 159 (1980).

22. Narain, M. et al., contributed paper to the 1987 International Symposium on Lepton Photon Interaction, Hamburg, July, 1987.

23. Jones, D.R.T. and Petkov, S.T., Phys. Lett 84B, 400 (1979); Cahn, R.N. and Dawson, S., Phys Lett. 136B , 196 (1984).

24. Kane, G.L. and Scanio, J., Nucl. Phys., B291 , 221, 1987.

25. Gunion, J.F. et al., to be published; Altarelli, G., Mele, B., and Pitolli, F., University of Rome preprint 531.

26. Gunion, J.F., Haber, H.E., Paige, F.E., Tung, Wu-ki, and Willenbrock, S.S.D., Davis preprint UCD-86-15.

27. Chanowitz, M.S. and Gaillard, M.K., Phys. Lett. 142B , 85 (1984); Nucl. Phys., B261 , 379 (1985).

28. Duncan, M.J., Kane, G.L., and Repko, W.W., Nucl. Phys. B272 , 517 (1986).

29. Chanowitz, M.S. and Gaillard, M.K., Ref. 27; Dawson, S., Nucl. Phys. B249 , 42 (1985); Kane, G.L., Repko, W.W., and Rolnick, W.B., Phys. Lett. 148B , 367 (1984).

30. Dicus, D.A. and Vega, R., Phys. Rev. Lett. 57 , 1110 (1986); Gunion, J.F., Kalinowski, J., and Tofighi-Niaki, A., UCD-86-19, August (1986).

31. Kane, G.L., talk at the Tucson meeting on the Physics of the 21st century, Dec. 1983, and ref. 29.

32. Johnson, P.W., Olness, F.I., Tung, W.-K., Phys. Rev. D36 , 291 (1987); Soper, D. and Kunszt, Z., Oregon preprint; Repko, W.W. and Tung, W.-K., reference 10.

33. Gunion, J.F., Kalyniak, P., Soldate, M., and Galison, P., Phys. Rev. D34 , 101 (1986).

34. Gunion, J.F., Kane, G.L., and Wudka, J., Davis preprint, UCD-87-28.

35. Ellis, J., Gaillard, M.K., and Nanopoulos, D.V., Nucl. Phys. B106 , 292 (1976).

36. Atwood, D. and Gunion, J.F., ref. 9; Ellis, K., Hinchliffe, I., Soldate, M., and VanderBij, J., FNAL preprint, PUB-87/100-T.

37. Paige, F., private communication.

38. Atwood, D., private communication.

39. Wang, E. et al., reference 11.

40. Cahn, R.N. and Chanowitz, M.S., Phys. Rev. Lett. 56 , 1327 (1986).

41. Duncan, M.J., Phys. Lett. 179B , 373 (1986).

42. Ellis, S.D., Kleiss, R., and W.J. Stirling, Phys. Lett. 163B 261 (1985); Gunion, J.F., Kunszt, Z., and Soldate, M., Phys. Lett. 163B , 389 (1985); (E: 168B 427).

43. Gunion, J.F. and Soldate, M., Phys. Rev. D34 , 826 (1986).

44. See. Alverson, G., et al., ref 10.

45. ref 39; and Yuan, C.-P., and Kane, G.L., in progress.

46. Chanowitz, M., Golden, M., and Georgi, H., LBL-22119; Chanowitz, M., Proceedings of the XXIII Conference on High Energy Physics, Berkeley, 1986, ed. S. Loken.

47. Seiden, A.; Gunion, J.F.; and Yuan, C.-P., and Kane, G.L., work in progress.

48. Gunion, J.F., Kalinowski, J., and Tofighi-Niaki, A., Davis preprint UCD-86-19.

49. For extensive detail about supersymmetric Higgs sectors, see Gunion, J.F. and Haber, H.E., Nucl. Phys $\underline{B272}$, 1 (1986); Nucl. Phys. $\underline{B278}$, 449 (1986). See also Kane, G.L. and Haber, H.E., Phys. Rep. $\underline{117}$, 75, (1985).

50. I appreciate conversations with H. Haber and D. Dominici on this subject.

51. Gunion, J.F., private communication.

52. Kamamiya, S. and Yamomata, Y., references 9 and 11.

53. For quarks see Dawson, S., and Haggerty, J., reference 11, and reference 12; for leptons see reference 18.

54. Collins, J.C. and Tung, Wu-ki, FERMILAB-PUB-86/39-T.

55. See Kane, G.L., ref 7 and 10; and Gunion, J.F. and Soper, D., Phys. Rev. $\underline{D35}$, 179, (1987).

56. Kane, G.L., Proceedings of the Conference "50 Years of Weak Interactions", Wingspread Foundation, May 1984, ed. D. Cline.

57. See also the talks of Chanowitz, M., in ref. 8 and Kane, G.L., in ref 10.

58. Gilman, F. and Price, L., ref 10; Brau, J., ref. 11.

59. See Barnett, M. et al., ref. 10.

60. del Aguila, F., Quiros, M., and Zwirner, F., Nucl. Phys. $\underline{B284}$, 530, (1987); $\underline{B287}$, 419, (1987).